GROUP THEORY:
essays for Philip Hall

1984

ACADEMIC PRESS

(Harcourt Brace Jovanovich, Publishers)

London   Orlando   San Diego   San Francisco   New York
Toronto   Montreal   Sydney   Tokyo

Published for the London Mathematical Society
by Academic Press Inc. (London) Ltd.

by   F. J. Grunewald
          B. Hartley
          T. O. Hawkes
          I. G. Macdonald
          D. S. Passman
          D. J. S. Robinson
          D. Segal
          R. Strebel
          J. G. Thompson

Commissioned by
the London Mathematical Society

and edited by
K. W. Gruenberg *and* J. E. Roseblade

Academic Press Inc. (London) Ltd.
24–28 Oval Road
London NW1 7DX

US edition published by
Academic Press Inc.
Orlando, Florida 32887

British Library Cataloguing in Publication Data

Group theory.
  1. Groups, Theory of
  I. Gruenberg, K. W.   II. Roseblade, J. E.
  III. Hall, Philip
  512'.22   QA171

  ISBN 0-12-304880-X   LCCCN 83-73506

Typeset by Advanced Filmsetters (Glasgow) Ltd.
Printed in Great Britain by Thomson Litho Ltd., East Kilbride

Preface

This book was to have been an eightieth birthday present for Philip Hall. In the summer of 1980 the Council of the London Mathematical Society asked us to edit a volume to mark Hall's 80th birthday on the eleventh of April 1984. We decided to produce a book in two parts: the first to consist of commissioned survey articles and the second of submitted research papers. Because we intended to invite research articles by advertisement, we had to tell Hall something of our plans; this we did at a pub lunch outside Cambridge in May 1981. At the same time we asked him if he would agree to take part in a birthday celebration in his honour which had been proposed by the Society. Characteristically he said that he would prefer no public festivity; but he liked the idea of a book, especially the surveys.

Our idea was that each survey would give a reasonably self-contained, up-to-date and forward-looking account of an area in which Hall had made important contributions. In view of Hall's considerable impact on modern group theory, we hoped that the essays would together form a fairly coherent picture of the subject. So as to avoid too much over-lap, we suggested to each author the area we should like him to cover, but only in broad terms; the choice of material within the suggested area was left entirely to him. It was inevitable, perhaps, that gaps would remain.

When Hall died on 30th December 1982, we felt that the second half of the planned book was no longer appropriate, but that the essays should still be published. We offer them here not as a memorial volume, since they were largely written while Philip Hall was alive and well, but as a tribute to his genius.

Our sincere thanks go to all the authors for their contributions, and for their understanding response to our editorial suggestions. We are also very grateful for the help and courtesy shown us by everyone involved in the production of the book at Academic Press.

Queen Mary College, London   K. W. Gruenberg
Jesus College, Cambridge   J. E. Roseblade

February 1984

Contents

Preface   v

Finite non-solvable groups   by John G. Thompson   1

Finite soluble groups   by T. O. Hawkes   13

**Topics in the theory of   by B. Hartley   61
nilpotent groups**

**Reflections on the classifica-   by Fritz Grunewald and   121
tion of torsion-free nilpotent   Dan Segal
groups**

**Finiteness, solubility and   by Derek J. S. Robinson   159
nilpotence**

**Group rings of polycyclic   by Donald S. Passman   207
groups**

**Finitely presented soluble   by Ralph Strebel   257
groups**

The algebra of partitions   by I. G. Macdonald   315

Index of notation   335

Subject index   341

Contributors   353

Finite Non-Solvable Groups

JOHN G. THOMPSON

1. Introduction 1
2. Weak closure and the J-subgroup 3
3. E-theorems and signalizers 8
4. Uniqueness theorems and characters 10
5. p-groups of symplectic type 11
References 12

1. Introduction

I welcome the greater latitude of a survey paper to present a view of finite group theory freed from the indispensable Satz-Beweis of a research paper in which the contours of solid reasoning are unbroken by any local color. From lack of competence, I restrict my attention mainly to those portions of the theory with which my own work has been concerned, giving to it some coloration which, if not unique, is probably distinctive.

A solvable group consists of Abelian groups laid on top of one another. The Fitting subgroup includes all the deep layers, which in various ways influence the surface. There may be canting so that, for example, some of the more superficial layers fit appropriately into the Fitting group.

A solvable group also consists of a family of pairwise permutable Sylow subgroups. It is a collection of core samples through the Abelian layers. The various shafts penetrate to different depths. Upheavals can be studied locally.

To extend this imagery to non-solvable groups, we need to capture the non-Abelian composition factors. They may be thought of as fossils. By this I have in mind that their discovery and study generate interest, that in some partially understood way they are evolved from the more inert solvable groups, and that they are rare.

GROUP THEORY: essays for Philip Hall
ISBN 0-12-304880-X

As for interest, Burnside ([2], Preface) stated his view: 'The subject is one which has hitherto attracted but little attention in this country; it will afford me much satisfaction if, by means of this book, I shall succeed in arousing interest among English mathematicians in a branch of pure mathematics which becomes the more fascinating the more it is studied.' This quotation betrays a certain insularity, but the open enthusiasm is still refreshing and my own experience confirms the appeal of finite group theory.

As for the evolution of simple groups from solvable groups, this is seen most vividly in my work on N-groups [11], but is amply present in work of others. A source of wonder to me in characterizing simple groups lies in the arduous case analysis which drills away at the local subgroups until finally only the very limited and integrated structure of an actual simple group survives. Many lines become extinct, but not without struggle. It is in this loose and suggestive way that I picture a sort of evolution.

As for rarity, this depends on our choice of a measure. If $G = G_1 \supset G_2 \supset \ldots \supset G_n = 1$ is a composition series for G, let

$$\frac{1}{|G|} \cdot \prod_{i \in I} |G_i : G_{i+1}| = s(G),$$

where I is the set of indices i such that G_i/G_{i+1} is non-Abelian. By the Jordan–Hölder theorem, $s(G)$ is a numerical invariant of G; and $0 \leqslant s(G) \leqslant 1$. If x and ε are real numbers and $x \geqslant 1$, let

$$f(x, \varepsilon) = \frac{\operatorname{card} \{G| \ |G| \leqslant x, \text{ and } s(G) \geqslant \varepsilon\}}{\operatorname{card} \{G| \ |G| \leqslant x\}}.$$

Then it seems plausible that

$$\lim_{x \to \infty} f(x, \varepsilon) = 0 \text{ for all } \varepsilon > 0.$$

This assertion is unproved, but if it is true, it gives a measure of rarity; most finite groups consist largely of solvable chunks.

The paleontological image of simple group theory lends itself to another aspect of the subject. For the classification of finite simple groups is an exercise in taxonomy. This is obvious to the expert and to the uninitiated alike. To be sure, the exercise is of colossal length, but length is a concomitant of taxonomy. Those of us who have been engaged in this work are the intellectual confrères of Linnaeus. Not surprisingly, I wonder if a future Darwin will conceptualize and unify our hard won theorems. The great sticking point, though there are several, concerns the sporadic groups. I find it aesthetically repugnant to accept that these groups are mere anomalies. In the full flood of proving characterization theorems, it is natural to press on to the 'general case', but when the proofs are complete an enigma remains. Baldly, we do not know where the sporadic simple groups come from, except in the sense that every group is a realization of the axioms which define a group. Nor do we have more than a very dim awareness of the relationship of group theory to other branches of mathematics and science. This lack is especially acute in regard to the sporadic groups. Attempts have been made to understand the Fischer–Griess group F_1 [3], but so far they have only scratched the rich numerological surface. To go below this surface is not possible with existing group-theoretical techniques. To take a pessimistic view, perhaps the surface encloses nothing more than hot air. But this has not been demonstrated, and is too bleak for my taste. Possibly, too, on the other hand, *The Origin of Groups* remains to be written, along lines foreign to those of Linnean outlook. Independently of these two possibilities, it is of archival importance to record the developments of an earlier period.

2.   Weak Closure and the J-Subgroup

My gratitude and sense of obligation outweigh my reticence in placing on record a letter which improved my spirits and my mathematics.

'King's College,
Cambridge, Eng.
Dec 6, 1958

Dear Dr. Thompson,

Thank you very much for the copy of your paper: "A Condition for a Finite Group to be p-Normal". It reached me yesterday; but being a slow reader, I've not yet got further than the slightly alarming Diagram V on page 30. The rest I hope to read this evening or tomorrow. But I feel that I must immediately congratulate you on an extremely fine piece of work. Not only are the results first rate, but your lucid exposition makes it a pleasure to read: and the problem of setting it down in detail cannot have been an easy one. In your covering letter, you are good enough to say that you await my comments "most anxiously"! As I should not like your rest to be broken through any fault or omission of mine, I hasten to append a few random thoughts which have occurred to me as a result of reading your very stimulating paper—though, of course, I have not yet had time to study it properly in detail.

I will concentrate on your latest & best result: Given a finite group G and a group $\mathfrak{A}$ of automorphisms of G, a subgroup K of G, a p-Sylow P of K which is $\mathfrak{A}$-invariant, p being an odd prime, if, for all $\mathfrak{A}$-invariant normal subgroups H of P, $N_K(H)/C_K(H)$ is always a p-group, then K is p'-closed.

Let me call that Theorem B. The beauty of it seems to me to lie among other things in the fact that notions like "p-normal" or "weakly closed" do not appear at all in the hypotheses; that it seems also to be genuinely an "odd-p" theorem—at least, I can see no way of twisting the enunciation so as to make it valid for $p = 2$ as well; and, of course, by comparison with your Theorem A, in the fact that K does not have to be $\mathfrak{A}$-invariant. That is clearly a most important advance.

In connection with this last point, it occurs to me, as it has probably done already to you, that G and $\mathfrak{A}$ are now really irrelevant (in the statement of the theorem, I mean; the proof, naturally, I do not know about.) For convenience, let me say that a p-subgroup H of a finite group K is *fertile* in K (I pick the word more or less at random from the dictionary & I'm not meaning to recommend it) if $N_K(H)/C_K(H)$ is not a p-group. So H is fertile in K if & only if it is fertile in $N = N_K(H)$; or again, if & only if there is an element $t \in K$ of order prime to p such that $H \lhd \{H, t\}$ and $\{H, t\}$ is not nilpotent. Then, the formulation of your theorem that I suggest is the following:

Theorem B*: Let K be a finite group, p an odd prime, P a p-Sylow subgroup of K. Suppose K is not p'-closed. Then there exists a subgroup H of P such that (i) H is fertile in K, (ii) H is characteristic in P and (iii) H/H' is elementary & H is of class (of nilpotency) at most 2.

For let $o(K) = n$; $(K : P) = m$; $K = s_1 P \cup s_2 P \cup \ldots \cup s_m P$ $(s_1 = 1)$; A the group of automorphisms of P; for $b \in K$, let $\rho(b) : x \to xb$ be the right translation of K by b; for $a \in A$, let $\tau(a)$ be the permutation of K defined by $s_i y \to s_i y^a$ for $y \in P$, $i = 1, \ldots, m$. Take G to be the group generated by the $\rho(b)$ and $\tau(a)$, $b \in K$, $a \in A$. We can regard K as a subgroup of G, by identifying it with its regular representation. Take $\mathfrak{A}$ to be the group of automorphisms of G induced by transformation by the elements $\tau(a)$, $a \in A$. Then P (identified with the group of all $\rho(y)$, $y \in P$) is $\mathfrak{A}$-invariant, owing to $\tau(a)^{-1}\rho(y)\tau(a) = \rho(y^a)$. The $\mathfrak{A}$-invariant subgroups of P are the characteristic subgroups. Applying your theorem B gives an H satisfying (i) & (ii). To get (iii), choose such an H which is as small as possible.

Let t be of order prime to p, $t \in K$, $H \lhd \{H, t\}$, $\{H, t\}$ not nilpotent. If H/H' were not elementary, the group H_1 of all $x \in H$ such that $x^p \in H'$ would be smaller than H & would satisfy (i) and (ii). (i) because $\{H_1, t\}$ is not nilpotent, (ii) because H_1 is characteristic in H. This contradicts minimality of H, so H/H' is elementary. If the class of H were > 2, let H_2 be the centralizer of H' in H. Then $H_2 < H$, H_2 satisfies (ii). By the minimality of H, every p'-element t of $N = N_K(H)$ is contained in $C_N(H')$, which is normal in N. Hence $[H, t] \leqslant H \cap C_N(H') = H_2$. If H_2 were infertile in K, we should have $[H_2, t] = 1$, and hence $[H, t] = 1$ which would make H infertile too. Hence H_2 must satisfy (i), contrary to the minimality of H. Consequently, H has class at most 2. Thus H also satisfies (iii).

Perhaps (iii) is not so important (and no doubt, it could also be made more precise by an analysis of the possible kinds of class 2 group which might be permissible). But obviously your use of an $\mathfrak{A}$ is the vital thing in obtaining a *characteristic* fertile subgroup of P. If I am not mistaken, this (or rather the $\mathfrak{A}$-invariance which it implies) is at the root of the beautiful application you make to Frobenius groups & solvability problems.

I wonder if you have yet considered the possibility (theoretical—I don't mean to say that I would know how to do it) of transforming your theorems, considered as criteria for K to be p'-closed, into results about the subgroup, I will call it $\eta_p(K)$, which is generated by all p'-elements of K. I have long been intrigued by Wielandt's Theorems of 1940 (Crelle) in this connection. Often as I have lectured on them, I've never felt that I understood their why or wherefore. It would certainly interest me greatly if you could bring your results into relation with Wielandt's. Although it is not, I think, stated quite explicitly in his paper, one possible version of his result would be the following

Theorem W: Let K be a finite group, p any prime, P a p Sylow subgroup of K, H_1 a weakly closed subgroup of P (with respect to K), $E_1 = E(H_1, P)$ the group generated by all commutators $[x, \underbrace{y, y, \ldots, y}_{p-1}]$ with $x \in P$ and $y \in H_1$, E_1^* the strong closure of E_1 in P (with respect to K) i.e. the subgroup generated by the intersection of P with the conjugates of E_1 in K, and $N = N_K(H)$. Then

$$N \cap \eta_p(K) \leqslant E_1^* \cdot \eta_p(N).$$

As a corollary, either (i) K is p'-closed or else (ii) $E_1 \neq 1$ or else (iii) N is not p'-closed.

This might be more useful if one knew more about weakly closed

subgroups. But at least it implies that either K is p'-closed or else P is fertile in K or else the class of P is at least p (better: $[x, y, \ldots, y] \not\equiv 1$ in P).
$$\underset{p-1}{}$$
Seen in this light, the great advantage of your theorem is that it involves no assumptions about the class of P or the like. For instance, can one always be sure that there exists even a single weakly closed $H_1 \neq 1$ in P such that $E_1 = 1$? I do not know. It seems possible that one might be able to infer such an H_1 by using the fact that a subgroup of P which is the only one of its isomorphism type in P is necessarily weakly closed. However it scarcely seems obvious.

One could attempt to combine your theorem with Wielandt's by saying that, if K is not p'-closed, then for every weakly closed H_1 in P with $E_1 = 1$, there must be an H characteristic in P which is already fertile in $N_K(H) \cap N_K(H_1)$. But I don't feel much the wiser.

I've now exhausted my ideas, such as they are. Thank you once more for sending me your paper, which I shall look forward eagerly to seeing in print. I hope no one is going to distract you from group theory: there are so many difficult problems, as you must know, which still defy solution.

 With all good wishes,
  Yours sincerely,
   Philip Hall.

P.S. Reading over what I have written, I wonder if the policy represented by (iii) of Theorem B* is really the best one. Why should we not choose H (characteristic in P & fertile in K) as *large* as possible? What would happen then? Assume K is not p'-closed & let H be maximal, subject to being fertile in K & characteristic in P. Then we can think about N/H where $N = N_K(H)$ and about its p-Sylow subgroup P/H (P is in N because H is characteristic in P). If N/H were not p'-closed, then, applying B*, we get a characteristic subgroup $\bar{H}/H$ of P/H which is fertile in N/H. But this is impossible by the maximality of H, for it would make $\bar{H} > H$, $\bar{H}$ characteristic in P & fertile in K. So N/H is p'-closed and if $M/H = \eta_p(N/H)$ we have $M \lhd N$, $M \cap P = H$, $N = PM$. So N is p-soluble of p-length at most 2. If the p-length $l_p(N)$ were equal to 1, it is easy to see that P itself would be fertile in K (whence $H = P$ by maximality of H). So we get

 Theorem B†: K, p and P as in B*. Assume K is not p'-closed & P is not fertile in K. Then there is an H satisfying (i) & (ii) of B* and such that $N = N_K(H)$ is p-soluble of p-length 2.

 Applying the corollary of Theorem 3.4.1 on page 29 of the paper on p-lengths, it follows that P is of class at least $p-1$ & even at least p, if p is not a Fermat prime. This comes very near to the result from (W)

above. It is true that (W) gives rather more in some respects viz. $[x, y, \ldots, y] \not\equiv 1$ in P but the above argument is very rough & could

$p-1$

no doubt be used to give much more. And (W) says nothing about characteristic subgroups of P. That your theorem should yield so much with little trouble seems to me a striking testimony to its usefulness. I think it very probable that you have already got a long way beyond this point. I should certainly be very glad to hear of anything you find throwing light on Wielandt's theorem, which has bothered me for a long time.'

Although this letter speaks for itself quite adequately, several comments can be made in the light of later results.

The idea of the postscript to take H as large as possible is incorporated in the induction step of [12], which itself goes back to the partial ordering of my thesis. This makes it possible to answer certain questions by studying solvable groups of p-length 2, an extremely valuable aid, since in a tight spot, one can calculate.

Lemma 8.2 of the Odd Order Paper (an obvious acronym is OOP) [5] is an outgrowth of Theorem B* of the letter. It is a fact of life that a nilpotent group need not contain a self-centralizing characteristic subgroup, so we salvage what we can.

In the fourth paragraph, '... it occurs to me ... that G and $\mathfrak{A}$ are now irrelevant ...'. This had not occurred to me, so Theorem B* with its elegant proof, was new to me. This seems strange now, but the lacunae of inexperience are gaping. Professor Hall's observation spurred me on to the discovery of the J-subgroup, and to the short paper [12], from which Theorem B* follows with little effort. My thesis with its triples $(G, P, \mathfrak{A})$ had left me with the nagging feeling that I should try to 'get rid' of $\mathfrak{A}$. The second paragraph of the letter contains a statement of my first attempt. I have not kept a proof of this result, and I have my doubts that my proof was correct. If it was, it was surely complicated. Theorem B* set me on the right track.

In the third paragraph from the end occurs another salient point: '... It seems possible ... obvious.' From here, it is only a small step to note that we obtain weakly closed subgroups of P by choosing an isomorphism class and forming the subgroup of P generated by all the subgroups of P in this class. This point is already implicit in the use of the word 'closure', since weak closure is an idempotent map. Choosing the class to consist of Abelian groups carries one to the threshold of the J-subgroup, and recognition of the importance of one aspect of the Hall–Higman paper [9] carries one across. A stumbling block here is that the relevant module to consider is not the Frattini quotient group of $O_p(G)$ but rather $\Omega_1(Z(O_p(G)))$, or equally well, the p-reducible p-core [11]. It is better, perhaps even imperative, to work from the bottom upwards than from the top downwards. This approach indicates in particular that to understand the transfer map, one must look far afield, as Wielandt did in his paper [13].

When I met Professor Wielandt in Tübingen in the summer of 1958, I told him that his paper interested me greatly, and explained that the Frobenius kernel would be nilpotent if we could find a group S of order p^2 whose normal closure in an S_p-subgroup P of G was Abelian and such that $C_G(s) = C_G(S)$ for all $s \in S \backslash \{1\}$. Here G is assumed to have a fixed point free automorphism of prime order. I thought that Wielandt's 'schwache Abschliessung' was relevant. In this I was not mistaken. I never found my group S, but I continued to be intrigued with what I thought of as the 'elements at infinity', which are contained in Grün's complex [7]. I made no progress here either and never have subsequently, but could not banish Wielandt's work. So it was extremely encouraging to read a letter which for the first time put me in touch with a resonating mind equally aware of the technical nuances which gripped me.

Few mathematicians, novices or veterans, have received a letter as graceful and stimulating as the one preserved here. Even fewer of us have written such letters.

3. E-Theorems and Signalizers

The underpinning here is Hall's work on solvable groups and in particular [8]. This work suggests that a group is solvable if and only if it satisfies $E_{p,q}$ for all primes p, q. This conjecture has recently been proved using taxonomy.‡

The starting point for Sylow systems is Burnside's $p^a q^b$-theorem. This theorem rests on an elementary result. The center C of the integral group ring of a finite group is of finite $\mathbb{Z}$-rank, and so from first principles, every homomorphism $\omega : C \to \mathbb{C}$ carries C to a ring of algebraic integers. The nature of the omegas was known to Frobenius and so Burnside's theorem emerges. It is one of several examples in which self evident properties of commutative algebra yield results on non-Abelian groups.

In beginning to work on the odd order theorem, there were two

‡ See Hawkes' essay, p. 54.

strategies which intrigued me. The first was to prove E-theorems. The second strategy, seemingly untouchable, was to construct, for each prime p, a proper subgroup H_p of our minimal simple G such that $p \nmid |G:H_p'|$. If one has the H_p in hand, then one is certainly in the presence of a perfect group, and Micawberish optimism suggested that something would turn up.

The E-theorems seemed more promising, since each proper subgroup contributes its mite (might?) to a possible S-subgroup. From my work on the Frobenius kernel, I wanted to work upwards from the bottom, and the bottom is the Fitting group. So, in trying to prove $E_{p,q}$, I could see the value in trying to locate q-subgroups Q which are normalized by a Sylow p-subgroup of G. In order to feel that I was not simply treading water, I decided to keep my eye on a G such that every Sylow subgroup was Abelian, an A-group. Here things went well and Walter Feit and I did a large, but worthwhile, amount of work on A-groups. We proved that a minimal simple A-group G of odd order did not contain an $S_{p'}$-subgroup provided p is a prime dividing $|G|$. This weak result required much work, both Sylow-theoretic and character-theoretic, and was a good warming-up exercise.

Pressing on, I was led to study the connection between q-groups Q_1, Q_2 which are normalized by the same Sylow p-subgroup P. When P is Abelian, I proved a transitivity theorem, but only on the hypothesis that P contains an elementary Abelian subgroup of order p^3. This was a moment of some self-doubt, since if one is obliged to make restrictive hypotheses in the 'trivial' case where $P' = 1$, what can one expect for the general Sylow subgroup? Furthermore, when $P' = 1$, or even when P is regular, $H = N_G(P)$ handles the second strategy and one is not much wiser. However, at the time, *faute de mieux*, I pursued these elementary Abelian subgroups of order p^3 and gradually came to believe that the resulting case analysis was worth the effort. The results of section 14 of OOP introduced signalizers into the theory. The transition from A-groups to the case of a Sylow p-subgroup P which contains a self-centralizing normal subgroup A which in its turn contains an elementary Abelian subgroup of order p^3 involved carrying over to P the transitivity theorem which I knew to hold when $A = P$. However, when this was accomplished, I discovered to my surprise and with considerable excitement that I could locate a suitable H_p. This is carried out in section 17 of OOP. So by focussing on the 'easier' E-theorems, I found that I had constructed the 'harder' H_p's. From my point of view, this construction is one of the pivots of OOP.

Having a rich sample of H_p's was a burden removed, like discharging an obligation to the literature on the transfer map. Bit by bit, the

various E- and C-theorems emerged, and the Sylow p-subgroups P with
$И(P) = \{1\}$ were squeezed into shape. The heuristics of these arguments
are invariable: what would be the case if we were playing the game in
a solvable group? This forced me to look at solvable groups in a
somewhat novel way. The so-called '3 against 2' argument appears
here, and began a long string of factorization theorems.

From the various E-theorems of OOP it became clear that a barrier
to the solvability of G was the equivalence relation $\sim$, which had
built into it the existence of elementary Abelian subgroups of order
p^3. Almost all of this work on E-theorems has been superseded by
Glauberman's ZJ-theorem [6], but it served its purpose.

4.   Uniqueness Theorems and Characters

A third strategy (or was it a tactic?) in OOP attempted to build a
bridge from Sylow theory to character theory. The far shore was
marked· by the granite of Suzuki's theorem on CA-groups [10], flanked
by [4]. The bridge was built of tamely embedded subsets with their
supporting subgroups and the associated tau isometry. The near shore
was dotted with the E-theorems and the uniqueness theorems.

Walter Feit and I went through a period of checking back and forth
with each other regarding what was provable and what was exploit-
able. This was immense fun, but demanding. It was clear from the
E-theorems and some uniqueness theorems that a maximal subgroup M
contained many special classes in the sense of [10]. It was also clear
that we understood fairly well when it could occur that two conjugacy
classes of M became fused in G. There was some murkiness and some
problem with 'obvious' results which still resisted complete proofs. Out
of this activity emerged the self-normalizing cyclic subgroups, which
immediately produced certain irreducible characters of G, the etas. It
was impossible from any known arguments to show that such cyclic
subgroups were impossible. We found no neat new arguments either. I
think that taxonomy may allow one to show that no perfect group
contains a self-normalizing cyclic subgroup. But in the context of the
odd order theorem, these cyclic subgroups were a terrible thorn.

To take but one important paper [1], there are cases when character
theory has wonderful success with fragments. But we need much more!
It seems to me, for example, that Brauer's characterization of characters
is still available for interesting applications. An underused theorem is
a waste of resources.

Suzuki's CA-theorem is a marvel of cunning. In order to have a genuinely satisfying proof of the odd order theorem, it is necessary, it seems to me, not to assume this theorem. Once one accepts this theorem as a step in a general proof, one seems irresistibly drawn along the path which was followed. To my colleagues who have grumbled about the tortuous proofs in the classification of simple groups, I have a ready answer: find another proof of Suzuki's theorem.

5. p-Groups of Symplectic Type

In unpublished notes, Hall classified those p-groups in which every characteristic Abelian subgroup is cyclic. Such groups are the central product of an extra special group and a group which is either cyclic or is a 2-group of maximal class. For this reason, they are said to be of symplectic type.

Already in OOP, the groups of symplectic type gave rise to difficulties. More precisely, one is led to consider, for a given group G, the following set of primes:

$\sigma(G) = \{p \mid G$ has a subgroup Z of order p such that the generalized Fitting subgroup of $C_G(Z)$ is of symplectic type$\}$.

There are many simple groups, both classical and sporadic ones, for which $\sigma(G)$ is non-empty. Perhaps the most extreme example concerning $\sigma(G)$ is F_1, and I make a conjecture, which, as with other conjectures, is probably decidable by using taxonomy.

(∗)   If G is a finite group, and $\sigma(G)$ has cardinality $\geqslant 5$, then $G \cong F_1$.

In fact, $\sigma(F_1) = \{2, 3, 5, 7, 13\}$, and this is the set of primes p for which Conway's group $.O$ contains an element of order p which has no fixed points on the Leech lattice. As far as I know, no explanation has been given for the equality of these two sets of primes. This is just another example of the numerological nature of our observations.

When G is a minimal simple group of odd order, the set $\sigma(G)$ appears when trying to handle those primes p such that G contains an elementary Abelian subgroup of order p^3, and in addition, 1 is the only p-signalizer of G. In this case, it follows that if M is a maximal subgroup of G, if $O_p(M) \neq 1$, and if M contains an elementary Abelian subgroup of order p^3, then the Fitting subgroup of M is a p-group. The game then turns to studying the weak closure in M of a normal elementary Abelian subgroup A of M. If A is of order at least p^3, various factorizations lead to the desired uniqueness theorem. When A is of order p or p^2, auxiliary arguments are needed, the most trying being the case when every normal Abelian subgroup of M is cyclic. This situation implies in particular that $F(M)$ is of symplectic type. The difficulties encountered here all evaporate with Glauberman's ZJ-theorem.

It seems to me that there are still some mysteries surrounding p-groups of symplectic type. The crucial role of extra special groups in the p-length paper [9] is no accident. At present, it would appear that extra special 2-groups and sporadic simple groups are connected only by the fact that certain arguments break down in the presence of extra special 2-groups, thereby leaving a 'gap' which a sporadic group occupies. Although I have no ideas on the matter, I hope that the mathematical community is not content to let this be the final word.

References

[1] Alperin, J., Brauer, R., Gorenstein, D., Finite groups with quasi-dihedral and wreathed Sylow 2-subgroups, *Trans. Amer. Math. Soc.* **151** (1970), 1–261.

[2] Burnside, W., 'Theory of Groups of Finite Order', Cambridge (1897).

[3] Conway, J. H., Norton, S. P., Monstrous moonshine, *Bull. London Math. Soc.* **2** (1979), 308–339.

[4] Feit, W., Hall, M., Jr., Thompson, J., Finite groups in which the centralizer of any non-identity element is nilpotent, *Math. Z.* **74** (1960), 1–17.

[5] Feit, W., Thompson, J., Solvability of groups of odd order, *Pacific J. Math.* **13** (1963), 775–1029.

[6] Glauberman, G., A characteristic subgroup of a p-stable group, *Canad. J. Math.* **20** (1968), 1101–1135.

[7] Grün, O., Beiträge zur Gruppentheorie, III, *Math. Nachr.* **1** (1948), 1–24.

[8] Hall, P., Theorems like Sylow's, *Proc. London Math. Soc.* (3) **6** (1956), 286–304.

[9] Hall, P., Higman, G. On the p-length of p-soluble groups and reduction theorems for Burnside's problem, *Proc. London Math. Soc.* (3) **6** (1956), 1–42.

[10] Suzuki, M., The nonexistence of a certain type of simple groups of odd order, *Proc. Amer. Math. Soc.* **8** (1957), 686–695.

[11] Thompson, J., Nonsolvable finite groups all of whose local subgroups are solvable, *Bull. Amer. Math. Soc.* **74** (1968), 383–437; *Pacific J. Math.* **33** (1970), 451–536; *Pacific J. Math.* **39** (1971), 483–534; *Pacific J. Math.* **48** (1973), 511–592; *Pacific J. Math.* **50** (1974), 215–297; *Pacific J. Math.* **51** (1974), 573–630.

[12] Thompson, J., Normal p-complements for finite groups, *J. Algebra* **1** (1964), 43–46.

[13] Wielandt, H., p-Sylowgruppen und p-Faktorgruppen, *J. reine angew. Math.* **182** (1940), 180–193.

Finite Soluble Groups

T. O. HAWKES

1. The early years of finite soluble group theory   13
2. Projectors   16
3. Injectors   ...   ... ·   32
4. Recent developments   54
References   58

This essay is a small selection of topics from the theory of finite soluble groups. Generalizations of Hall subgroups are the dominating theme. We begin with a short summary of Philip Hall's early work. In section 2 we discuss the theory of projectors, which began with W. Gaschütz's description of Hall and Carter subgroups by means of saturated formations. We devote section 3 to aspects of a dual theory of injectors, first studied by B. Fischer. Finally in section 4 we look at some recent theorems about Hall subgroups which have been proved by reference to the known list of finite simple groups. *Throughout the essay 'group' will mean 'finite group'.*

1.   The Early Years of Finite Soluble Group Theory

In 1928, in his first published work on group theory [32], Philip Hall proved that, if p is a prime, each finite soluble group G has a Sylow p-complement, that is a subgroup S whose index in G is the order of a Sylow p-subgroup of G. Nine years later he brought to light in [33] the remarkable fact that this complementary theorem to Sylow's is true only in soluble groups; if G is finite and insoluble, then there is at least one prime p for which G has no Sylow p-complement.

An interesting way to prove Hall's theorem that the existence of Sylow complements is a sufficient condition for solubility is to cite a theorem of H. Wielandt [64] proved in 1960. This states that a group

GROUP THEORY: essays for Philip Hall
ISBN 0-12-304880-X

with three soluble subgroups whose indices are coprime in pairs is soluble. It can be proved by elementary arguments. Now assume G has a Sylow p-complement for each $p \in \pi(G)$, the set of distinct primes dividing the order of G. If $|\pi(G)| \leqslant 2$, then G is soluble by Burnside's $p^a q^b$-theorem. Otherwise G has Sylow p-complements for at least 3 different primes p in $\pi(G)$, and hence three proper subgroups whose indices are coprime in pairs. Since these subgroups inherit the characteristic property from G, they are soluble by induction, and therefore G is soluble by Wielandt's theorem.

If $\mathbb{P}$ denotes the set of all prime numbers, the existence in a group of Sylow p-complements for all p in $\mathbb{P}$ implies the existence of Hall π-subgroups for all subsets π of $\mathbb{P}$. For let $\pi(G) = \{p_1, p_2, \ldots, p_n\}$, and for $1 \leqslant i \leqslant n$ let S_i be a Sylow p_i-complement of G. (A set of the form

$$K = \{S_i : 1 \leqslant i \leqslant n\}$$

is called a *complement basis* of G.) If π is a set of primes, then the subgroup

$$G_\pi = \cap \{S_i : 1 \leqslant i \leqslant n \quad \text{and} \quad p_i \notin \pi\}$$

has the property that its order is a π-number and its index a π'-number; in other words, G_π is a *Hall π-subgroup* of G. Then a set of the form

$$\Sigma = \{G_\pi : \pi \subseteq \pi(G)\},$$

comprising the 2^n Hall subgroups of G which can be obtained in this way from a given complement basis K is called a *Hall system* of G (Hall called such a set a Sylow system). Note that $K \subseteq \Sigma$.

In addition to being a representative set of the Hall subgroups of G (i.e. containing one member of each conjugacy class), a Hall system Σ has the property that if $H, K \in \Sigma$, then $HK = KH \in \Sigma$. In particular, if P_i denotes the Sylow p_i-subgroup of G in Σ, then $P_i P_j = P_j P_i$ for $1 \leqslant i, j \leqslant n$. (A representative set $B = \{P_1, P_2, \ldots, P_n\}$ of Sylow subgroups of G which permute in pairs is called a *Sylow basis* of G.) A product of the form $P_{i_1} P_{i_2} \ldots P_{i_t}$ where the order of the terms is immaterial is clearly a Hall subgroup in Σ, and so the Sylow subgroups in a Hall system form a Sylow basis, whose products generate that system.

The group G itself, therefore, can be written as a product $G = P_1 P_2 \ldots P_n$ of nilpotent subgroups which permute in pairs. This property is characteristic of soluble groups (cf. Wielandt [63] and O. H. Kegel [48]). Wielandt showed in [62] that it would be enough to deal with the case $n = 2$: Let $G = G_1 G_2 \ldots G_n$, where (i) $G_i G_j = G_j G_i$ for $1 \leqslant i, j \leqslant n$, and (ii) each G_i is nilpotent. If each product $G_i G_j$ is soluble, then so is G.

A representative set of Sylow p-complements, the result of selecting at random just one p-complement of G for each $p \in \pi(G)$, is always a complement basis. This procedure applied instead to Sylow subgroups, however, does not necessarily lead to a Sylow basis, because the condition of permutability in pairs is not automatically respected. For groups G with $|\pi(G)| \leqslant 2$ this is no restriction, since then each Sylow subgroup is simultaneously a Sylow complement. The class of groups in which every representative set of Sylow subgroups yields a Sylow basis has been studied by B. Huppert [46]; he shows it comprises those groups which are epimorphic images of direct products $G_1 \times G_2 \times \ldots \times G_m$ of groups G_i with $|\pi(G_i)| \leqslant 2$.

The set of all Hall systems of a group G is clearly invariant under the action of $\text{Aut}(G)$, the full group of automorphisms of G, and in [34] Hall shows that it forms a single orbit, even under the action of the inner automorphism group $\text{Inn}(G)$ of G; in other words, any two Hall systems can be transformed, one to another, by conjugating by a suitable element of G. Thus a transitive permutation representation of G itself arises from conjugation on the set of Hall systems of G. The stabilizer in this representation of a given Hall system Σ is called its *system normalizer* and is denoted by $N_G(\Sigma)$. The number of Hall systems of G is therefore $|G : N_G(\Sigma)|$. A group is nilpotent if and only if it has precisely one Hall system and then it coincides with the normalizer of this system. In [35] Hall shows that a system normalizer covers the central chief factors of a group and avoids the rest. (A chief factor H/K of G is *central* if $H/K \leqslant Z(G/K)$, and a subgroup D *covers* H/K if $H \leqslant DK$ and *avoids* it if $H \cap DK = K$.) It follows that a system normalizer of G is nilpotent and that its order is the product of the orders of the central chief factors in a given chief series of G. Hall's characterization of soluble groups suggests an intimate connection between the commutator or normal structure of a group on the one hand, and its Sylow structure on the other, a connection which is reinforced by the well known characterizations of nilpotence. A system normalizer connects these two aspects of structure in a soluble group and, at the same time, provides a measure of a group's nilpotence. Since G acts transitively by conjugation on its Hall systems, which form an $\text{Aut}(G)$-invariant set, the system normalizers form a characteristic conjugacy class of subgroups of G. Furthermore, the join of the system normalizers is G itself, and their intersection is $Z_\infty(G)$, the hypercentre of G. System normalizers are preserved under epimorphisms, but they do not persist in intermediate groups: J. Shamash [59] constructs a group G with a system normalizer D and a subgroup H containing D such that D normalizes no Hall system of H. Nevertheless, we shall see in section 2 that if a system normalizer D is self-normalizing, then D is a system normalizer of each intermediate group H, and furthermore, if the Hall system of G normalized by D *reduces* into H (i.e. intersects H in a Hall system of H), then D is at least contained in, if not necessarily equal to, a system normalizer of H. We shall have more to say on the way the reducibility of a Hall system can be used to control properties in the subgroup lattice of a soluble group in section 3.

2.   Projectors

In 1961 R. W. Carter published a striking theorem about the nilpotent subgroups of a finite soluble group.

THEOREM 2.1 (Carter [12])   *If G is a soluble group then*

(E) *G has a self-normalizing nilpotent subgroup, and*
(C) *if C_1 and C_2 are self-normalizing nilpotent subgroups of G, then $C_1 = C_2^g$ for some $g \in G$.*

We shall prove a more general version of this theorem shortly. Self-normalizing nilpotent subgroups are now called *Carter subgroups*. Their existence and conjugacy match the existence and conjugacy of self-idealizing nilpotent subalgebras in the classical theory of finite-dimensional Lie algebras. Cartan subalgebras play an important part in the classification of semi-simple Lie algebras. Carter's discovery aroused considerable interest, although it was clear from the start that it could not be extended in an obvious way to arbitrary finite groups. The alternating group A_5 shows that Carter subgroups need not exist in insoluble groups, and the symmetric group S_5 shows that possessing a single conjugacy class of Carter subgroups is not characteristic of soluble groups.

Carter subgroups may be seen as analogues (for the class of nilpotent groups) of Sylow p-subgroups (for the class of p-groups) and Hall π-subgroups (for the class of π-groups). All are maximal subgroups of their class, are preserved under epimorphisms, and satisfy an E-theorem (Existence) and a C-theorem (Conjugacy). One important component of the theorems of Sylow and Hall is missing from Carter's theorem, however, and that is a D-theorem (Dominance), viz.

(D) *Every p-subgroup of a group is contained in a Sylow p-subgroup; every π-subgroup of a soluble group is contained in a Hall π-subgroup.*

Some years were to elapse before the discovery of new conjugacy classes of generalized Sylow subgroups satisfying a D-theorem.

2.1 Covering subgroups

In 1963 Gaschütz published a far-reaching generalization of Hall and Carter subgroups. First we give some definitions and introduce notation.

Definitions. Let $\mathfrak{X}$ be a class of groups. A subgroup E of a group G is an $\mathfrak{X}$-*covering subgroup* of G if

(CS1) $E \in \mathfrak{X}$, and
(CS2) if $E \leqslant H \leqslant G$ and $K \triangleleft H$ with $H/K \in \mathfrak{X}$, then $H = EK$.

The set of $\mathfrak{X}$-covering subgroups of G is denoted by $\mathrm{Cov}_{\mathfrak{X}}(G)$.

The class of soluble groups will be denoted by $\mathfrak{S}$. If π is a set of prime numbers, then $\mathfrak{E}_\pi$ will denote the class of all finite π-groups, and $\mathfrak{S}_\pi = \mathfrak{S} \cap \mathfrak{E}_\pi$. If $\mathfrak{X}$ is a class of groups, then $\pi(\mathfrak{X})$ will be the set of all prime divisors of the orders of the groups in $\mathfrak{X}$; in particular, $\pi(G)$ is the set of primes dividing $|G|$.

A *formation* $\mathfrak{F}$ is a $\langle \mathrm{Q}, \mathrm{R}_0 \rangle$-closed class of groups. We note that a class $\mathfrak{X}$ is R_0-closed if and only if each group G has a smallest normal subgroup K such G/K is in $\mathfrak{X}$. This is called the $\mathfrak{X}$-*residual* and is denoted by $G^{\mathfrak{X}}$. Thus when $\mathfrak{X}$ is a formation, Property (CS2) is equivalent to the statement that E covers the $\mathfrak{X}$-residual of each intermediate group. The class $\mathfrak{A}$ of finite abelian groups is a formation (with the derived subgroup as its residual). A formation $\mathfrak{F}$ is *saturated* if $\mathfrak{F}$ contains G whenever it contains $G/\Phi(G)$. The class $\mathfrak{N}$ of finite nilpotent groups is the smallest saturated formation containing $\mathfrak{A}$. The class $\mathfrak{S}$ of all finite soluble groups is also a saturated formation.

THEOREM 2.2 (Gaschütz [24]) *Let $\mathfrak{F}$ be a formation of soluble groups. Every soluble group possesses an $\mathfrak{F}$-covering subgroup if and only if $\mathfrak{F}$ is saturated, and in this case the $\mathfrak{F}$-covering subgroups of a group form a conjugacy class of subgroups.*

It is clear from the definition that an $\mathfrak{X}$-covering subgroup E of a group G is also an $\mathfrak{X}$-covering subgroup of each intermediate group H.

Moreover, if $\mathfrak{X}$ is a formation, an $\mathfrak{X}$-covering subgroup of G is mapped by an epimorphism θ to an $\mathfrak{X}$-covering subgroup of $\theta(G)$. The following lemma, which lies at the heart of the inductive proof of Theorem 2.2 is a sort of converse of these two facts.

LEMMA 2.3 *Let $\mathfrak{F}$ be a formation and N a normal subgroup of a group G. If E is a subgroup of G such that*

(1) $EN/N \in \operatorname{Cov}_{\mathfrak{F}}(G/N)$ *and*
(2) $E \in \operatorname{Cov}_{\mathfrak{F}}(EN)$,

then $E \in \operatorname{Cov}_{\mathfrak{F}}(G)$.

Proof. Let $E \leqslant H \leqslant G$, and put $K = H^{\mathfrak{F}}$. We have to show that $EK = H$. If θ denotes the natural isomorphism from HN/N to $H/(H \cap N)$, its application to (1) gives $\theta(EN/N) \in \operatorname{Cov}_{\mathfrak{F}}(H/(H \cap N))$, where

$$\theta(EN/N) = (H \cap EN)/(H \cap N) = E(H \cap N)/(H \cap N).$$

Now $(H/(H \cap N))^{\mathfrak{F}} = K(H \cap N)/(H \cap N)$ since $\mathfrak{F}$ is a formation, and so by the definition of a covering subgroup we have

(2.α) $H = E(H \cap N)K(H \cap N) = E(H \cap N)K.$

Consequently $E(H \cap N)/(E(H \cap N) \cap K) \cong E(H \cap N)K/K = H/K \in \mathfrak{F}$, and because $E \in \operatorname{Cov}_{\mathfrak{F}}(E(H \cap N))$ by (2), the definition implies that $E(E(H \cap N) \cap K) = E(H \cap N)$. On substituting in (2.$\alpha$), we obtain $H = E(E(H \cap N) \cap K)K = EK$, as required.

To prove Theorem 2.2, first suppose that $\mathfrak{F}$ is a saturated formation, let G be a soluble group, and assume inductively that every group of order less than $|G|$ has a unique conjugacy class of $\mathfrak{F}$-covering subgroups. Let N be a minimal normal subgroup of G, and let $E^*/N \in \operatorname{Cov}_{\mathfrak{F}}(G/N)$. If $E^* < G$, induction yields an $E \in \operatorname{Cov}_{\mathfrak{F}}(E^*)$. Since $E^*/N \in \mathfrak{F}$, we have $E^* = EN$, and therefore $E \in \operatorname{Cov}_{\mathfrak{F}}(G)$ by Lemma 2.3. Now suppose that $E^* = G$, so that $G/N \in \mathfrak{F}$. If $G \in \mathfrak{F}$, then $G \in \operatorname{Cov}_{\mathfrak{F}}(G)$. If $G \notin \mathfrak{F}$, then $N \not\leqslant \Phi(G)$ because $\mathfrak{F}$ is saturated (this is the only point in the proof where saturation is used), and in this case N is complemented in G by a maximal subgroup, E say. Since $E \cong G/N \in \mathfrak{F}$ and since $N = G^{\mathfrak{F}}$, it is clear that E fulfils the requirements of an $\mathfrak{F}$-covering subgroup of G.

Thus we have shown that $\operatorname{Cov}_{\mathfrak{F}}(G)$ is non-empty, and it remains to prove that it forms a single conjugacy class. Let $E, \bar{E} \in \operatorname{Cov}_{\mathfrak{F}}(G)$. For a minimal normal subgroup N of G we have $EN/N = (\bar{E}N/N)^{xN} = \bar{E}^x N/N$ for some x in G, and if $EN < G$, by induction $E = \bar{E}^{xy}$ for some $y \in EN$. Therefore suppose that $EN = G$. If $G \in \mathfrak{F}$, then $E = \bar{E} = G$. If not, then E is a maximal subgroup of G complementing N. If $\mathrm{Core}_G(E)$ $(= \bigcap_{g \in G} E^g) \neq 1$, we could rechoose N in $\mathrm{Core}_G(E)$, and the above induction argument would again yield the conjugacy of E and $\bar{E}$. On the other hand, if $\mathrm{Core}_G(E) = 1$, then G is primitive (having a faithful permutation representation on the cosets of the maximal subgroup E), and it is a well known property of a primitive soluble group that any two complements of its unique minimal normal subgroup are conjugate; in particular, the complements E and $\bar{E}$ of N are conjugate in G.

Conversely, let G be a soluble group with $G/\Phi(G)$ in $\mathfrak{F}$ and assume G has an $\mathfrak{F}$-covering subgroup, E say. By definition E covers $G/\Phi(G)$, and so $E = G$. Hence G is in $\mathfrak{F}$ and $\mathfrak{F}$ is saturated.

We now deduce

THEOREM 2.4   *If G is a soluble group, then $\mathrm{Cov}_\mathfrak{N}(G)$ is the set of Carter subgroups of G.*

Proof.   Let $E \in \mathrm{Cov}_\mathfrak{N}(G)$, and $x \in N_G(E)$. If $H = E\langle x \rangle$, then $H/E \in \mathfrak{N}$, and consequently $H = EE = E$. Hence $E = N_G(E)$, and therefore E is a Carter subgroup of G.

Conversely, let C be a Carter subgroup of G. We show by induction on $|G|$ that $C \in \mathrm{Cov}_\mathfrak{N}(G)$. Let $C \leqslant H \leqslant G$, and put $K = H^\mathfrak{N}$. We suppose that $CK < H$ and derive a contradiction. Since $H/K \in \mathfrak{N}$, there is an element g in $N_H(CK) \setminus CK$. By induction C, $C^g \in \mathrm{Cov}_\mathfrak{N}(CK)$, and therefore $C = C^{gx}$ for some $x \in CK$ by Theorem 2.2. Hence $gx \in N_G(C) = C$, and $g \in Cx^{-1} \subseteq CK$, which is the contradiction we sought. Thus $CK = H$, and so C is an $\mathfrak{N}$-covering subgroup of G.

Clearly Theorems 2.2 and 2.4 imply Theorem 2.1, and thus the theory of $\mathfrak{F}$-covering subgroups includes that of Carter subgroups. It also includes Hall subgroups of soluble groups, for the E-, C- and D-properties of Hall π-subgroups show that they are just the $\mathfrak{S}_\pi$-covering subgroups.

As we shall see below, the class $\mathfrak{U}$ of supersoluble groups is also a saturated formation and therefore Theorem 2.2 applies to $\mathfrak{U}$. Gaschütz discovered an interesting characterization of $\mathfrak{U}$-covering subgroups.

THEOREM 2.5 (Gaschütz [24])   *A subgroup E of a soluble group G is a $\mathfrak{U}$-covering subgroup if and only if it satisfies Property 2.6.*

PROPERTY 2.6   *Whenever* $1 \leqslant D <_{\max} E \leqslant R <_{\max} S \leqslant G$, *then*

(1) $|E:D|$ *is a prime, and*
(2) $|S:R|$ *is composite.*

2.2   Saturated formations

Gaschütz demonstrated the considerable scope of Theorem 2.2 with the following method.

Definitions

(a) If H/K is a chief factor of a group G, then G acts on H/K by conjugation. This action yields a homomorphism from G to Aut(H/K), whose image is denoted by Aut$_G(H/K)$, and by the first isomorphism theorem Aut$_G(H/K) \cong G/C_G(H/K)$.

(b) A *formation function f* is a map from the set of prime numbers to classes of groups such that $f(p)$ is either empty or a formation for all p.

(c) If f is a formation function, we define an associated class $LF(f)$ as the class of all finite groups G satisfying

(2.β)   Aut$_G(H/K) \in f(p)$ *whenever* H/K *is a chief factor of* G *and* $p \mid |H/K|$.

If $f(p) = \emptyset$, the interpretation of this condition is that $p \nmid |G|$. The class $LF(f)$ is called the *local formation defined by f*. A formation is called *local* if it has the form $LF(f)$.

THEOREM 2.7 (Gaschütz [24])   *If f is a formation function, then $LF(f)$ is a saturated formation.*

Proof.   Let $\mathcal{H}(G)$ denote the set of pairs $(H/K, \text{Aut}_G(H/K))$ where H/K runs through the chief factors of G. If $N, N_1, \ldots, N_t$ are normal subgroups of G, then up to isomorphism we have $\mathcal{H}(G/N) \subseteq \mathcal{H}(G)$, and $\mathcal{H}(G/(\cap_{i=1}^{t} N_i)) \subseteq \cup_{i=1}^{t} \mathcal{H}(G/N_i)$. Therefore the class $LF(f)$ is a formation.

To see that it is saturated, one needs the well known fact that $O_{p'p}(G)$ is the intersection of the centralizers of those chief factors whose orders are divisible by p. Hence $X \in LF(f)$ implies

$$X/O_{p'p}(X) \in \text{R}_0 f(p) = f(p)$$

for all p. Consequently, if $G/\Phi(G) \in LF(f)$, then

$$(G/\Phi(G))/O_{p'p}(G/\Phi(G)) \in f(p),$$

and because $O_{p'p}(G/\Phi(G)) = O_{p'p}(G)/\Phi(G)$, it follows that $G/O_{p'p}(G) \in f(p)$ for all primes p. Thus, if H/K is a chief factor of G and $p \in \pi(H/K)$, then $\mathrm{Aut}_G(H/K) \cong G/C_G(H/K) \in \mathsf{Q}(G/O_{p'p}(G)) \subseteq \mathsf{Q}f(p) = f(p)$. Hence $G \in LF(f)$, as required.

Examples 2.8

(a) Let $\pi \subseteq \mathbb{P}$, and set $f(p) = \mathfrak{S}$ if $p \in \pi$ and $f(p) = \emptyset$ if $p \in \pi'$. Then $LF(f)$ is the class $\mathfrak{S}_\pi$ of soluble π-groups.

(b) If $f(p) = (1)$ for all $p \in \mathbb{P}$, then $LF(f) = \mathfrak{N}$.

(c) If $f(p)$ is the class of abelian groups of exponent dividing $p-1$, then $LF(f) = \mathfrak{U}$, the class of supersoluble groups.

(d) If $f(p) = \mathfrak{N}^{r-1}$ for all $p \in \mathbb{P}$, then $LF(f) = \mathfrak{N}^r$, the class of groups of nilpotent length at most r.

(e) If $f(p) = \mathfrak{S}_{p'}$ and $f(q) = \mathfrak{S}$ for all $q \in \mathbb{P} \setminus \{p\}$, then $LF(f)$ is the class of groups of p-length 1.

Gaschütz's procedure can obviously be used to generate saturated formations of ever-increasing complexity. If $\mathscr{F}_r$ denotes the saturated formations constructed at stage r, then at stage $r+1$ we put into $\mathscr{F}_{r+1}$ all classes $LF(f)$ with the $f(p)$'s chosen from $\mathscr{F}_r$ (cf. section 3).

In 1963 U. Lubeseder [53] proved the converse of Theorem 2.7 for soluble groups. Thus we have

THEOREM 2.9 (Gaschütz and Lubeseder, 1963) *A formation of soluble groups is saturated if and only if it is local.*

Lubeseder's proof requires elementary ideas from the theory of modular representations, which are dispensed with in the account of the theorem in Huppert's book [47]. In 1978 Schmid [57] showed that solubility is not necessary for Lubeseder's result, although his proof reinstates the facts about blocks used by Lubeseder and also makes essential use of a theorem of Gaschütz, about the existence of certain non-split extensions. We sketch the outlines of this proof.

Let $\mathfrak{F}$ be a saturated formation. The following two elementary facts will be used repeatedly in the proof.

(a) Let A be a group of operators for a group B, and write $\Phi_A(B)$ for the intersection of the maximal A-subgroups of B (so that if A is empty, then $\Phi_A(B)$ is merely $\Phi(B)$). Suppose that L is a normal A-subgroup of

B contained in $\Phi_A(B)$. Then because $\mathfrak{F}$ is saturated, $\mathfrak{F}$ contains the semi-direct product $B \rtimes A$ whenever it contains $(B/L) \rtimes A$ ($\cong (B \rtimes A)/L$).

(b) If R/S is an abelian chief factor of a group $G \in \mathfrak{F}$, then $(R/S) \rtimes (G/K) \in \mathfrak{F}$ for any $K \lhd G$ such that $K \leqslant C_G(R/S)$. (This is a general property of formations—see D. W. Barnes and O. H. Kegel [3].)

Step 1. Let X be a group, and let M, N be $\mathbb{F}_p X$-modules with N irreducible and X acting faithfully on M. If $M \rtimes X \in \mathfrak{F}$, then $N \rtimes X \in \mathfrak{F}$.

By a theorem of R. Steinberg [61] N appears as a quotient of the tensor power $M^{\otimes r} = M \otimes \ldots \otimes M$ (r copies) for some r. A construction due to B. Hartley (see [42]) yields a p-group P, admitting X as a group of operators, with $P/\Phi(P) \cong M^{(r)} = M \oplus \ldots \oplus M$ (r copies) and containing an X-invariant abelian normal subgroup L, isomorphic with $M^{\otimes r}$ as an $\mathbb{F}_p X$-module. If $G = P \rtimes X$, then $G/\Phi(P) \cong M^{(r)} \rtimes X \in \mathrm{R}_0 \mathfrak{F} = \mathfrak{F}$. Hence $G \in \mathfrak{F}$ by (a), and it follows from (b) that $N \rtimes X \in \mathfrak{F}$.

Step 2.   If a prime p divides the order of a group G in $\mathfrak{F}$, then $\mathfrak{F}$ contains a cyclic group of order p.

If $p \mid |G|$, then by a theorem of Gaschütz [23] there is a group H containing a minimal normal p-subgroup M satisfying (i) $M \leqslant \Phi(H)$ and (ii) $H/M \cong G$. If $G \in \mathfrak{F}$, then $H \in \mathfrak{F}$ by saturation, and putting $X = H/C_H(M)$, we have $M \rtimes X \in \mathfrak{F}$ by (b). The conclusion of Step 2 follows by taking the trivial $\mathbb{F}_p X$-module for N in Step 1.

Step 3 (P. Förster, unpublished).   *Let $p \in \mathbb{P}$, $G \in \mathfrak{F}$, and let M and N be non-zero $\mathbb{F}_p G$-modules all of whose composition factors are in the same block B of $\mathbb{F}_p G$. If $N \rtimes G \in \mathfrak{F}$, then $M \rtimes G \in \mathfrak{F}$.*

Since $\mathfrak{F} = \mathrm{Q}\mathfrak{F}$, nothing is lost by supposing that N is irreducible. Let P be the projective indecomposable $\mathbb{F}_p G$-module with $P/\Phi(P) \cong N$. Then $P \rtimes G \in \mathfrak{F}$ by (a). Let W be a composition factor of P, and note that $W \rtimes G \in \mathfrak{F}$ by (b). Let P^* be a projective indecomposable module which contains W as a composition factor, and let K be a submodule of P^*, minimal subject to W in $\mathrm{Q}(K)$. Then clearly $K/\Phi(K) \cong W$, and so $K \rtimes G \in \mathfrak{F}$ by (a). If S is the socle of P^*, then S is simple and $S \leqslant K$, whence $S \rtimes G \in \mathfrak{F}$ by (b). Now $P^*/\Phi(P^*) \cong S$, and therefore $P^* \rtimes G \in \mathfrak{F}$ by (a) once more. But the projective indecomposable modules of a block are connected by common composition factors, and consequently we have shown that $Q \rtimes G \in \mathfrak{F}$ for all projective indecomposable modules Q

in B. Finally, since M is an epimorphic image of a direct sum of such modules Q, we conclude that $M \lhd G \in \mathrm{QR}_0 \mathfrak{F} = \mathfrak{F}$, as claimed.

Step 4. Let $G \in \mathfrak{F}$, and let p divide $|G|$. If N is an irreducible module over $\mathbb{F}_p$ for the group $X = G/O_{p'p}(G)$, then $N \lhd X \in \mathfrak{F}$.

Let M be the direct sum of the irreducible modules in the first block B_1 of $\mathbb{F}_p G$. Then $\mathrm{Ker}(G \text{ on } M) = O_{p'p}(G)$ by a theorem of Brauer (see G. O. Michler [55], Theorem 2) and so M may be viewed as a faithful X-module. Let T be the trivial irreducible $\mathbb{F}_p G$-module. Then $T \lhd G = T \times G \in \mathfrak{F}$ by Step 2, and so an application of Step 3 to B_1 (with T as N) yields $M \lhd G \in \mathfrak{F}$. Consequently $M \lhd X \cong M \lhd (G/\mathrm{Ker}(G \text{ on } M)) \in \mathrm{Q}\mathfrak{F} = \mathfrak{F}$, and from Step 1 we conclude that $N \lhd X \in \mathfrak{F}$.

To complete the proof of Theorem 2.9, let $\pi = \pi(\mathfrak{F})$ (p. 17) and set

$$f(p) = \begin{cases} \mathrm{Q}(G \in \mathfrak{F} \mid O_{p'}(G) = 1) & \text{if } p \in \pi, \text{ and} \\ \emptyset & \text{if } p \notin \pi. \end{cases}$$

It can be verified that f is a formation function and that $\mathfrak{F} \subseteq LF(f)$. To prove the reverse inclusion, one first verifies from Step 4 that $f(p) = \mathfrak{S}_p f(p)$ (the class of $\mathfrak{S}_p$-by-$f(p)$ groups) for all $p \in \pi$. It then follows that $LF(f)$ consists of the π-groups in the class

$$\bigcap_{p \in \pi} \mathfrak{E}_{p'} \mathfrak{S}_p f(p) = \bigcap_{p \in \pi} \mathfrak{E}_{p'} f(p) \subseteq \bigcap_{p \in \pi} \mathfrak{E}_{p'} \mathfrak{F} = \mathfrak{F}.$$

This completes the proof.

In an unpublished manuscript R. Baer has explored a different definition of local formation. It is more flexible than the one discussed above in that the simple components, rather than the primes dividing its order, are used to label a chief factor and its permitted automorphism group. Hence the actions allowed on the insoluble chief factors can be independent of those on the abelian chief factors.

Definitions. Let $\mathfrak{K}$ denote the class of (finite) simple groups.

(a) A map $f: \mathfrak{K} \to \{\text{classes of groups}\}$ is called a *Baer function* provided that $f(J)$ is a formation whenever J is abelian and $f(J)$ is non-empty. (Thus no restrictions are placed on $f(J)$ when $J \notin \mathfrak{A}$.)

(b) If f is a Baer function, let $BLF(f)$ denote the class of all finite groups G whose chief factors H/K satisfy $\mathrm{Aut}_G(H/K) \in f(J)$, where J is a composition factor of H/K. The class $BLF(f)$ is called the *Baer local formation defined by f.*

If f is a formation function in the sense of Definition (b) p. 20, and if we set

$$g(J) = \begin{cases} f(J) \text{ if } J \in \mathfrak{A}, \\ \bigcap_{p \,|\, |J|} f(p) \text{ if } J \in \mathfrak{R} \setminus \mathfrak{A}, \end{cases}$$

then it is easy to verify that $LF(f) = BLF(g)$. Thus local formations in the sense of Definition (c) p. 20 are a special case of Baer local formations. Moreover, in the universe of soluble groups the two definitions evidently coincide. The price to be paid for the greater generality of Baer's approach is that the Baer local formations are no longer saturated. However, there is a suitable substitute for saturation.

Definition.   A class $\mathfrak{F}$ is *solubly-saturated* if the condition $G/\Phi(N) \in \mathfrak{F}$ for a soluble normal subgroup N of G implies that $G \in \mathfrak{F}$.

As one would expect, this concept is weaker than saturation. But it evidently coincides with saturation for classes of finite soluble groups, and it plays a precisely analogous role in Baer's generalization.

THEOREM 2.10 (Baer)   *A formation is solubly-saturated if and only if it is a Baer local formation.*

Although the soluble version of Theorem 2.9 is included in Baer's theorem, it is not clear that the theorem itself can be deduced from it.

2.3   Schunck classes

We turn to the question of which classes $\mathfrak{X}$ give rise to $\mathfrak{X}$-covering subgroups. To describe these classes we need the class $\mathfrak{P}$ of primitive soluble groups: these are the groups which possess a core-free maximal subgroup. They have a unique minimal normal subgroup, and this is complemented and coincides with the Fitting subgroup. Equivalently they are groups expressible as semi-direct products $N {\rtimes} H$ of an $\mathbb{F}_p H$-module N by H, where N is irreducible and faithful for H.

Definition.   A *Schunck class* is a class $\mathfrak{H}$ of soluble groups with the property that

$$\mathsf{Q}(G) \cap \mathfrak{P} \subseteq \mathfrak{H} \Leftrightarrow G \in \mathfrak{H}.$$

Thus a Schunck class must contain the group G whenever it contains all the primitive epimorphic images of G.

It is not hard to see that a Schunck class is Q-closed, D_0-closed, i.e.

closed under forming finite direct products, and saturated, and that every saturated formation is a Schunck class. Simple examples of Schunck classes which are not saturated formations are the classes $\mathfrak{Q}^\pi$ of π-perfect groups, comprising all soluble groups G with $G = O^\pi(G)$.

THEOREM 2.11 (H. Schunck [58])   *Every soluble group G has $\mathfrak{X}$-covering subgroups if and only if $\mathfrak{X}$ is a Schunck class. In this case* $\mathrm{Cov}_\mathfrak{X}(G)$ *is a conjugacy class of G.*

This theorem was published in 1967 in the same year that the basic ideas for a dual theory of Fitting classes appeared in print (see section 3). Covering subgroups underwent a further mutation two years later, when Gaschütz [27] presented Theorem 2.11 in a generalized form, basing his treatment on the following idea.

Definition.   Let $\mathfrak{X}$ be a class of groups. An $\mathfrak{X}$-projector X of a group G is a subgroup with the property that $\theta(X)$ is a maximal $\mathfrak{X}$-subgroup of $\theta(G)$ for all homomorphisms θ of G.

It is easy to check that X is an $\mathfrak{X}$-covering subgroup of G if and only if X is an $\mathfrak{X}$-projector of H for all $X \leqslant H \leqslant G$. Gaschütz proves Theorem 2.11 for $\mathfrak{X}$-projectors as well as for $\mathfrak{X}$-covering subgroups. He also shows for a Schunck class $\mathfrak{X}$ that the $\mathfrak{X}$-projectors coincide with the $\mathfrak{X}$-covering subgroups. The advantage of projectors over covering subgroups, apart from their greater generality, lies in the fact that they possess good duals (namely the injectors (cf. section 3)).

Let $\mathfrak{F}$ be a saturated formation and set $\pi = \pi(\mathfrak{F})$. Then by Step 2 in the proof of Theorem 2.9,

$$(2.\gamma) \qquad\qquad \mathfrak{N} \cap \mathfrak{E}_\pi \subseteq \mathfrak{F} \subseteq \mathfrak{E}_\pi.$$

Therefore, if G is a group in which every $\mathfrak{F}$-subgroup is contained in some $\mathfrak{F}$-projector of G, then each $\mathfrak{F}$-projector must contain a Sylow p-subgroup of G for each p in π and must therefore be a Hall π-subgroup of G.

Definition.   A Schunck class $\mathfrak{H}$ is called a *D-class* if the $\mathfrak{H}$-projectors satisfy a *D*-theorem, i.e. if in each soluble group each $\mathfrak{H}$-subgroup is contained in some $\mathfrak{H}$-projector of G.

The preceding remarks imply that the classes $\mathfrak{S}_\pi$ are the only *D*-classes among saturated formations. However, for Schunck classes which are not formations the situation is quite different. Observe that $(2.\gamma)$ need no longer hold. An example is $\mathfrak{F} = \mathfrak{Q}^p$, the class of p-perfect soluble groups: certainly $p \in \pi = \pi(\mathfrak{Q}^p)$, whence $\mathfrak{S}_p \subseteq \mathfrak{N} \cap \mathfrak{E}_{\pi}$, but $\mathfrak{S}_p \cap \mathfrak{Q}^p = (1)$. The class $\mathfrak{Q}^p$ is a D-class; for the unique $\mathfrak{Q}^p$-projector of a group G is $O^p(G)$, and since $G/O^p(G) \in \mathfrak{S}_p$, we have $HO^p(G)/O^p(G) = 1$ for any p-perfect subgroup H of G, in other words $H \leqslant O^p(G)$.

That most useful tool for the study of saturated formations, the local definition, is no longer available for general Schunck classes. The following concept, due to Doerk [15], is a valuable substitute.

Definition.   Let $\mathfrak{H}$ be a Schunck class. The *boundary* $b(\mathfrak{H})$ of $\mathfrak{H}$ consists of all soluble groups G satisfying

(1) $G \notin \mathfrak{H}$, and
(2) $G/N \in \mathfrak{H}$ for all $1 \neq N \lhd G$.

Note that groups in $b(\mathfrak{H})$ are non-trivial by (1). It is easy to see that $b(\mathfrak{H})$ consists of all primitive groups in which the socle is complemented by an $\mathfrak{H}$-projector. In particular, on setting $\mathfrak{B} = b(\mathfrak{H})$, we have:

(BDY1) $\mathfrak{B} \subseteq \mathfrak{P}$ (the class of primitive groups), and
(BDY2)   no group in $\mathfrak{B}$ is a proper epimorphic image of another group in $\mathfrak{B}$.

It can be shown that any class $\mathfrak{B}$ of soluble groups satisfying (BDY1) and (BDY2) is the boundary of some Schunck class $\mathfrak{H}$, and $\mathfrak{H}$ is characterized as the class of all soluble $\mathfrak{B}$-perfect groups (i.e. groups with no epimorphic images in $\mathfrak{B}$). For example, the boundary of $\mathfrak{Q}^p$ consists of the cyclic groups of order p. There is a bijection between the set of Schunck classes and the set of Schunck boundaries, and a question about Schunck classes can sometimes be resolved more easily by translating it into a question about their boundaries. As an illustration, we cite the following criterion for a Schunck class to be a D-class.

THEOREM 2.12 (Förster [22])   *Let $\mathfrak{H}$ be a Schunck class. Then $\mathfrak{H}$ is a D-class if and only if each group G in $b(\mathfrak{H})$ satisfies the following condition:*

$(2.\delta)$ *Denote by H a complement in the primitive group G to the socle N. If L is an $\mathfrak{H}$-subgroup of H, and if U is a composition factor of N, viewed as an L-module, then $B = U \rtimes (L/C_L(U))$ is in $b(\mathfrak{H})$.*

It is easy to construct examples of D-classes with the help of this criterion. For example, let $\pi \subseteq \mathbb{P}$, and let $\mathfrak{F}$ be an $\langle s, q \rangle$-closed class contained in $\mathfrak{S}_{\pi'}$. Then the class $\mathfrak{B}$ of all primitive groups B such that

(1) $B/\mathrm{Soc}(B) \in \mathfrak{F}$, and
(2) $\mathrm{Soc}(B) \in \mathfrak{S}_{\pi}$

is obviously a Schunck boundary satisfying $(2.\delta)$, and so the Schunck class of $\mathfrak{B}$-perfect groups is a D-class. If $\pi = \{p\}$ and $\mathfrak{F} = (1)$, we obtain the class $\mathfrak{Q}^p$ by this means.

Proof of Theorem 2.12.   First suppose that $\mathfrak{H}$ is a D-class. Let $G \in b(\mathfrak{H})$, and let N, L, H and B be as in $(2.\delta)$. Then the D-property implies that every $\mathfrak{H}$-subgroup of G has trivial intersection with N and hence that L is $\mathfrak{H}$-maximal in LN. Therefore by a theorem of Gaschütz (see Lemma II.11 of [27]), L is an $\mathfrak{H}$-projector of LN, and since B is isomorphic with a quotient of a subgroup of LN containing L, it follows that $L/C_L(U)$ is an $\mathfrak{H}$-projector of B. Because B is primitive with socle U, we conclude that $B \in b(\mathfrak{H})$.

Conversely, let $\mathfrak{H}$ be a Schunck class whose boundary groups all satisfy $(2.\delta)$. We suppose that $\mathfrak{H}$ is not a D-class and derive a contradiction. Then there exists a group G of smallest order having an $\mathfrak{H}$-subgroup X not contained in any $\mathfrak{H}$-projector of G. Let N be a minimal normal subgroup of G. Then $XN/N \in \mathfrak{H}$, and by the choice of G there is an $\mathfrak{H}$-projector H^*/N of G/N such that $XN \leqslant H^*$. Furthermore, $H^* = HN$ for some $H \in \mathrm{Proj}_{\mathfrak{H}}(G)$. If $HN < G$, then X is contained in some $\mathfrak{H}$-projector of HN and this is a conjugate of H because $H \in \mathrm{Proj}_{\mathfrak{H}}(HN)$. This contradiction shows that $HN = G$ for all minimal normal subgroups N of G. Since clearly $H \neq G$, it follows that G is a primitive group with its socle N complemented in G by H, and therefore $G \in b(\mathfrak{H})$.

If θ denotes the natural isomorphism from H to HN/N, let $L = \theta^{-1}(XN/N)$. Then $LN = XN$, and L is an $\mathfrak{H}$-subgroup of the $\mathfrak{H}$-projector H of G. We claim the Condition $(2.\delta)$ implies that L is $\mathfrak{H}$-maximal in LN. If $L \leqslant K \in \mathfrak{H}$, then $K = L(K \cap N)$, and we assert that $K \cap N = 1$. If not, there is an L-composition factor U of N of the form $U = (K \cap N)/T$. But then $U \rtimes C_L(U) \cong K/(TC_L(U)) \in \mathsf{Q}\mathfrak{H} = \mathfrak{H}$, which contradicts $(2.\delta)$ because $\mathfrak{H} \cap b(\mathfrak{H}) = \emptyset$. Therefore $K \cap N = 1$, and L is $\mathfrak{H}$-maximal in LN, as claimed. By the theorem of Gaschütz cited earlier in the proof it follows that L is an $\mathfrak{H}$-projector of LN. If $L < H$, then $LN < G$, and the minimal choice of G implies that X is contained in some conjugate of L and hence in some conjugate of H. Therefore $L = H$, and $HN = XN = G$. Consequently H and X are complements in G of its socle N, and are therefore conjugate in G by a well known property of primitive groups. This final contradiction completes the proof.

Definition. If $\mathfrak{H}$ and $\mathfrak{K}$ are Schunck classes, their *join* (or *composition*) $\mathfrak{H} \vee \mathfrak{K}$ is defined thus:

$$\mathfrak{H} \vee \mathfrak{K} = (G \in \mathfrak{S} \mid G = \langle H, K \rangle \text{ whenever } H \in \mathrm{Proj}_{\mathfrak{H}}(G) \text{ and } K \in \mathrm{Proj}_{\mathfrak{K}}(G)).$$

THEOREM 2.13 (Gaschütz [27], Chapter VII, Theorem 12) *Let $\mathfrak{H}$ and $\mathfrak{K}$ be Schunck classes, let G be a soluble group, and let H and K be $\mathfrak{H}$- and $\mathfrak{K}$-projectors of G chosen so as to minimize the order of $\langle H, K \rangle$. Then $\langle H, K \rangle$ is an $(\mathfrak{H} \vee \mathfrak{K})$-projector of G. In particular (cf. Theorem 2.11), $\mathfrak{H} \vee \mathfrak{K}$ is a Schunck class.*

It turns out that if $\mathfrak{H}$ and $\mathfrak{K}$ are D-classes, then so is $\mathfrak{H} \vee \mathfrak{K}$, and that the set of all D-classes forms a lattice with respect to the operators $\vee$ and $\cap$. G. J. Wood [67] was the first to investigate this lattice. He showed *inter alia* that it is complemented but not modular. A more recent discussion of this lattice and related notions appears in §9 of Förster [22].

If $\mathfrak{H}$ and $\mathfrak{K}$ are D-classes and if $\mathfrak{H} \subseteq \mathfrak{K}$, then each $\mathfrak{H}$-projector of a soluble group is contained in some $\mathfrak{K}$-projector of it. This is not the case for Schunck classes in general; for example, nilpotent groups form a subclass of supersoluble groups, but in the symmetric group S_4 the $\mathfrak{N}$-projectors have order 8 whereas the $\mathfrak{U}$-projectors have order 6. Thus the partial order of class inclusion in general yields no information about inclusions between corresponding projectors. However, inclusion between projectors (up to conjugacy) can be used to define another partial order, denoted by $<<$, on the set of Schunck classes, as follows:

$$\mathfrak{H} << \mathfrak{K} \text{ if and only if } \mathrm{Proj}_{\mathfrak{H}}(G) \lesssim \mathrm{Proj}_{\mathfrak{K}}(G) \text{ for all } G \in \mathfrak{S}.$$

(If $\mathscr{C}_1$ and $\mathscr{C}_2$ are conjugacy classes of subgroups, the notation $\mathscr{C}_1 \lesssim \mathscr{C}_2$ means that there exist $C_1 \in \mathscr{C}_1$ and $C_2 \in \mathscr{C}_2$ with $C_1 \leqslant C_2$.) The extension of the definition of composition to an arbitrary set of Schunck classes yields a supremum on $<<$, and so $<<$ induces on the set $\mathscr{H}$ of all Schunck classes the structure of a complete lattice, in which the set $\mathscr{D}$ of all D-classes forms a sublattice. Although the definition of $\vee$ is the same as in $\mathscr{D}$, the definition of $\wedge$ in $\mathscr{H}$ is more complicated than $\cap$. The structure of $(\mathscr{H}, \vee, \wedge)$ is reflected in the inclusion relationships between the corresponding conjugacy classes of projectors in a given group and thus yields structural information about the group. The boundary concept has proved helpful in analysing this lattice (cf. Doerk [15], Förster [22] and Hawkes [41]).

2.4 Subgroups defined by numbers

Interest in the theory of projectors has recently centred round their arithmetical properties. In our discussion we follow Gaschütz.

Let $\mathbb{P}^* = \{p^n \mid p \in \mathbb{P}, n \geq 1\}$, the set of all prime powers. Given $\Omega \subseteq \mathbb{P}^*$, let $\mathfrak{H}_\Omega$ denote the class of all finite soluble groups H with the following property:

$$(\Omega 1) \qquad\qquad M <_{\max} H \Rightarrow |H:M| \in \Omega.$$

There is a well known bijection $\{M^H\} \to H/\mathrm{Core}_H(M)$ from the conjugacy classes of maximal subgroups M of H to the primitive epimorphic images of H, and so $(\Omega 1)$ is equivalent to the condition that each primitive epimorphic image of H should belong to the class $\mathfrak{P}_\Omega$ of primitive groups of degrees belonging to Ω; this is because $|H:M|$ is the degree of $H/\mathrm{Core}_H(M)$ as a primitive permutation group, or equivalently, the order of its socle. Therefore $\mathfrak{H}_\Omega$ is evidently a Schunck class; for example, when $\Omega = \mathbb{P}$, the class $\mathfrak{H}_\Omega$ coincides with the class $\mathfrak{U}$ of supersoluble groups, which is even a saturated formation.

For a given set of prime powers Ω we now describe a recursive procedure for constructing a set $\Omega(G)$ of subgroups of a finite soluble group G: If $G \in \mathfrak{H}_\Omega$, set $\Omega(G) = \{G\}$. If not, let M_1 be a maximal subgroup of G chosen to have index as small as possible in $\mathbb{P}^* \setminus \Omega$. If $(\Omega 1)$ is satisfied with $H = M_1$, in other words if $M_1 \in \mathfrak{H}_\Omega$, then put M_1 into $\Omega(G)$. If not, let M_2 be a maximal subgroup of M_1 chosen in the same way as M_1, and continue in this way until a subgroup M_r is reached,

$$(2.\varepsilon) \qquad\qquad H = M_r <_{\max} \cdots <_{\max} M_1 <_{\max} G,$$

such that $H = M_r$ satisfies $(\Omega 1)$. The set $\Omega(G)$ is defined to comprise all terminal subgroups of such maximal chains.

THEOREM 2.14 (Gaschütz [28]) *For any $\Omega \subseteq \mathbb{P}^*$ and for any finite soluble group G, the set $\Omega(G)$ defined above is a conjugacy class of G.*

The reason for this surprising fact is that $\Omega(G)$ turns out to be precisely the set of $\mathfrak{H}_\Omega$-projectors of G. To understand why, suppose $G \notin \mathfrak{H}_\Omega$ and consider the first member $M_1 <_{\max} G$ of the chain $(2.\varepsilon)$. Put $T = \mathrm{Core}_G(M_1)$, and let R/T denote the socle of the primitive group G/T; further, let p^m be the common value of $|R/T| = |G:M_1|$. Since R/T can be regarded as a faithful irreducible module for M_1/T over $\mathbb{F}_p$, a theorem of Gaschütz ([25], Satz 4) implies that the degrees of the chief factors of M_1/T are less than p^m, and the choice of M_1 then implies that all the primitive epimorphic images of M_1/T ($\cong G/R$) have degrees in Ω. Thus $G/R \in \mathfrak{H}_\Omega$ and $G/T \in b(\mathfrak{H}_\Omega)$. If H is an $\mathfrak{H}_\Omega$-projector of G, then HT/T and M_1/T are complements of the socle of the primitive group G/T and as such are conjugate; in particular, M_1 contains a conjugate of H, which is an $\mathfrak{H}_\Omega$-projector of G, and also of M_1 by the persistence of projectors in intermediate groups. Similarly M_2 contains an $\mathfrak{H}_\Omega$-projector of M_1, which is also a conjugate of H. Continuing down the chain, we conclude that $H^x \leqslant M_r$ for some $x \in G$, and since $M_r \in \mathfrak{H}_\Omega$, it follows from the definition of a projector that $H^x = M_r \in \Omega(G)$.

The subgroups H in $\Omega(G)$ satisfy Condition $(\Omega 1)$ and also the chain condition $(2.\varepsilon)$ with $|M_{i-1}:M_i| \notin \Omega$ for $1 \leqslant i \leqslant r$. This last can be strengthened as follows:

$(\Omega 2)$ If $H \leqslant R <_{\max} S \leqslant G$, then $|S:R| \notin \Omega$.

Subgroups H of a group G which satisfy $(\Omega 1)$ and $(\Omega 2)$ were first discussed by Gaschütz in [25]; we shall call them *Gaschütz Ω-subgroups* of G. The arguments used above can be easily adapted to show that if H is a Gaschütz Ω-subgroup of G, then $H \in \Omega(G)$; thus, when it is non-empty, the set $\text{Gasch}_\Omega(G)$ of Gaschütz Ω-subgroups is a conjugacy class. The question of which sets Ω of prime powers guarantee the existence of Gaschütz Ω-subgroups in each soluble group is an interesting one and has yet to be satisfactorily answered. A representation-theoretic criterion is described in Hawkes [43], but no purely numerical necessary and sufficient conditions seem to be known. There are, however, many known examples of such sets Ω; one such is $\mathbb{P}$, for by Theorem 2.5 the supersoluble projectors are Gaschütz $\mathbb{P}$-subgroups. Further examples are discussed in 2.15 and 2.16 below.

Examples 2.15

(a) Let $\Omega = \mathbb{P}^* \setminus \{2\}$. Then $\text{Gasch}_\Omega(G) = \{O^2(G)\}$ for all finite groups G.

(b) Let M be a set of natural numbers coprime with a given prime p, and set
$$\Omega = \mathbb{P}^* \setminus \{p^n \mid n \notin M\}.$$

Then $\text{Gasch}_\Omega(G) \neq \emptyset$ for all soluble G. (See [43] for details).

Let us now strengthen Condition $(\Omega 2)$ still further:

$(\Omega 2^*)$ If $H \leqslant R < S \leqslant G$, then $|S:R| \notin \Omega$.

Subgroups H of G which satisfy the requirements of ($\Omega1$) and ($\Omega2^*$) are called *generalized Sylow Ω-subgroups* of G (Gaschütz [29]). It is clear that a Sylow Ω-subgroup is a Gaschütz Ω-subgroup, but Example 2.15(a) above shows that they do not always coincide; for example, a group of order 4 clearly has no Sylow ($\mathbb{P}^* \setminus \{2\}$)-subgroups. The following are examples of sets Ω for which $\mathrm{Syl}_\Omega(G)$ is non-empty (and therefore a conjugacy class) in each finite soluble group G.

Examples 2.16

(a) If $\pi \subseteq \mathbb{P}$ and $\Omega = \{p^n \mid p \in \pi, n \geqslant 1\}$, then the Sylow Ω-subgroups of G are its Hall π-subgroups. Thus the Gaschütz and Sylow Ω-subgroups are natural generalizations of Hall subgroups.

(b) If $\Omega = \mathbb{P}$, the supersoluble projectors are Sylow Ω-subgroups.

(c) (Gaschütz [25]). Let n be a natural number, and set $\Omega = \{q \in \mathbb{P}^* \mid q \leqslant n\}$. Then $\mathrm{Syl}_\Omega(G) \neq \varnothing$ for all soluble G (cf. Theorem 2.17 below). Thus for each natural number n a finite soluble group G has precisely one conjugacy class of subgroups H satisfying

$$1 \leqslant U <_{\max} H \leqslant R < S \leqslant G \Rightarrow |H\!:\!U| \leqslant n \text{ and } |S\!:\!R| > n.$$

An elegant criterion for the universal existence of Sylow Ω-subgroups is given by Gaschütz [27], Chapter III, Theorem 10 *et seq.* In stating it we say that a subset Ω of $\mathbb{P}^*$ is prime if its complement is multiplicatively closed.

THEOREM 2.17   *If Ω is a prime subset of $\mathbb{P}^*$ then every finite soluble group has a unique conjugacy class of Sylow Ω-subgroups. Conversely, if every finite soluble group has Sylow Ω-subgroups, then there exists a prime subset Γ of $\mathbb{P}^*$ such that the Sylow Γ- and Sylow Ω-subgroups of each soluble group coincide.*

The only Gaschütz Ω-subgroups which satisfy a D-theorem are the Hall subgroups and the Gaschütz ($\mathbb{P}^* \setminus \{2\}$)-subgroup $O^2(G)$ (see [44]). Replacing ($\Omega1$) by the condition

($\Omega1^*$) $1 \leqslant U <_{\max} V \leqslant H \Rightarrow |V\!:\!U| \in \Omega$

is equivalent to requiring that $\mathfrak{H}_\Omega$ should be subgroup-closed. The study of this condition led to an unexpected connection with work of H. Heineken [45]. If d denotes a function which associates with each prime p a set $d(p)$ of natural numbers, an associated formation $\mathfrak{F}_d$ can be defined to comprise all finite soluble groups with the property that for each prime p the ranks (i.e. $\mathbb{F}_p$-dimensions) of its p-chief factors belong to $d(p)$. If $\Omega(d) = \{p^n \,|\, p \in \mathbb{P}, n \in d(p)\}$, this is equivalent to asking that the orders of its chief factors belong to $\Omega(d)$. Heineken was interested in the question of when the formation $\mathfrak{F}_d$ is saturated, and when each $d(p)$ is non-empty he discovered a set of necessary conditions for this to be so; these are in the form of a list of arithmetical requirements to be satisfied by the sets $d(p)$, some of them quite complicated.

D. Harman shows that Heineken's necessary conditions are also sufficient for $\mathfrak{F}_d$ to be saturated. With the help of this fact he proves

THEOREM 2.18 (Harman [39])   *Assume that $d(p)$ is non-empty for all p in $\mathbb{P}$, and set $\Omega = \Omega(d)$. Then the following statements are equivalent:*

(1) $\mathfrak{H}_\Omega$ *is subgroup-closed;*

(2) $\mathfrak{H}_\Omega$ *is a formation;*

(3) $\mathfrak{H}_\Omega = \mathfrak{F}_d$;

(4) $\mathfrak{F}_d$ *is saturated.*

Moreover, Harman extends Heineken's work in two directions. He modifies Heineken's list of arithmetical conditions so as to remove the restriction $d(p) \neq \varnothing$; and he solves the analogous problem for the formations $\widetilde{\mathfrak{F}}_d$ consisting of all finite soluble groups G with the property that for all primes p the absolutely irreducible constituents of the p-chief factors of G, regarded as $\mathbb{F}_p G$-modules, have dimensions in $d(p)$.

3. Injectors

3.1 Fitting sets

For a group G and class of groups $\mathfrak{X}$ we write

$$G_\mathfrak{X} = \langle K \,|\, K \lhd\lhd G, K \in \mathfrak{X} \rangle;$$

this is the $\mathfrak{X}$-*radical* of G. If $\mathfrak{X} = \mathrm{N}_0 \mathfrak{X}$, then $G_\mathfrak{X}$ is the unique maximal normal $\mathfrak{X}$-subgroup; if $\mathfrak{X} = \langle \mathrm{s_n}, \mathrm{N_0} \rangle \mathfrak{X}$, then $K \cap G_\mathfrak{X} = K_\mathfrak{X}$ if $K \lhd\lhd G$. H. Fitting [21] showed that the class $\mathfrak{N}$ of nilpotent groups is $\mathrm{N_0}$-closed; $G_\mathfrak{N}$ is the Fitting subgroup $F(G)$. In [18] Fischer studied $\langle \mathrm{s_n}, \mathrm{N_0} \rangle$-

closed classes of soluble groups, which he called *Fitting classes*. There are abundant examples. The local formation $LF(f)$ is a Fitting class if each of the non-empty formations $f(p)$ is a Fitting class. Thus, starting from the trivial Fitting classes (1) and $\mathfrak{S}$, one can use this fact repeatedly to construct Fitting classes of increasing diversity, including the classes $\mathfrak{S}_\pi$ ($\pi \subseteq \mathbb{P}$), the class of groups of nilpotent length (or of p-length) at most n, the class of Sylow tower groups of a given complexion, and so on. The classes $(G \in \mathfrak{S} \,|\, \text{Soc}(G) \leqslant Z(G))$ and $(G \in \mathfrak{S} \,|\, O_p(G) \leqslant Z_\infty(G))$ are examples of Fitting classes which are not formations. The class of supersoluble groups is an example of a saturated formation which is not a Fitting class.

Fischer's aim was to see how far it was possible to dualize the theory of saturated formations and projectors by interchanging the roles of normal subgroups and quotient groups. From this point of view the closure operations s_n and N_0 are the natural duals of Q and R_0, and so a Fitting class should be regarded as the dual of a formation. In fact, it turns out that Fitting classes parallel Schunck classes more closely in the dual theory because they are precisely the classes for which a theory of injectors, dual to projectors, is possible. At the time of Fischer's initial investigation the projectors were still known by the 'covering-subgroup' definition (see p. 17) and by close analogy the dual concept chosen by Fischer, now called a Fischer subgroup, is defined thus: Let $\mathfrak{X}$ be a class of groups. A subgroup E of a group G is a *Fischer $\mathfrak{X}$-subgroup* if

(1) $E \in \mathfrak{X}$, and
(2) whenever $E \leqslant H \leqslant G$, then $H_\mathfrak{X} \leqslant E$.

Thus a Fischer $\mathfrak{X}$-subgroup belongs to $\mathfrak{X}$ and contains each $\mathfrak{X}$-subgroup that it normalizes. Fischer proved that if $\mathfrak{F}$ is a Fitting class, then every finite soluble group possesses Fischer $\mathfrak{F}$-subgroups. However, he was not able to prove that the Fischer subgroups of a soluble group are all conjugate. Some years later, indeed, R. S. Dark [14] gave an example of a Fitting class $\mathfrak{F}$ and a soluble group which has two conjugacy classes of Fischer $\mathfrak{F}$-subgroups.

As it turned out, the definition of projector, rather than covering-subgroup, is the right thing to dualize in order to guarantee conjugacy. Let $\mathfrak{X}$ be a class of groups. An *$\mathfrak{X}$-injector* of a group G is an $\mathfrak{X}$-subgroup E of G such that $E \cap K$ is $\mathfrak{X}$-maximal in K for all subnormal subgroups K of G. This was the course taken in 1967 by Fischer, Gaschütz, and Hartley in [19], where they prove the following result.

THEOREM 3.1 *Let $\mathfrak{F}$ be a Fitting class and G a finite soluble group. Then*

(E) G has $\mathfrak{F}$-injectors, and

(C) any two $\mathfrak{F}$-injectors of G are conjugate.

When $\mathfrak{F}$ is the Fitting class of π-groups, the $\mathfrak{F}$-injectors of a finite soluble group, like its $\mathfrak{F}$-projectors, turn out to be the Hall π-subgroups. This is the only situation in which the injectors and projectors coincide, and so the two theories are quite independent generalizations of the classical theory of Sylow and Hall subgroups. The question here of a D-theorem is easily resolved. Let $\mathfrak{F}$ be a Fitting class with the property that for every soluble group G each $\mathfrak{F}$-subgroup of G is contained in an $\mathfrak{F}$-injector. If $p \in \pi = \pi(\mathfrak{F})$, it is easy to show that $\mathfrak{F}$ contains a cyclic group of order p and hence contains all p-groups. Consequently an $\mathfrak{F}$-injector of G must contain a Sylow p-subgroup of G for each $p \in \pi$, and therefore $\mathfrak{F} = \mathfrak{S}_\pi$. In this respect Fitting classes are closer to saturated formations than to Schunck classes.

We shall prove a more general version of Theorem 3.1. The proof is no harder, but the rewards in terms of information about the subgroup lattice of the group are greater. This 'local' approach to injectors is due to W. Anderson [1].

Definition.   A *Fitting set* $\mathscr{F}$ of a group G is a set of subgroups of G satisfying

(FS1) if $F \in \mathscr{F}$ and $g \in G$, then $F^g \in \mathscr{F}$,

(FS2) if $K \vartriangleleft\!\vartriangleleft F \in \mathscr{F}$, then $K \in \mathscr{F}$, and

(FS3) if $F_1, F_2 \in \mathscr{F}$ and each normalizes the other, then $F_1 F_2 \in \mathscr{F}$.

Examples 3.2

(a) If $\mathfrak{F}$ is a Fitting class, then $\text{Trace}(\mathfrak{F}) = \{F \leqslant G \,|\, F \in \mathfrak{F}\}$ is a Fitting set of G.

(b) If $N \vartriangleleft G$, then $\{F \leqslant N \,|\, F \vartriangleleft\!\vartriangleleft N\}$ is a Fitting set of G.

(c) If $\mathscr{F}$ is a Fitting set of G and $H \leqslant G$, then $\mathscr{F}_H = \{F \,|\, F \in \mathscr{F}$ and $F \leqslant H\}$ is a Fitting set of H. (By convention, $\mathscr{F}_H$ is called the Fitting set $\mathscr{F}$ of H.)

The definitions of a radical and an injector relative to a Fitting set $\mathscr{F}$ are the same as for a Fitting class provided one observes the convention of Example (c) above and bears in mind that the $\mathscr{F}$-radical is no longer necessarily a characteristic subgroup. The key to the proof is

HARTLEY'S LEMMA *Let $\mathscr{F}$ be a Fitting set of a finite soluble group G, and K a normal subgroup of G with G/K nilpotent. Let W be an $\mathscr{F}$-maximal subgroup of K, and let V_1 and V_2 be $\mathscr{F}$-maximal subgroups of G which contain W.*

(a) *If $W \lhd K$, then there exists a Carter subgroup C_i of G such that $V_i = (WC_i)_\mathscr{F}$ for $i = 1, 2$;*
(b) *V_1 and V_2 are conjugate in $\langle V_1, V_2 \rangle$.*

Proof. (a) Write V for V_i. If $W \lhd K$, then $W = K_\mathscr{F}$, and therefore $W \lhd G$ by conditions FS1 and FS3. Since $V \cap K$ is normal in the $\mathscr{F}$-group V, it is an $\mathscr{F}$-subgroup of K; because it contains W and W is $\mathscr{F}$-maximal, $V \cap K = W$. Let $N = N_G(V)$. Since G/K is nilpotent, we have $[V,_r N] \leqslant V \cap K$ for suitably large r. Hence $V/W \leqslant Z_\infty(N/W)$. Let C^*/W be a Carter subgroup of N/W. Since Carter subgroups contain the hypercentre of a group, we have $V/W \leqslant C^*/W$, and hence $V \lhd\lhd C^*$. Thus $V \leqslant (C^*)_\mathscr{F}$, and the $\mathscr{F}$-maximality of V forces equality. Let $N^* = N_G(C^*)$. Since $V = (C^*)_\mathscr{F} \leqslant C^* \lhd N^*$, it follows that $N^* \leqslant N = N_G(V)$. Thus $N^* \leqslant N_N(C^*) = C^*$, since by definition C^*/W is self-normalizing in N/W. We conclude that $N^* = C^*$ and hence that C^*/W is a Carter subgroup of G/W. But because Carter subgroups are invariant under epimorphisms, we can write $C^* = WC$ for some Carter subgroup C of G, and so (a) is justified.

(b) Let $G^* = \langle V_1, V_2 \rangle$ and $K^* = K \cap G^*$. Clearly $G^*/K^* \in \mathfrak{N}$ and $K^* \cap V_i = K \cap V_i = W$; therefore $W \lhd \langle V_1, V_2 \rangle = G^*$. Let $i \in \{1, 2\}$. By (a) we have $V_i = (WC_i)_\mathscr{F}$ for a suitable Carter subgroup C_i of G^*. Now the conjugacy of Carter subgroups implies that $(WC_2)^x = WC_1$ for some $x \in G^*$, and hence that V_2^x is a normal $\mathscr{F}$-subgroup of WC_1. Thus $V_2^x \leqslant (WC_1)_\mathscr{F} = V_1$, and equality holds by the $\mathscr{F}$-maximality of V_2.

Proof of Theorem 3.1 (for a Fitting set $\mathscr{F}$). Assume inductively that all proper subgroups of G possess a single conjugacy class of $\mathscr{F}$-injectors, and let K be the nilpotent residual of G. Let W be an $\mathscr{F}$-injector of K and V an $\mathscr{F}$-maximal subgroup of G containing W. If we can show that $V \cap M$ is an $\mathscr{F}$-injector of M for all maximal normal subgroups M of G, it will follow that V is an $\mathscr{F}$-injector of G. Since $K \lhd M$, there is an $\mathscr{F}$-injector V_0 of M such that $V_0 \cap K = W$, and if V^* is an $\mathscr{F}$-maximal subgroup of G containing V_0, then by Hartley's Lemma $V = (V^*)^g$ for some $g \in G$. Thus $V_0^g \leqslant (V^*)^g \cap M = V \cap M$, and the $\mathscr{F}$-maximality of V_0^g in M implies that $V \cap M = V_0^g$, which is an $\mathscr{F}$-injector of M, as desired. This proves the existence of $\mathscr{F}$-injectors in G, and their conjugacy now follows easily by induction and Hartley's Lemma.

There are some good reasons for studying injectors in the framework of Fitting sets in preference to Fitting classes. One such is that an injector is mapped by an epimorphism of a group to an injector of its image. This is useful in proofs by induction.

PROPOSITION 3.3   *Let $\mathscr{F}$ be a Fitting set of a soluble group G, and let $V \in \mathrm{Inj}_{\mathscr{F}}(G)$. If $N \lhd G$, define*

$$\mathscr{F}_{G/N} = \{SN/N \mid S \in \mathrm{Inj}_{\mathscr{F}}(SN)\}.$$

Then $\mathscr{F}_{G/N}$ is a Fitting set of G/N, and VN/N is an $\mathscr{F}_{G/N}$-injector of G/N.

Another reason is that Fitting sets focus attention on the subgroup lattice of a group. We call a subgroup of a group an injector if it is an $\mathscr{F}$-injector for some Fitting set $\mathscr{F}$ of the group. Of course, there is usually more than one Fitting set giving rise to a particular conjugacy class of injectors. However, there is a smallest one:

THEOREM 3.4   *A subgroup H is an injector of a soluble group G if and only if the set*

$$\mathscr{H} = \{S \mid S \lhd\lhd H^g \text{ for some } g \in G\}$$

is a Fitting set of G.

Proof.   If $\mathscr{H}$ is a Fitting set, then it is straightforward to verify that H is an $\mathscr{H}$-injector of G. We now suppose that H is an $\mathscr{F}$-injector of G for some Fitting class $\mathscr{F}$ and proceed by induction on $|G|$ to show that $\mathscr{H}$ is a Fitting set of G. Since H clearly satisfies (FS1) and (FS2), it remains to show that if $N_1, N_2 \in \mathscr{H}$ with N_1, N_2 normal in $N = N_1 N_2$, then $N \in \mathscr{H}$. Let K be a minimal normal subgroup of G, and set

$$\mathscr{H}^* = \{S/K \leqslant G/K \mid S/K \lhd\lhd (HK/K)^x \text{ for some } x \in G/K\}.$$

By Proposition 3.3 and induction we know that $\mathscr{H}^*$ is a Fitting set of G/K and hence that $NK/K = (N_1 K/K)(N_2 K/K) \in \mathscr{H}^*$. Thus $NK/K \lhd\lhd H^g K/K$ for some $g \in G$. If $K \leqslant H$, then

$$N_i \lhd\lhd N_i K \lhd NK \lhd\lhd H^g K = H^g \quad \text{for } i = 1, 2.$$

Hence $N = N_1 N_2 \lhd\lhd H^g$, and then $N \in \mathscr{H}$, as required.

If, on the other hand, $K \not\leqslant H$, it follows from the fact that injectors either cover or avoid chief factors that $H \cap K = 1$, and then, since H is an $\mathscr{F}$-injector of G for a Fitting set $\mathscr{F}$ of G, that 1 is an $\mathscr{F}$-maximal subgroup of K. Since $N_i \in \mathscr{H} \subseteq \mathscr{F}$, we have $N = N_1 N_2 \in \mathscr{F}$ and hence $N \cap K = 1$. In this situation it is not hard to prove that N and H^g are respectively $\mathscr{F}$-injectors of NK and $H^g K$. Since H^g meets the subnormal subgroup NK of $H^g K$ in an $\mathscr{F}$-injector of NK, it follows from the conjugacy of injectors that $N = (NK \cap H^g)^h$ for some $h \in NK$. Thus $N = (NK \cap H^{gh}) \lhd\lhd H^{gh}$, and again $N \in \mathscr{H}$, as required.

Some interesting dualizations of Fitting sets, extending the notion of a projector, can be found in Prentice [56], Wielandt [65], and Wright [68].

3.2 Pronormal subgroups

Injectors, like projectors, persist in intermediate groups:

THEOREM 3.5 (Fischer, Gaschütz and Hartley [19]) *Let $\mathscr{F}$ be a Fitting set of a finite soluble group G and let E be an $\mathscr{F}$-injector of G. If $E \leqslant H \leqslant G$, then E is an $\mathscr{F}$-injector of H.*

We may take $H = \langle E, E^g \rangle$ here and deduce from Theorem 3.1 that for g in G, the subgroups E and E^g are conjugate in their join.

Definition. A subgroup H of a group G is *pronormal* if H and H^g are conjugate in $\langle H, H^g \rangle$ for all $g \in G$.

Pronormality is one of the most important of a hierarchy of subgroup embedding properties explored by Philip Hall in his Cambridge lecture courses. The archetypes of a pronormal subgroup are a Sylow subgroup and a normal subgroup. Thus properties of conjugacy and persistence enjoyed by the projectors and injectors of a group ensure that they too are pronormal subgroups.

Pronormality is precisely the property that makes the Frattini argument work: Assume that H is a pronormal subgroup of a group G and that $H \leqslant K \lhd G$. If $g \in G$, then $H^g = H^k$ for some $k \in \langle H, H^g \rangle \leqslant K$. Hence $gk^{-1} \in N = N_G(H)$ and it follows that $G = NK$. Thus, if a normal subgroup K of a group G contains a non-normal pronormal subgroup of G, then K is supplemented in G by a proper subgroup of G. This fact can be used as an inductive step in the search for a complement to K in G. The idea of using supplements (or partial complements, as he called them) in the construction of soluble groups was developed by Hall [37] as a way of avoiding, or at least minimizing, the difficulties presented by the complicated nature of factor systems. Hall's idea is based on the following observation: If K is a soluble normal subgroup of a finite group G and if Σ is a Hall system of K, then

$$G = NK,$$

where N is the *relative system normalizer* $N_G(\Sigma)$. This follows from the Frattini argument because the set of Hall systems of K is invariant under $\mathrm{Aut}(K)$ and permuted transitively by $\mathrm{Inn}(K)$. Moreover, $D = N \cap K$ is a system normalizer of K and avoids the non-central chief factors of K. Thus, if K is not nilpotent, N is a proper subgroup of G, and therefore the possible structures for N, which is an extension of D by G/K, may be supposed known by induction. Hall states the conditions which given groups N, K and D must satisfy to be embeddable in a group G in the way described: Attention can be restricted to those automorphisms of K which leave a system normalizer D invariant, and they must be made compatible with the double role of D, as a subgroup of K on the one hand, and a normal subgroup of N on the other. As Hall points out, the drawback of his procedure is the problem of getting started. For groups G of large nilpotent length there are many candidates for K; for example the nilpotent residual of G is a good choice for K because D is relatively small and the automorphism group of K fixing D must act without fixed points on K/K' since $[K, G] = K$. But if every proper normal subgroup of G is nilpotent, we have $N_G(\Sigma) = G$ for all choices of K and the procedure cannot be used. However, Hall had already developed other methods for dealing with extensions of nilpotent groups in [36].

As we pointed out above, a pronormal subgroup H of G contained in K also yields a supplement, viz. $N_G(H)$, to K in G. When K is nilpotent, however, this offers no better service than a relative system normalizer for the construction of a partially complemented extension. For then H is subnormal in G, and a subgroup which is both pronormal and subnormal is easily seen to be normal; in this case therefore, $N_G(H) = G$. Furthermore, a pronormal subgroup can do no better than a relative system normalizer in providing a supplement whose intersection with K is small. Indeed, if H is a pronormal subgroup of K, then there is a Hall system Σ with

$(3.\alpha)$ $N_G(\Sigma) \cap K \leqslant N_G(H) \cap K.$

To formulate the reason for this we need the following concept.

Definition.   A subgroup U of G is *abnormal* in G if $g \in \langle U, U^g \rangle$ for all $g \in G$.

Clearly a maximal subgroup of a group is either normal or abnormal, and G itself is the only subgroup of G which is both normal and abnormal. For a finite soluble group G it can be shown that

$(3.\beta)$ U is abnormal in G if and only if whenever $U \leqslant R <_{\max} S \leqslant G$, then R is not normal in S.

LEMMA 3.6   *If H is a pronormal subgroup of K, then $N_K(H)$ is abnormal in K. Hence a subgroup is abnormal if and only if it is pronormal and self-normalizing.*

Proof.   Let $k \in K$. Then $H^k = H^x$ for some $x \in \langle H, H^k \rangle$. Therefore $kx^{-1} \in N_K(H)$, and consequently $k \in N_K(H)\langle H, H^k \rangle \leqslant \langle N_K(H), N_K(H)^k \rangle$, as required.

Returning to why $(3.\alpha)$ is true, let H be a pronormal subgroup of K. By Lemma 3.6 the supplement $N = N_G(H)$ meets K in an abnormal subgroup $N_K(H)$ of K. However, Hall showed that the system normalizers of a soluble group K are precisely the minimal elements of the set of all subgroups U of K which can be joined to K by a chain of the form

$$U = U_0 < U_1 < \ldots < U_n = K,$$

where U_{i-1} is abnormal in U_i for $i = 1, \ldots, n$; in fact, in view of $(3.\beta)$ the links in this chain can be taken to be maximal. It follows from this remarkable characterization that an abnormal subgroup always contains a system normalizer. Therefore $(3.\alpha)$ holds.

The pronormal subgroups of a group do not form a sublattice of the subgroup lattice. However, there is a theorem (3.8 below) of Fischer which clearly indicates how to form the join of a set of pronormal conjugacy classes and obtain thereby a new pronormal conjugacy class. Essential to the proof of this theorem is a characterization of pronormality which was discovered by Fischer and independently by A. Mann. It allows one to compare different pronormal conjugacy classes by using a fixed Hall system to select and label distinguished representatives of the classes. A Hall system Σ of G is said to *reduce into a subgroup H of G* (in symbols: $\Sigma \searrow H$) if for all $\pi \subseteq \mathbb{P}$ the Hall π-subgroup in Σ meets H in a Hall π-subgroup of H. When this happens, the set $\Sigma \cap H = \{S \cap H \mid S \in \Sigma\}$ is a Hall system of H. Recall that a complement basis of H extends to some complement basis of G, and it follows from the conjugacy of Hall systems that $\Sigma^g \searrow H$ for some g in G.

THEOREM 3.7 (Fischer, unpublished; Mann [54])   *A subgroup H of a soluble group G is pronormal in G if and only if each Hall system of G reduces into a unique conjugate of H.*

Proof.   Suppose that H is pronormal in G, and let Σ be a Hall system of G reducing into H and H^g. We have to show that $H = H^g$ and may suppose H is not normal in G. Let N be a minimal normal subgroup of G. Since the Hall system $\Sigma N/N$ of G/N reduces into the pronormal subgroups HN/N and $H^g N/N$, by induction we have $HN = H^g N$, $= L$ say. Since $\Sigma \searrow H$ and $\Sigma \searrow N$ (because $N \lhd G$), then $\Sigma \searrow HN$. Thus $\Sigma \cap L$ is a Hall system of L reducing into the conjugate pronormal subgroups H and H^g of L. Therefore, if $L < G$ then $H = H^g$ by induction. Thus we suppose that $HN = G$ for all minimal normal subgroups N of G. Hence H and H^g are non-normal maximal subgroups of G and have p-power index for some prime p. Let S be the Sylow p-complement in Σ. Since Σ reduces into a maximal subgroup of p-power index if and only if that subgroup contains the p-complement of Σ, we conclude that $S \leqslant H \cap H^g$. Therefore $S^{g^{-1}} = S^h$ for some $h \in H$, and hence $g \in H N_G(S)$. However, by a well known property of non-normal maximal subgroups established by Hall in the proof of Theorem 3.5 of [35], we have $N_G(S) \leqslant H$. Thus $g \in H$ and $H = H^g$, as required.

   To prove that the condition is also sufficient, let H be a subgroup such that any Hall system of G reduces into just one conjugate of H. Let $g \in G$, and let Σ_0 be a Hall system of H. Extend Σ_0 to a Hall system Σ_1 of $J = \langle H, H^g \rangle$ and extend Σ_1 to a Hall system Σ of G. Then $\Sigma \cap H = \Sigma_0$.

   Now there is an x in J such that $\Sigma_1^x \searrow H^g$. Then $\Sigma^x \searrow H^g$ and so $\Sigma \searrow H^{gx^{-1}}$. By hypothesis $gx^{-1} \in N_G(H)$, and therefore $H^g = H^x$ with $x \in J$. Therefore H is pronormal in G.

   There are easy examples to show that when H and L are pronormal subgroups of a group G, it does not necessarily follow that either $H \cap L$ or $\langle H, L \rangle$ is pronormal in G. However, if G has a Hall system Σ reducing into both H and L, then their join $\langle H, L \rangle$ is indeed pronormal in G. This is a consequence of the following promised theorem of Fischer (unpublished).

THEOREM 3.8   *Let Σ be a Hall system of a finite soluble group G, and let $\{H_\lambda \mid \lambda \in \Lambda\}$ be a set of pronormal subgroups of G such that $\Sigma \searrow H_\lambda$ for all $\lambda \in \Lambda$. Let $\mathscr{S}$ denote the set of all subgroups of G of the form*

$$\langle H_\lambda^{g_\lambda} \mid \lambda \in \Lambda, \, g_\lambda \in G \rangle.$$

Then the minimal elements of $\mathscr{S}$, partially ordered by inclusion, form a conjugacy class of pronormal subgroups of G; furthermore, the join $J = \langle H_\lambda \mid \lambda \in \Lambda \rangle$ is the unique member of this class into which Σ reduces and, in particular, is pronormal.

Proof.   Since G is finite, we may suppose without loss of generality that $\Lambda = \{1, 2, \ldots, n\}$. Let $L \in \mathscr{S}$, say $L = \langle H_1^{g_1}, \ldots, H_n^{g_n} \rangle$. There exists a g in G such that $\Sigma^g \searrow L$. We assert that

$(3.\gamma)$ *if $\Sigma^g \searrow L$, then $J^g \leqslant L$.*

Let $i \in \{1, 2, \ldots, n\}$. There is an element $l_i \in L$ such that the Hall system $\Sigma^g \cap L$ of L reduces into $H_i^{g_i l_i}$, and in consequence Σ^g reduces into both H_i^g and $H_i^{g_i l_i}$. By Theorem 3.7 we have $H_i^g = H_i^{g_i l_i} \leqslant L$, and therefore

$$J^g = \langle H_1^g, \ldots, H_n^g \rangle = \langle H_1^{g_1 l_1}, \ldots, H_n^{g_n l_n} \rangle \leqslant L,$$

as desired. Since the conjugates of J belong to $\mathscr{S}$, they must be the minimal elements of $\mathscr{S}$.

If $\Sigma \searrow J^g$, then $\Sigma^{g^{-1}} \searrow J$, and it follows from $(3.\gamma)$ that $J^{g^{-1}} \leqslant J$. Therefore $J = J^g$, as required. Thus J is pronormal in G by Theorem 3.7, and the theorem is proved.

We mention an important application of this result to the theory of projectors. Let $\mathfrak{H}$ and $\mathfrak{K}$ be Schunck classes. In a given group G choose a fixed Hall system Σ, and let H and K denote respectively the $\mathfrak{H}$- and $\mathfrak{K}$-projector of G into which Σ reduces. Then, using Theorem 3.8, one can show that $\langle H, K \rangle$ is the $(\mathfrak{H} \vee \mathfrak{K})$-projector of G into which Σ reduces (cf. Theorem 2.13).

3.3   S-closed Fitting sets

Definition.   A subgroup H of a group G is *strongly pronormal* in G if each of its Sylow subgroups is a Sylow subgroup of a normal subgroup of G, or equivalently, if for all primes p each P in $\mathrm{Syl}_p(H)$ belongs to $\mathrm{Syl}_p(\langle P^G \rangle)$. (Strongly pronormal subgroups are sometimes called normally embedded.)

This property was first studied by Fischer [18]. Let H be a strongly pronormal subgroup of G, and write $H = P_1 P_2 \dots P_t$ as a product of pairwise permutable Sylow subgroups P_i. Since $P_i \in \mathrm{Syl}_{p_i}(N)$ for some $N \lhd G$, the subgroup P_i is pronormal in G. By order considerations H is clearly a minimal element of the set $\langle P_1^{g_1}, P_2^{g_2}, \dots, P_t^{g_t} \rangle$ $(g_1, \dots, g_t \in G)$, and so from Theorem 3.8 we can deduce that a strongly pronormal subgroup of a finite soluble group is pronormal.

It follows from Fischer's work that an injector for a subgroup-closed Fitting set is strongly pronormal (see Hartley [40], Corollary to Lemma 3). Injectors which are strongly pronormal give rise to an especially tractable theory, and the following result explains why.

THEOREM 3.9 (Anderson [1]) *A strongly pronormal subgroup of a soluble group G is an injector for some subgroup-closed Fitting set of G.*

To prove Anderson's theorem, we first need a lemma.

LEMMA 3.10 *Let H be a strongly pronormal subgroup of G, and suppose that the set $\mathscr{F}$ of conjugates of subgroups of H is a Fitting set of G. Then H is an $\mathscr{F}$-injector of G.*

Proof. We proceed by induction on $|G|$. Let M be a maximal normal subgroup of G. Then the set $\mathscr{F} \cap M = \{U \,|\, U \leqslant M, U \leqslant H^g \text{ for some } g \in G\} = \{U \,|\, U \leqslant (H \cap M)^g \text{ for some } g \in G\}$ is a Fitting set of M. But $H \cap M$ is strongly pronormal in G (and therefore in M), and by the Frattini argument $G = N_G(H \cap M)M$. Hence $\mathscr{F} \cap M = \{U \,|\, U \leqslant (H \cap M)^m \text{ for some } m \in M\}$, and by induction $H \cap M$ is an $(F \cap M)$-injector of M. It follows that $H \cap K$ is $\mathscr{F}$-maximal in K for every *proper* subnormal subgroup K of G, and since H is clearly $\mathscr{F}$-maximal in G, we have shown that H is an $\mathscr{F}$-injector of G.

Proof of Theorem 3.9. Let H be a strongly pronormal subgroup of G, and let

$$\mathscr{F} = \{U \,|\, U \leqslant H^g \text{ for some } g \in G\}.$$

By Lemma 3.10 it will suffice to show that $\mathscr{F}$ is a Fitting set of G. Conditions (FS1) and (FS2) are obviously fulfilled. We must prove (FS3): if $R, S \lhd RS$ with R and S in $\mathscr{F}$, we must show that $RS \leqslant H^g$ for some $g \in G$.

Let Σ be a Hall system of G reducing into RS, and let H^x be the conjugate of H into which Σ reduces. Let p be a prime, let P^* be the

Sylow p-subgroup in Σ, and let P be a Sylow p-subgroup of H such that $P^x = H^x \cap P^*$. Then define

$$\mathscr{F}_p = \{U \,|\, U \leqslant P^g \text{ for some } g \in G\}.$$

Since by hypothesis $P \in \mathrm{Syl}_p(\langle P^G \rangle)$, by Sylow's theorem $\mathscr{F}_p$ consists of all p-subgroups of $\langle P^G \rangle$ and is evidently a Fitting set of G. The subgroup P^x is clearly $\mathscr{F}_p$-maximal and subnormal in P^* and is therefore the $\mathscr{F}_p$-radical of P^*. By choice Σ reduces into RS and therefore into $R \triangleleft RS$; in particular, $R \cap P^* \in \mathrm{Syl}_p(R)$. Since $R \in \mathscr{F}$, the p-subgroup $R \cap P^*$ is contained in some conjugate of H and hence belongs to $\mathscr{F}_p$. Thus $R \cap P^*$ is contained in the $\mathscr{F}_p$-radical P^x of P^*, and therefore $R \cap P^* \leqslant H^x$. Similarly $S \cap P^* \leqslant H^x$, and H^x contains the Sylow p-subgroup $(R \cap P^*)(S \cap P^*)$ of RS. Since p was chosen arbitrarily, we conclude that $RS \leqslant H^x$, as desired.

Subgroup-closed Fitting classes have received considerable attention and have recently been fully described by R. A. Bryce and J. Cossey. These authors first studied $\langle \mathrm{s}, \mathrm{N}_0, \mathrm{Q} \rangle$-closed classes of finite soluble groups, which are easily seen to be exactly the subgroup-closed Fitting formations. In [7] they show that these formations are saturated and describe them as follows. Let $\mathscr{F}_0 = \{\emptyset, (1), \mathfrak{S}\}$, and define $\mathscr{F}_i = \{LF(f) \,|\, \text{each } f(p) \in \mathscr{F}_{i-1}\} \cup \mathscr{F}_{i-1}$ for $i = 1, 2, \ldots$. Then $\mathfrak{F} = \langle \mathrm{s}, \mathrm{N}_0, \mathrm{Q} \rangle \mathfrak{F} \subseteq \mathfrak{S}$ if and only if $\mathfrak{F} \cap \mathfrak{N}^r \in \mathscr{F}_r$ for $r = 1, 2, \ldots$. Then in [8], [10] and [11] they develop and deploy some special techniques of representation theory to prove that an $\langle \mathrm{s}, \mathrm{N}_0 \rangle$-closed class of soluble groups is already Q-closed and thereby classify all subgroup-closed Fitting classes.

3.4 Normal Fitting classes
If $\mathfrak{F}$ is a formation and G and H are soluble groups, then

$$(G \times H)^{\mathfrak{F}} = G^{\mathfrak{F}} \times H^{\mathfrak{F}}.$$

Here $G^{\mathfrak{F}}$ denotes the $\mathfrak{F}$-residual of G described in subsection 2.1. (A proof of this property of residuals, together with an example to show that it can fail for insoluble groups, is to be found in [16].) An obvious question is whether a similar equation is true for radicals; in other words, if $\mathfrak{F}$ is a Fitting class, is $(G \times H)_{\mathfrak{F}} = G_{\mathfrak{F}} \times H_{\mathfrak{F}}$? The answer is no:

Examples 3.11 (Blessenohl and Gaschütz [5])

(1) Let p be a prime, and let

(3.δ) $\qquad\qquad\qquad 1 = G_0 \leqslant G_1 \leqslant \ldots \leqslant G_n = G$

be a chief series of a soluble group G. Each p-chief factor G_i/G_{i-1} can be viewed as a linear space over $\mathbb{F}_p$ on which each group element g induces a non-singular linear map $\lambda_i(g)$. Let

$$(3.\varepsilon) \qquad\qquad \Delta(g) = \prod \det(\lambda_i(g)),$$

where the product is taken over the p-chief factors of $(3.\delta)$. Then Δ is a homomorphism from G to $\mathbb{F}_p^\times$ and is independent of the chief series by the Jordan–Hölder theorem. The class of groups

$$\mathfrak{D}(p) = (G \in \mathfrak{S} \mid \Delta(g) = 1 \text{ for all } g \in G)$$

is then a Fitting class.

(2) Let G be a group. Each g in G induces on the elements of the odd order radical $O(G)$ of G a permutation $\tau(g)$ defined thus

$$\tau(g): x \to g^{-1}xg \quad (x \in O(G)).$$

Thus τ is a homomorphism from G to $\mathrm{Sym}(O(G))$. It can be shown that the class of groups

$$\mathfrak{X} = (G \mid \tau(g) \text{ is even for all } g \in G),$$

is $\langle s_n, N_0 \rangle$-closed, and so the class $\mathfrak{Y} = \mathfrak{X} \cap \mathfrak{S}$, is again a Fitting class.

Now let G denote the symmetric group of degree 3. Then clearly $|G : G_{\mathfrak{D}}| = |G : G_{\mathfrak{Y}}| = 2$, where $\mathfrak{D} = \mathfrak{D}(3)$. However, a simple calculation shows that the radicals $(G \times G)_{\mathfrak{D}}$ and $(G \times G)_{\mathfrak{Y}}$ each have index 2. Thus we have two examples of Fitting classes whose radicals fail to respect direct products.

One line taken in the study of Fitting classes has been their classification according to the embedding properties of their injectors in soluble groups, and in this direction the pursuit of those with normal injectors—*normal* Fitting classes, as they are known—has been especially fruitful; of course, for such classes the radical is the unique injector in every soluble group. The first investigation was carried out in 1970 by Blessenohl and Gaschütz in [5]. They quickly settle the question of which Schunck classes have normal projectors—these turn out to be exactly the classes of π-perfect groups (the projector in G being $O^\pi(G)$)—and then go on to lay the foundations for the fascinating and much more complex dual theory.

First they introduce the Fitting classes $\mathfrak{D}(p)$ and $\mathfrak{Y}$ cited above and show them to be normal; these were the first published non-trivial examples. (Obviously (1) and $\mathfrak{S}$ are examples, although, by convention and with good reason, (1) is excluded from the definition of a normal Fitting class.) They go on to show that an intersection of normal Fitting classes is again normal, whence there is a smallest one, but that nevertheless a normal Fitting class $\mathfrak{F}$ must be large, in the following sense: if $G \in \mathfrak{F}$ and C is cyclic of prime order, then $(G \times G) \setminus C \in \mathfrak{F}$, a fact which implies that any finite soluble group is a subgroup of an $\mathfrak{F}$-group, that is $s\mathfrak{F} = \mathfrak{S}$. Furthermore, they show that the $\mathfrak{F}$-radical contains G' so that $G/G_{\mathfrak{F}}$ is abelian for any soluble group G; this characterizes normal Fitting classes because for any non-normal Fitting class $\mathfrak{G}$ Gaschütz and Blessenohl prove that

$$s(G/G_{\mathfrak{G}} \mid G \in \mathfrak{S}) = \mathfrak{S}.$$

In 1971 F. P. Lockett [51] exploited the aberrant behaviour of radicals in direct products and showed how to associate with each Fitting class $\mathfrak{F}$ another containing it, called $\mathfrak{F}^*$, such that

$$(G \times H)_{\mathfrak{F}^*} = G_{\mathfrak{F}^*} \times H_{\mathfrak{F}^*}$$

for all G, $H \in \mathfrak{S}$. His construction characterizes the normal Fitting classes as just those for which $\mathfrak{F}^* = \mathfrak{S}$. Moreover, Lockett proved that among all Fitting classes $\mathfrak{G}$ with $\mathfrak{G}^* = \mathfrak{F}^*$, there is a smallest one $\mathfrak{F}_*$, and in each finite group G the section $G_{\mathfrak{F}^*}/G_{\mathfrak{F}_*}$ is a central factor of G. The set of all Fitting classes lying between $\mathfrak{F}_*$ and $\mathfrak{F}^*$ forms a lattice, known as the *Lockett section* of $\mathfrak{F}$, which turns out to be isomorphic with the subgroup lattice of a certain abelian group, discovered by H. Lausch [50]. Lockett's work deals only with the existence and theoretical properties of his section 'Locksec $(\mathfrak{F})$' and gives no indication of how to calculate $G_{\mathfrak{F}^*}$ or $G_{\mathfrak{F}_*}$ in a given group G. However, H. Lausch, H. Laue and G. R. Pain [49] discovered a transfer-like construction for certain members of Locksec $(\mathfrak{F})$, and T. R. Berger [4] subsequently showed that this construction yields a sufficiently varied and abundant supply to allow the $\mathfrak{F}_*$-radical of a group to be calculated entirely from knowledge of the internal structure of its $\mathfrak{F}$-radical. Berger's proof of this fact appears to lie deep and only works for Fitting classes $\mathfrak{F}$ which satisfy an extra condition (cf. (3.ζ) below). It is not yet known whether this condition is redundant, although it certainly holds when $\mathfrak{F}$ is subgroup-closed. We shall now explain the basis of Lockett's section, then construct Lausch's abelian group, and briefly describe Berger's revelation; finally we give two applications of this work to the structure of finite groups, which are due to Gaschütz.

3.5   The Lockett section of a Fitting class

Let $\mathfrak{F}$ be a Fitting class. For a group G let $G(\mathfrak{F})$ denote the subgroup of G consisting of all (second) components of $(G \times G)_{\mathfrak{F}}$. We write $G^2, G^3, \ldots$ for $G \times G$, $G \times G \times G, \ldots$ and note that if $(g_1, \ldots, g_n)$ is in $(G^n)_{\mathfrak{F}}$, then so is $(g_{\sigma(1)}, \ldots, g_{\sigma(n)})$ for any permutation σ of $1, \ldots, n$. We also note that $(G^2)_{\mathfrak{F}}$ embeds in $(G^n)_{\mathfrak{F}}$ in several ways; for example the maps $(x, y) \to (1, x, y)$, $(x, y) \to (x, 1, y)$, $(x, y) \to (x, y, 1)$ all embed $(G^2)_{\mathfrak{F}}$ in $(G^3)_{\mathfrak{F}}$.

LEMMA 3.12   *Let $\mathfrak{F}$ be a Fitting class and G, H groups:*

(a) *if $N \lhd G$, then $N(\mathfrak{F}) = N \cap G(\mathfrak{F})$;*
(b) *if $G = MN$ with M and N normal in G, then $G(\mathfrak{F}) \geqslant M(\mathfrak{F})N(\mathfrak{F})$;*
(c) *if $g \in G(\mathfrak{F})$, then $(g^{-1}, g) \in (G^2)_{\mathfrak{F}}$;*
(d) *$(G \times H)_{\mathfrak{F}} \leqslant G(\mathfrak{F}) \times H(\mathfrak{F})$.*

Proof.   Statements (a) and (b) are immediate. Statement (c) follows at once from the fact that if $(x, g) \in (G^2)_{\mathfrak{F}}$, then

$$(1, g^{-1}, g) = (x, 1, g)(x^{-1}, g^{-1}, 1)$$

is in $(G^3)_{\mathfrak{F}}$. For (d) note that the equation

$$(1, h^{-1}, h) = (g^{-1}, h^{-1}, 1)(g, 1, h)$$

in $G \times H \times H$ shows that if $(g, h) \in (G \times H)_{\mathfrak{F}}$, then $(h^{-1}, h) \in (H^2)_{\mathfrak{F}}$, and so $h \in H(\mathfrak{F})$; similarly $g \in G(\mathfrak{F})$.

Definition.   Following Lockett [52] we define $\mathfrak{F}^*$ to be the class of all groups G such that $G = G(\mathfrak{F})$.

By Lemma 3.12(a) (with $N = G(\mathfrak{F})$) we have $G(\mathfrak{F})(\mathfrak{F}) = G(\mathfrak{F})$; therefore $\mathfrak{F}^*$ consists of all groups $G(\mathfrak{F})$.

THEOREM 3.13 (Lockett [52])   *Let $\mathfrak{F}$ be a Fitting class. Then*

(i) *$\mathfrak{F}^*$ is a Fitting class;*
(ii) *$\mathfrak{F}^{**} = \mathfrak{F}^*$;*
(iii) *$\mathfrak{F} = \mathfrak{F}^*$ if and only if $(G \times H)_{\mathfrak{F}} = G_{\mathfrak{F}} \times H_{\mathfrak{F}}$ for all groups G and H.*

Proof.   Assertion (i) is immediate from (a) and (b) of Lemma 3.12. For (ii) let $G \in \mathfrak{F}^{**}$ and $g \in G$. We show that $g \in G(\mathfrak{F})$. Write $E = (G^2)_{\mathfrak{F}^*}$. Applying Lemma 3.12(c) first to G for the class $\mathfrak{F}^*$ we get $(g^{-1}, g) \in E$ and then to E ($= E(\mathfrak{F})$) for the class $\mathfrak{F}$ we get

$$((g,g^{-1}),(g^{-1},g)) \in (E^2)_{\mathfrak{F}} \leqslant (G^4)_{\mathfrak{F}}.$$

Therefore

$$(1,1,1,g^{-1},g) = (g,g^{-1},g^{-1},1,g)(g^{-1},g,g,g^{-1},1) \in (G^5)_{\mathfrak{F}},$$

which gives g in $G(\mathfrak{F})$.

The sufficiency in (iii) is clear. If $\mathfrak{F} = \mathfrak{F}^*$, then $G(\mathfrak{F}) = G_{\mathfrak{F}}$ for all groups G, and therefore $(G \times H)_{\mathfrak{F}} = G_{\mathfrak{F}} \times H_{\mathfrak{F}}$ by Lemma 3.12(d).

For Fitting classes of the form $\mathfrak{F}^*$, injectors as well as radicals respect direct products: an $\mathfrak{F}^*$-injector of $G \times H$ always has the form $U \times V$ with U and V $\mathfrak{F}^*$-injectors of G and H respectively.

Definition. The lower limit $\mathfrak{F}_*$ of the Lockett section of $\mathfrak{F}$ is the intersection of all Fitting classes $\mathfrak{X}$ such that $\mathfrak{X}^* = \mathfrak{F}^*$.

Obviously $\mathfrak{F}_*$ is a Fitting class contained in $\mathfrak{F}$. That $(\mathfrak{F}_*)^*$ is the same as $\mathfrak{F}^*$ follows directly from

LEMMA 3.14 *For Fitting classes $\mathfrak{X}_\lambda$, $\lambda \in \Lambda$, we have*

$$(\cap \mathfrak{X}_\lambda)^* = \cap \mathfrak{X}_\lambda^*.$$

Proof. From Lemma 3.12(c), $G \in (\cap \mathfrak{X}_\lambda)^*$ if and only if for all g in G we have $(g^{-1},g) \in (G^2)_{\cap \mathfrak{X}_\lambda} = \cap (G^2)_{\mathfrak{X}_\lambda}$, which is precisely the condition for G to be in $\cap \mathfrak{X}_\lambda^*$.

We have now proved most of the following

THEOREM 3.15 (Lockett [52]) *Let $\mathfrak{F}$ be a Fitting class. Then*

$$(\mathfrak{F}_*)_* = \mathfrak{F}_* = (\mathfrak{F}^*)_* \subseteq \mathfrak{F} \subseteq \mathfrak{F}^* = (\mathfrak{F}_*)^* \subseteq \mathfrak{F}_*\mathfrak{A}.$$

The missing inclusion $\mathfrak{F}^* \subseteq \mathfrak{F}_*\mathfrak{A}$ follows from the following observation:

(3.16) *If A is a group of operators of a group G in $\mathfrak{F}^*$, then $[G,A] \leqslant G_{\mathfrak{F}}$.*

Proof. Let $\alpha \in A$. If $(G^2)_{\mathfrak{F}}$ contains (x,g), then it contains (x,g^α) and hence also $(1,g^{-1}g^\alpha)$. Therefore, if $G \in \mathfrak{F}^*$, it follows that $g^{-1}g^\alpha \in G_{\mathfrak{F}}$ for all $g \in G$, as desired.

On taking $\mathfrak{F}_*$ for $\mathfrak{F}$ and G acting on G by conjugation for A, we get $G' \leqslant G_{\mathfrak{F}_*}$, and hence $\mathfrak{F}^* \subseteq \mathfrak{F}_* \mathfrak{A}$.

As remarked earlier, the *Lockett section* of $\mathfrak{F}$, written Locksec$(\mathfrak{F})$, consists of all Fitting classes $\mathfrak{G}$ with $\mathfrak{G}^* = \mathfrak{F}^*$, i.e. such that $\mathfrak{F}_* \subseteq \mathfrak{G} \subseteq \mathfrak{F}^*$. We now justify our earlier assertion that the condition $\mathfrak{F}^* = \mathfrak{S}$ characterizes normal Fitting classes.

PROPOSITION 3.17 $\mathfrak{F}$ *is a normal Fitting class if and only if* $\mathfrak{F} \in$ Locksec$(\mathfrak{S})$.

Proof. Let $\mathfrak{F} \in$ Locksec$(\mathfrak{S})$ and $G \in \mathfrak{S}$. By Theorem 3.15 we have $G' \leqslant G_{\mathfrak{F}}$, and since an $\mathfrak{F}$-injector of G contains $G_{\mathfrak{F}}$, it is therefore normal in G.

Conversely, let $\mathfrak{F}$ be a normal Fitting class. For a soluble group G we have $G' \leqslant G_{\mathfrak{F}}$ by Satz 5.3 of Blessenohl and Gaschütz [5]. Let α be the automorphism of $G \times G$ sending (g, g') to (g', g) and let $S = (G \times G) \rtimes \langle \alpha \rangle$. If $g \in G$, then

$$(g^{-1}, g) = (g, 1)^{-1}(g, 1)^{\alpha} \in (G \times G) \cap S' \leqslant (G \times G) \cap S_{\mathfrak{F}} = (G \times G)_{\mathfrak{F}},$$

and so $g \in G(\mathfrak{F})$. Thus $G \in \mathfrak{F}^*$, and it follows that $\mathfrak{F}^* = \mathfrak{S}$.

3.6 The Lausch group

Because we shall soon need to work with sets rather than classes of groups, we choose once for all a fixed set $\mathscr{E}$ containing exactly one representative of each isomorphism class of finite groups. Then for any class $\mathfrak{X}$, we put Set$(\mathfrak{X}) = \mathfrak{X} \cap \mathscr{E}$. For groups G, H, a monomorphism $\alpha \colon G \to H$ such that $\alpha(G)$ is normal in H is called a *normal embedding*. The following idea is useful for the construction of the members of the Lockett section of a given Fitting class $\mathfrak{F}$.

Definitions

(a) A pair (A, d) is called an $\mathfrak{F}$-*Fitting pair* if A is a (possibly infinite) abelian group and if for each $G \in$ Set$(\mathfrak{F})$ there is a homomorphism $d_G \colon G \to A$ such that

(FP1) $d_G = \alpha \circ d_H$ for all $G, H \in$ Set$(\mathfrak{F})$ and all normal embeddings $\alpha \colon G \to H$, and
(FP2) A is the union of the images $G d_G$, $G \in$ Set$(\mathfrak{F})$.

(It is straightforward to check that this definition is unchanged when 'normal' in (FP1) is replaced by 'subnormal'.)

(b) If R and S are characteristic subgroups of a group G, we say that R/S is *characteristically central* in G if and only if $[R, T] \leqslant S$ with $T = \mathrm{Aut}(G)$. We say that R/S is *characteristically hypercentral* in G if $[R, \underbrace{T, \ldots, T}_{r}] \leqslant S$ for some r.

If $\alpha \in \mathrm{Aut}(G)$, then α is a normal embedding, and so $g^{-1}g^{\alpha} \in \mathrm{Ker}(d_G)$ for all $g \in G$. Thus $G/\mathrm{Ker}(d_G)$ is characteristically central in G for all G in $\mathrm{Set}(\mathfrak{F})$.

PROPOSITION 3.18   *Let (A, d) be an $\mathfrak{F}$-Fitting pair. Then*

$$\mathfrak{K} = \mathrm{Ker}(A, d) = (G \in \mathrm{Set}(\mathfrak{F}) \,|\, Gd_G = 1)$$

is a Fitting class, and $\mathfrak{F}_* \subseteq \mathfrak{K} \subseteq \mathfrak{F}$.

(*Remark*: Brackets ( ) are used to denote the class generated by the groups inside; thus $\mathfrak{K}$ is a class of groups.)

Proof.   If $Gd_G = 1$ and $N \triangleleft G$, then $1 = Nd_G = Nd_N$. Similarly, if M and N belong to $\mathfrak{K}$ and are normal in $G = MN$, then $Gd_G = (Md_M)(Nd_N) = 1$. Thus $\mathfrak{K}$ is a Fitting class and $\mathfrak{K} \subseteq \mathfrak{F}$. To complete the proof it is enough to prove that $\mathfrak{F} \subseteq \mathfrak{K}^*$. Let $G \in \mathrm{Set}(\mathfrak{F})$. Since $G_{\mathfrak{K}} = \mathrm{Ker}(d_G)$, by the above observation $G/G_{\mathfrak{K}}$ is characteristically central. Therefore by the following lemma $G \in \mathfrak{K}^*$, as desired.

LEMMA 3.19   *Any two of the following statements about a Fitting class $\mathfrak{F}$ are equivalent:*

(i) $G \in \mathfrak{F}^*$;

(ii) $(G^n)/(G^n)_{\mathfrak{F}}$ *is characteristically central in the direct power G^n for $n \geqslant 1$;*

(iii) $(G^n)/(G^n)_{\mathfrak{F}}$ *is characteristically hypercentral in G^n for $n \geqslant 1$.*

Proof.   (i) $\Rightarrow$ (ii): If $G \in \mathfrak{F}^*$, then $G^n \in \mathfrak{F}^*$ by Theorem 3.13(i), and Statement (ii) follows from (3.16). It is clear that (ii) $\Rightarrow$ (iii), and so it remains to prove that (iii) $\Rightarrow$ (i): Let n be a prime not dividing $|G|$. The automorphism σ of G^n defined by $\sigma: (g_1, g_2, \ldots, g_n) \to (g_n, g_1, g_2, \ldots, g_{n-1})$ is an operator of coprime order acting hypercentrally and so trivially on $G^n/(G^n)_{\mathfrak{F}}$. Therefore, for any $g \in G$ we have

$$(g^{-1}, g, 1, \ldots, 1) = (g, 1, \ldots, 1)^{-1}(g, 1, \ldots, 1)^{\sigma} \in (G^n)_{\mathfrak{F}}.$$

Hence $(G \times G)_{\mathfrak{F}}$ contains (g^{-1}, g), and consequently $G \in \mathfrak{F}^*$.

As an illustration consider the Fitting class $\mathfrak{D}(p)$ of Examples 3.11. A simple application of Clifford's theorem shows that the map Δ defined in Equation (3.ε) satisfies the axioms of an $\mathfrak{S}$-Fitting pair with $A = \mathbb{F}_p^\times$. It follows that $\mathfrak{D}(p) = \mathrm{Ker}(\mathbb{F}_p^\times, \Delta)$ satisfies $(\mathfrak{D}(p))^* = \mathfrak{S}$; in other words, $\mathfrak{D}(p)$ is a normal Fitting class.

The Lausch group $\Lambda = \Lambda(\mathfrak{F})$, which we are about to construct, gives a *universal $\mathfrak{F}$-Fitting pair* (Λ, δ) in the following sense: for any Fitting pair (A, d) there exists a homomorphism $\phi \colon \Lambda \to A$ such that the diagram

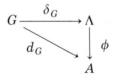

commutes for all $G \in \mathrm{Set}(\mathfrak{F})$.

Construction 3.20.   For any class $\mathfrak{X}$ of groups, let

$$\Delta(\mathfrak{X}) = \mathrm{Dr}\,\{G \mid G \in \mathrm{Set}(\mathfrak{X})\}$$

be the restricted direct product of the groups in $\mathrm{Set}(\mathfrak{X})$, and denote by $\Gamma(\mathfrak{X})$ the subgroup generated by all $(g^{-1}\varepsilon_G)(g\alpha\varepsilon_H)$ with G, H in $\mathrm{Set}(\mathfrak{X})$, g in G, and $\alpha \colon G \to H$ a normal embedding, and where ε_G and ε_H are the natural embeddings of G and H into $\Delta(\mathfrak{X})$.

It is clear that $[G, \mathrm{Aut}(G)]\varepsilon_G \leqslant \Gamma(\mathfrak{X})$; in particular $\Gamma(\mathfrak{X})$ contains $\Delta(\mathfrak{X})'$.

The *Lausch group* $\Lambda(\mathfrak{X})$ of $\mathfrak{X}$ is defined to be the abelian group

$$\Lambda(\mathfrak{X}) = \Delta(\mathfrak{X})/\Gamma(\mathfrak{X}),$$

and the *associated map* δ is defined by

$$g\delta_G = (g\varepsilon_G)\Gamma(\mathfrak{X})$$

whenever $g \in G \in \mathrm{Set}(\mathfrak{X})$.

It is not hard to verify the following basic result.

THEOREM 3.21 (Lausch [50])   *If $\mathfrak{F}$ is a Fitting class, then $(\Lambda(\mathfrak{F}), \delta)$ is a universal $\mathfrak{F}$-Fitting pair.*

The most important property of $\Lambda(\mathfrak{F})$ is that its subgroups are in one-to-one correspondence with $\mathrm{Locksub}(\mathfrak{F})$, the Fitting classes lying between $\mathfrak{F}_*$ and $\mathfrak{F}$. The exact connection is described in the following theorem.

THEOREM 3.22 (Lausch [50], Bryce and Cossey [9]) *For* $\mathfrak{X} \in \mathrm{Locksub}(\mathfrak{F})$, *define*

$$\Xi(\mathfrak{X}) = \{g\delta_G \,|\, g \in G \in \mathrm{Set}(\mathfrak{X})\}.$$

Then the map Ξ *is an isomorphism from the lattice* $\mathrm{Locksub}(\mathfrak{F})$ *to the subgroup lattice of* $\Lambda(\mathfrak{F})$.

A method for determining the lattice structure is implicit in a theorem of Berger to which we now turn.

We begin by describing the $\mathfrak{F}$-Fitting pairs of Laue, Lausch and Pain [49]. Each such pair is determined by a prime p and a group U in $\mathfrak{F}$. To verify the details of the construction is laborious and technically complicated; we therefore only sketch the procedure in outline.

Let $U \in \mathfrak{F}$ and $p \in \mathbb{P}$. A Sylow p-subgroup P^* of $\mathrm{Aut}(U)$ has a unique normal subgroup P such that UP is the $\mathfrak{F}$-radical of $U \rtimes P^*$. We define two subgroups of P:

$$A = \langle \alpha \in P \,|\, [U, \alpha] < U \rangle, \quad \text{and}$$
$$B = \langle [g, \alpha] \,|\, \alpha \in \mathrm{Aut}(U), g \text{ and } [g, \alpha] \in P \rangle.$$

Since B contains P', their product $P_0 = AB$ is a normal subgroup of P with an abelian quotient. Our object is to construct an $\mathfrak{F}$-Fitting pair $(P/P_0, d)$ and so we only need to consider those groups U in $\mathfrak{F}$ and primes p for which the associated p-group P/P_0 is non-trivial. The problem then is to define d_G for each G in $\mathrm{Set}(\mathfrak{F})$. Step 1 is to consider an $\mathfrak{F}$-group of the special form XS with a normal embedding $\psi: U \to XS$ such that $\psi(U) = X$ and where S is a p-group. One makes use of the action of S on the normal subgroup X to obtain a homomorphism $\eta: S \to P/P_0$ which is independent of both ψ and the choice of Sylow p-subgroup of $\mathrm{Aut}(U)$. This map then lifts to a homomorphism $\hat{\eta}: S/S' \to P/P_0$. Step 2 is to consider a general group G of $\mathrm{Set}(\mathfrak{F})$ and a subgroup X of G with X isomorphic with U. Assume that

(3.ζ)   if $S \in \mathrm{Syl}_p(N_G(X))$, then $XS \in \mathfrak{F}$.

On composing the transfer $V_{G \to S}: G \to S/S'$ with $\hat{\eta}$, we obtain a homomorphism

$$\theta_X: G \to P/P_0,$$

which turns out to be independent of the chosen representative of the conjugacy class of X.

The third and final step is to choose representatives $X_1, \ldots, X_k$, one from each of the set $\mathscr{S}$ of conjugacy classes of subnormal subgroups of G that are isomorphic with U. Let S_i be a Sylow p-subgroup of $N_G(X_i)$, and for $i = 1, \ldots, k$ let $t_i \equiv |N_G(X_i) : S_i|^{-1}$ modulo the exponent of P/P_0. Write θ_i for θ_{X_i}. If $g \in G$, where $G \in \mathrm{Set}(F)$, define

$$gd_G = \begin{cases} 1 & \text{if } \mathscr{S} = \varnothing, \\ \prod_{i=1}^{k} (g\theta_i)^{t_i} & \text{if } k \geqslant 1. \end{cases}$$

Then after considerable labour (cf. [49]) it can be proved that

$$(P/P_0, d) \text{ is an } \mathfrak{F}\text{-Fitting pair.}$$

For given $\mathfrak{F}$ the function d depends on U and p, and so we designate it $d^{U,p}$. If we now set $\mathfrak{K}(U,p) = \mathrm{Ker}(P/P_0, d^{U,p})$, we may formulate Berger's theorem [4] as follows:

THEOREM 3.23   *If G is in $\mathfrak{F}$, then*

$$G_{\mathfrak{F}_*} = \bigcap \{G_{\mathfrak{K}(U,p)} \mid U \vartriangleleft\vartriangleleft G, \quad p \mid |G|\}.$$

(In fact, in the right-hand intersection it is enough for U to run through just those subnormal subgroups U for which $|U| \leqslant |G^{\mathfrak{N}}|$ and for which P_0 is a proper subgroup of P; moreover, p may be restricted to the prime divisors of $|G/G^{\mathfrak{N}}|$.)

We make no attempt to prove this theorem. The reader may refer to Berger [4].

An alternative approach via the Lausch group is given by Brison [6]. Both Berger and Brison mention some interesting consequences of the result, including the fact that the Lausch group is a restricted direct product of finite cyclic groups.

For those Fitting classes $\mathfrak{F}$ for which Condition (3.ζ) holds for all subnormal subgroups X of all groups $G \in \mathfrak{F}$, Berger's theorem yields a finite procedure for deciding whether an $\mathfrak{F}$-group G belongs to $\mathfrak{F}_*$, and this algorithm depends only on knowledge of the subnormal subgroups of G and their automorphism groups. One of the more interesting outstanding questions in the theory of Fitting classes is whether Berger's theorem is true for all Fitting classes $\mathfrak{F}$. Certainly the significance of Condition (3.ζ) has not yet been fully understood. Brison has shown that $\mathfrak{S}_*$ fails to satisfy (3.ζ), but this, of course, does not invalidate the theorem.

3.7 Two applications

We conclude with two applications to the structure of a general finite group. They are both due to Gaschütz [26].

THEOREM 3.24 *Let G be a finite group, and let $\mathfrak{F}$ be a normal Fitting class. Then the section $(G_{\mathfrak{S}}G')/(G_{\mathfrak{F}}G')$ of G/G' is isomorphic with $G_{\mathfrak{S}}/G_{\mathfrak{F}}$.*

Proof. Set $R = G_{\mathfrak{F}}$, and note that $G_{\mathfrak{S}}/R \leqslant Z(G/R)$ by (3.16) since $\mathfrak{F}^* = \mathfrak{S}$. We assert that

$$(3.\eta) \qquad\qquad (G_{\mathfrak{S}}/R) \cap (G/R)' = 1.$$

Suppose this is false, and let x be an element of $(G_{\mathfrak{S}}/R) \cap (G/R)'$ of prime order, p say. Let H/R be a Sylow p-subgroup of G/R. Since H is soluble and $\mathfrak{F}^* = \mathfrak{S}$, we have $H' \leqslant H_{\mathfrak{F}}$ by (3.16) again, and therefore $H' \cap G_{\mathfrak{S}}$ lies in the unique maximal $\mathfrak{F}$-subgroup $(G_{\mathfrak{S}})_{\mathfrak{F}} \,(= R)$ of $G_{\mathfrak{S}}$. Consequently $(H'R/R) \cap (G_{\mathfrak{S}}/R) = 1$. Let v denote the transfer homomorphism $G/R \to H/R$, and let $n = |G:H|$. Since $x \in Z(G/R)$, we have $v(x) = x^n(H/R)'$, and since $x \in (G/R)' \leqslant \mathrm{Ker}(v)$, it follows that $x^n \in (H/R)' \cap (G_{\mathfrak{S}}/R) = 1$. Consequently $x = 1$ because $p \nmid n$. This contradiction justifies $(3.\eta)$ and shows that $G_{\mathfrak{S}} \cap RG' = R$, from which we conclude that $G_{\mathfrak{S}}G'/RG' \cong G_{\mathfrak{S}}/(G_{\mathfrak{S}} \cap RG') = G_{\mathfrak{S}}/R$, as required.

COROLLARY *If G is a finite perfect group, then $G_{\mathfrak{S}} \in \mathfrak{S}_*$.*

If S denotes the symmetric group of degree 3, then $S_{\mathfrak{D}} = G'$, where $\mathfrak{D} = \mathfrak{D}(3)$ is the normal Fitting class in the first of Examples 3.11. Since $\mathfrak{S}^* \subseteq \mathfrak{D} \subseteq \mathfrak{S}$, it follows that $S' = S_{\mathfrak{S}_*}$, and therefore, if $S \lhd\lhd G$, we have $S \cap G_{\mathfrak{S}_*} = S'$. Thus we obtain a result of Dark [13], that S_3 cannot appear as a subnormal subgroup of a perfect group.

The second application is a splitting theorem.

THEOREM 3.25 (Gaschütz [26]) *Let $\mathfrak{F}$ be a normal Fitting class with the property that $S/S_{\mathfrak{F}}$ is elementary abelian for all soluble groups S. Then, for any group G, $G_{\mathfrak{S}}/G_{\mathfrak{F}}$ is a direct factor of $G/G_{\mathfrak{F}}$.*

Proof. Let $R = G_{\mathfrak{F}}$, let $H/R \in \mathrm{Syl}_p(G/R)$, and let $P/R \in \mathrm{Syl}_p(G_{\mathfrak{S}}/R)$ for some prime p. Since $G_{\mathfrak{S}}$ and H are soluble groups, by hypothesis P/R and $H/H_{\mathfrak{F}}$ are elementary abelian groups. Because R is $\mathfrak{F}$-maximal in $G_{\mathfrak{S}}$ and because $R \leqslant H_{\mathfrak{F}} \cap G_{\mathfrak{S}} \in \mathrm{s_n}\,\mathfrak{F} = \mathfrak{F}$, we have $R = H_{\mathfrak{F}} \cap G_{\mathfrak{S}} = H_{\mathfrak{F}} \cap P$. Let $L/H_{\mathfrak{F}}$ be a subgroup complementary to the subgroup $PH_{\mathfrak{F}}/H_{\mathfrak{F}}$ of the elementary abelian group $H/H_{\mathfrak{F}}$. Then $L \cap P \leqslant H_{\mathfrak{F}} \cap P = R$, furthermore $PL = H$, and therefore L/R complements P/R in H/R. Thus the extension G/R of the abelian normal subgroup $G_{\mathfrak{S}}/R$ splits over its Sylow subgroups, and by a well known theorem of Gaschütz G/R splits over $G_{\mathfrak{S}}/R$. Since $G_{\mathfrak{S}}/R$ is in the centre of G/R, it is therefore a direct factor.

In conclusion, we remark that the definitions of the Lockett section and the Lausch group of a Fitting class do not require solubility, and that in consequence the fundamental Theorems 3.15 and 3.22 are true for arbitrary $\langle s_n, N_0 \rangle$-closed classes $\mathfrak{F}$ of finite groups.

4.   Recent Developments

The most important achievement in finite group theory in the last twenty years is undoubtedly the classification of the finite simple groups. Some long-standing open questions about finite groups can now be resolved by systematically checking the simple groups; that is the alternating, Chevalley and sporadic groups. The following theorem is an example. It may be viewed as a generalization of the theorem of Wielandt on the solubility of the product of two nilpotent groups.

THEOREM 4.1 (E. Fisman [20])   *Let $G = AB$, where A and B are soluble subgroups of relatively prime orders. Then each insoluble chief factor of G is isomorphic with one of the following groups:*

$$PSL(2, q) \text{ with } q = 2^n \ (n \geqslant 2) \text{ or } q \equiv -1 (\text{mod } 4); \quad PSL(3, 3); \quad M_{11}.$$

Another conjecture that has been resolved in this way is mentioned by Hall in his 1956 paper entitled 'Theorems like Sylow's' [38], where he uses the following terminology:

A finite group G is said to satisfy

E_π *if it has at least one conjugacy class of Hall π-subgroups,*
C_π *if it has exactly one such class, and*
D_π *if it satisfies C_π and every π-subgroup of G is contained in some Hall π-subgroup of G.*

In this language his characterization of finite soluble groups may be stated thus: a soluble group satisfies D_π for all $\pi \subseteq \mathbb{P}$, and if a group satisfies $E_{p'}$ for all $p \in \mathbb{P}$, then it is soluble. The conjecture referred to can now be formulated as follows:

CONJECTURE   *If G satisfies $E_{p,q}$ for all $\{p,q\} \subseteq \mathbb{P}$, then G is soluble.*

Hall established that for $n \geqslant 5$, the alternating group A_n does not satisfy $E_{p,q}$ for any pair of odd primes p,q. Z. Arad and M. B. Ward [2], building on earlier work of E. L. Spitznagel [60], prove

THEOREM 4.2   *Let G be a simple Chevalley group of characteristic p which satisfies $E_{p,q}$ for all odd primes q. Then $G \cong PSL(2,p)$, where p is a Mersenne prime > 3, and in this case G fails to satisfy $E_{2,p}$.*

For each of the 26 sporadic groups Arad and Ward find a pair of odd primes $\{p,q\}$ for which $E_{p,q}$ fails to be satisfied, and since a minimum counterexample to the conjecture is clearly simple, with this the conjecture becomes a theorem.

In [2] Arad and Ward also consider groups which satisfy $E_{2'}$. Such a group contains a Sylow 2-complement, which is soluble by the Feit–Thompson theorem, and therefore by Hall's theorem it satisfies $E_{p,q}$ for all pairs of odd primes p and q. But by Theorem 4.2 a simple group of this type must be isomorphic with $PSL(2,p)$ (p a Mersenne prime > 3), and since these groups do not satisfy $E_{3'}$, they obtain the following improvement of Hall's characteristic property of soluble groups.

THEOREM 4.3 (Arad and Ward [2])   *If a group G satisfies $E_{2'}$ and $E_{3'}$, then G is soluble.*

On p. 288 of [38] Hall also refers to the following conjecture:

(4.4)   *If $2 \notin \pi$ and G satisfies E_π, then G satisfies D_π.*

We show below that in general this is false; but in the special case $\pi = \mathbb{P} \setminus \{2\}$ it is true.

THEOREM 4.5 (Arad and Ward [2])   *A group which satisfies $E_{2'}$ also satisfies $D_{2'}$.*

We now indicate how this theorem can be deduced from the other results. This first step is to prove

PROPOSITION 4.6 [2]   *If G satisfies $E_{2'}$, then it satisfies $C_{2'}$.*

Proof.   A non-abelian simple group which satisfies $E_{2'}$ is isomorphic with $PSL(2,p)$ for some Mersenne prime $p > 3$ and satisfies $C_{2'}$. A minimal counterexample G to the Proposition therefore has a proper normal subgroup $N \neq 1$. Since G/N and N inherit $E_{2'}$ from G, both satisfy $C_{2'}$, and by the Feit–Thompson theorem [17] the Hall $2'$-subgroups of G/N are soluble. Therefore G satisfies $C_{2'}$ by Theorem C1 of Hall [38].

We now prove Theorem 4.5. Let G be a group with a Hall $2'$-subgroup. If G is simple, by Theorem 4.2 there is a Mersenne prime $p > 3$ such that $G \cong PSL(2,p)$, and scrutiny of its list of subgroups (found by L. E. Dickson) confirms that G satisfies $D_{2'}$. Therefore suppose that G has a proper minimal normal subgroup K. Then G/K satisfies $D_{2'}$ by induction, and if K is soluble, then G satisfies $D_{2'}$ by Theorem D5 of Hall [38] and the Feit–Thompson theorem. Hence we can suppose that $F(G) = 1$. Then K, which inherits $E_{2'}$ from G, must be a direct product of copies of a $PSL(2,p)$ for some Mersenne prime $p > 3$; the structure of K is therefore known and the properties of it used below can be checked. In particular, K satisfies $D_{2'}$.

Let H be a fixed Hall $2'$-subgroup and M a maximal $2'$-subgroup of G. We must show that M is contained in some conjugate of H. If $P \in \mathrm{Syl}_p(K)$, it is a property of K that $N_K(P) \in \mathrm{Hall}_{2'}(K)$, and by the Frattini argument $N_{KM}(P) \in \mathrm{Hall}_{2'}(KM)$. If $KM < G$, by induction $M \leqslant N_{KM}(P)$ for a suitable choice of P. Thus we have $M \leqslant N_G(P) < G$ since $F(G) = 1$. Because $N_G(P)/N_K(P)$ ($\cong G/K$) satisfies $E_{2'}$ and $|N_K(P)|$ is odd, $N_G(P)$ satisfies $E_{2'}$, and therefore by induction M is contained in a Hall $2'$-subgroup H^* of $N_G(P)$. But H^* is a Hall $2'$-subgroup of G and so is conjugate to H by Proposition 4.6.

Consequently we can suppose that $KM = G$. Since $K \cap M$ has odd order and K satisfies $D_{2'}$, we have $K \cap M \leqslant N_K(P)$ for some P. If $M \leqslant N_G(P)$ ($\in \mathrm{Hall}_{2'}(G)$), the proof is complete. If not, the set $M \setminus N_G(P)$ contains an element x of odd order, and then

$$K \cap M \leqslant N_K(P) \cap N_K(P^x) \cap N_K(P^{x^2}).$$

In K, however, only the identity element normalizes 3 distinct Sylow subgroups of K; thus $K \cap M = 1$.

Since G/K has odd order, it is soluble, and a minimal normal subgroup T/K of G/K is an elementary abelian r-group for some odd prime r. Let $R = T \cap M$, and put $N = N_G(R)$. Then $M \leqslant N$, and $N < G$ because $F(G) = 1$. Analysis of the possible action of an elementary abelian r-group on K shows that $C = C_K(R)$ ($= N \cap K$) satisfies $C_{2'}$, and since N/C ($\cong M$) has odd order, we conclude from Theorem E2 of Hall [38] that N satisfies $E_{2'}$ and therefore satisfies $D_{2'}$ by induction. Hence M is a Hall $2'$-subgroup of N, and C is a 2-group. Let

$R \leqslant R^* \in \mathrm{Syl}_r(T)$. If $R < R^*$, then $R < N_{R^*}(R)$, and it follows that $r \mid |C|$, which is not the case. Consequently $R \in \mathrm{Syl}_r(T)$. Since $H \cap T \in \mathrm{Hall}_{2'}(T)$, by Sylow's theorem we may suppose that $R \leqslant H \cap T$. By the Frattini argument applied to the normal subgroup $H \cap T \, (= (H \cap K)R)$ of H, we have $H = (H \cap K)N_H(R)$, and so $G = KH = KN_H(R)$. Hence $N_H(R)$ is a $2'$-subgroup of N of order at least $|G:K| = |M|$. Since M is a Hall $2'$-subgroup of N, which satisfies $D_{2'}$, we conclude that $M = N_H(R)^n \leqslant H^n$ for some $n \in N$, as required.

Fletcher Gross has proved the following theorem, which shows decisively that (4.4) is false.

THEOREM 4.7 *Let π be a finite set of odd primes with $|\pi| > 1$. Then there exists a finite group G such that*

(1) *every prime in π divides $|G|$;*
(2) *G satisfies C_π; and*
(3) *G does not satisfy D_π.*

I am grateful to him for the following example, which concerns the case $\pi = \{3, 5\}$. If $G = SL(3, 11)$, and if α is a primitive 5th root of unity in $\mathbb{F}_{11}$, then the matrices

$$\begin{pmatrix} \alpha & 0 & 0 \\ 0 & \alpha^{-1} & 0 \\ 0 & 0 & 1 \end{pmatrix} \quad \text{and} \quad \begin{pmatrix} 0 & 1 & 0 \\ 0 & 0 & 1 \\ 1 & 0 & 0 \end{pmatrix}$$

generate a subgroup H of G of order $3 . 5^2$, and since

$$|G| = 2^4 . 3 . 5^2 . 7 . 11^3 . 19,$$

we have $H \in \mathrm{Hall}_{\{3, 5\}}(G)$. Furthermore, because any group of order $3 . 5^2$ has a normal Sylow 5-subgroup, it follows that G satisfies $C_{\{3,5\}}$. Now the subgroup H is a Frobenius group and therefore contains no subgroup of order 15. On the other hand, G contains a subgroup isomorphic with $GL(2, 11)$, which in turn has a subgroup isomorphic with $\mathbb{F}_{11^2}^\times \cong C_{11^2-1}$. Therefore G fails to satisfy $D_{\{3,5\}}$.

Gross has also informed me that he knows of no counterexample to the

CONJECTURE *If $2 \notin \pi$ and G satisfies E_π, then G satisfies C_π.*

We conclude with another result recently announced by Gross.

THEOREM 4.8 (Gross [31]) *If the automorphism groups of the non-abelian composition factors of a group G satisfy E_π, then G does; if they satisfy D_π and the factors themselves satisfy C_π, then G satisfies D_π.*

It is disappointing that the generalized Sylow theory of soluble groups has not played a more significant part in the classification of simple groups. It may find a role in the revision (cf. Gross [30]), but even if not, I believe that it has a contribution to make to the study of composite groups (cf. Wielandt [66]).

References

[1] Anderson, W., Injectors in finite soluble groups, *J. Algebra* **36** (1975), 333–338.

[2] Arad, Z., Ward, M. B., New criteria for the solvability of finite groups, *J. Algebra* **77** (1982), 234–246.

[3] Barnes, D. W., Kegel, O. H., Gaschütz functors on finite soluble groups, *Math. Z.* **94** (1966), 134–142.

[4] Berger, T. R., The smallest normal Fitting class revealed, *Proc. London Math. Soc.* (3) **42** (1981), 59–86.

[5] Blessenohl, D., Gaschütz, W., Über normale Schunck- und Fittingklassen, *Math. Z.* **118** (1970), 1–8.

[6] Brison, O. J., Relevant groups for Fitting classes, *J. Algebra* **68** (1981), 31–53.

[7] Bryce, R. A., Cossey, J., Fitting formations of finite soluble groups, *Math. Z.* **127** (1972), 217–223.

[8] Bryce, R. A., Cossey, J., Metanilpotent Fitting classes, *J. Austral. Math. Soc.* **17** (1974), 285–304.

[9] Bryce, R. A., Cossey, J., A problem in the theory of normal Fitting classes, *Math. Z.* **141** (1975), 99–110.

[10] Bryce, R. A., Cossey, J., Subgroup closed Fitting classes, *Math. Proc. Cambridge Phil. Soc.* **83** (1978), 195–204.

[11] Bryce, R. A., Cossey, J., Subgroup closed Fitting classes are formations, *Math. Proc. Cambridge Phil. Soc.* **91** (1982), 225–258.

[12] Carter, R. W., Nilpotent self-normalizing subgroups of soluble groups, *Math. Z.* **75** (1961), 136–139.

[13] Dark, R. S., On subnormal embedding theorems for groups, *J. London Math. Soc.* **43** (1968), 387–390.

[14] Dark, R. S., Some examples in the theory of injectors of finite soluble groups, *Math. Z.* **127** (1972), 145–156.

[15] Doerk, K., Über Homomorphe endlicher auflösbarer Gruppen, *J. Algebra* **30** (1974), 12–30.

[16] Doerk, K., Hawkes, T. O., On the residual of a direct product, *Arch. Math.* **30** (1978), 458–468.

[17] Feit, W., Thompson, J. G., Solvability of groups of odd order, *Pacific J. Math.* **13** (1963), 755–1029.

[18] Fischer, B., Klassen konjugierter Untergruppen in endlichen auflösbaren Gruppen, Habilitationsschrift, Universität Frankfurt am Main (1966).

[19] Fischer, B., Gaschütz, W., Hartley, B., Injektoren endlicher auflösbarer Gruppen, *Math. Z.* **80** (1963), 300–305.

[20] Fisman, E., On the product of two finite solvable groups, *J. Algebra* **80** (1983), 517–536.

[21] Fitting, H., Beiträge zur Theorie der endlichen Gruppen, *Jber. Dtsch. Math.-Ver.* **48** (1938), 77–141.

[22] Förster, P., Charakterisierungen einiger Schunckklassen endlicher auflösbarer Gruppen III, *J. Algebra* **62** (1980), 124–153.

[23] Gaschütz, W., Über modulare Darstellungen endlicher Gruppen, die von freien Gruppen induziert werden, *Math. Z.* **60** (1954), 274–286.

[24] Gaschütz, W., Zur Theorie der endlichen auflösbaren Gruppen, *Math. Z.* **80** (1963), 300–305.

[25] Gaschütz, W., Existenz und Konjugiertsein von Untergruppen, die in endlichen auflösbaren Gruppen durch gewisse Indexschranken definiert sind, *J. Algebra* **53** (1978), 389–394.

[26] Gaschütz, W., Zwei Bemerkungen über normale Fittingklassen, *J. Algebra* **30** (1974), 277–278.

[27] Gaschütz, W., Lectures on subgroups of Sylow type in finite soluble groups, *Notes on Pure Mathematics* **11**, Australian National University, Canberra (1979).

[28] Gaschütz, W., Eine Kennzeichnung der Projektoren endlicher auflösbarer Gruppen, *Arch. Math.* **33** (1979), 401–403.

[29] Gaschütz, W., Ein allgemeiner Sylowsatz in endlichen auflösbaren Gruppen, *Math. Z.* **170** (1980), 217–220.

[30] Gross, F., Elementary abelian operator groups and admissible formations, *J. Austral. Math. Soc.* (series A) **34** (1983), 71–91.

[31] Gross, F., On the existence of Hall subgroups (preprint), University of Utah, Salt Lake City.

[32] Hall, P., A note on soluble groups, *J. London Math. Soc.* **3** (1928), 98–105.

[33] Hall, P., A characteristic property of soluble groups, *J. London Math. Soc.* **12** (1937), 198–200.

[34] Hall, P., On the Sylow systems of a soluble group, *Proc. London Math. Soc.* (2) **43** (1937), 316–323.

[35] Hall, P., On the system normalizers of a soluble group, *Proc. London Math. Soc.* (2) **43** (1937), 507–528.

[36] Hall, P., The classification of prime-power groups, *J. reine angew. Math.* **182** (1940), 130–141.

[37] Hall, P., The construction of soluble groups, *J. reine angew. Math.* **182** (1940), 206–214.

[38] Hall, P., Theorems like Sylow's, *Proc. London Math. Soc.* (3) **6** (1956), 287–304.

[39] Harman, D., Characterizations of some classes of finite soluble groups, Ph.D. Thesis, University of Warwick (1981).

[40] Hartley, B., On Fischer's dualization of formation theory, *Proc. London Math. Soc.* (3) **19** (1969), 193–207.

[41] Hawkes, T. O., The family of Schunck classes as a lattice, *J. Algebra* **39** (1976), 527–550.

[42] Hawkes, T. O., On metanilpotent Fitting classes, *J. Algebra* **63** (1980), 459–483.

[43] Hawkes, T. O., On Gaschütz's theory of generalized Sylow subgroups, *Arch. Math.* **35** (1980), 15–22.

[44] Hawkes, T. O., Parker, D., On subgroups like Hall's, *Bull. London Math. Soc.* **13** (1981), 385–391.

[45] Heineken, H., Group classes defined by chief factor ranks, *Bollettino U.M.I.* (5) **16-B** (1979), 754–764.

[46] Huppert, B., Zur Sylowstruktur auflösbarer Gruppen II, *Arch. Math.* **15** (1964), 251–257.

[47] Huppert, B., 'Endliche Gruppen I', Springer-Verlag, Berlin, Heidelberg, New York (1967).

[48] Kegel, O. H., Produkte nilpotenter Gruppen, *Arch. Math.* **12** (1961), 90–93.

[49] Laue, H., Lausch, H., Pain, G. R., Verlagerung und normale Fittingklassen endlicher Gruppen, *Math. Z.* **154** (1977), 257–260.

[50] Lausch, H., On normal Fitting classes, *Math. Z.* **130** (1973), 67–72.

[51] Lockett, F. P., On the theory of Fitting classes of finite soluble groups, Ph.D. Thesis, University of Warwick (1971).

[52] Lockett, F. P., The Fitting class 𝔉*, *Math. Z.* **137** (1974), 131–136.

[53] Lubeseder, U., Formationsbildung in endlichen auflösbaren Gruppen, Dissertation, Christian-Albrechts-Universität Kiel (1963).

[54] Mann, A., A criterion for pronormality, *J. London Math. Soc.* **44** (1969), 175–176.

[55] Michler, G. O., The kernel of a block of a group algebra, *Proc. Amer. Math. Soc.* **37** (1973), 47–49.

[56] Prentice, M. J., 𝔛-normalizers and 𝔛-covering subgroups, *Proc. Camb. Phil. Soc.* **66** (1969), 215–230.

[57] Schmid, P., Every saturated formation is a local formation, *J. Algebra* **51** (1978), 144–148.

[58] Schunck, H., 𝔥-Untergruppen in endlichen auflösbaren Gruppen, *Math. Z.* **97** (1967), 326–330.

[59] Shamash, J., On the Carter subgroup of a solvable group, *Math. Z.* **109** (1969), 288–310.

[60] Spitznagel, E. L., Jr., Hall subgroups of certain families of finite groups, *Math. Z.* **97** (1967), 259–290.

[61] Steinberg, R., Complete sets of representations of algebras, *Proc. Amer. Math. Soc.* **13** (1962), 746–747.

[62] Wielandt, H., Über das Produkt von paarweise vertauschbaren nilpotenten Gruppen, *Math. Z.* **55** (1951), 1–7.

[63] Wielandt, H., Über Produkte von nilpotenten Gruppen, *Illinois J. Math.* **2** (1958), 611–618.

[64] Wielandt, H., Über die Normalstruktur von mehrfach faktorisierbaren Gruppen, *J. Austral. Math. Soc.* **1** (1960), 143–146.

[65] Wielandt, H., Topics in the theory of composite groups, *Mimeographed Lecture Notes*, Department of Mathematics, University of Wisconsin, Madison (1967).

[66] Wielandt, H., Zusammengesetzte Gruppen: Hölders Programm heute, *Proceedings of Symposia in Pure Mathematics* **37**, 161–173, American Mathematical Society, Providence (1980).

[67] Wood, G. J., A lattice of homomorphs, *Math. Z.* **130** (1973), 31–37.

[68] Wright, C. R. B., On internal formation theory, *Math. Z.* **134** (1973), 1–9.

[69] Yen, T., Permutable pronormal subgroups, *Proc. Amer. Math. Soc.* **34** (1972), 340–342.

Topics in the Theory of Nilpotent Groups

B. HARTLEY

1. Finite p-groups 62
2. Augmentation ideals and the lower central series 83
3. Some residual nilpotence questions 99
4. Actions on the lower central factors of a free group 110
References 116

This article is not intended to be an encyclopaedic account of the theory of nilpotent groups, nor even of any particular part of that theory. It is an enormous area, and I would not be competent to write a full account of it, even if space were available. Rather than covering a large area somewhat superficially, with statements of results and brief comments only, I have elected to consider a fairly small number of topics in greater depth, with outlines of some of the more important arguments, to convey a better impression of how these areas have developed and are developing. Thus, wide areas have been neglected. The choice of topics has been guided largely by my own personal preferences, but I have tried to select fields where important recent developments have occurred, and which, I believe, offer interesting challenges for further investigation.

We begin with a section on finite p-groups, referring particularly to recent work of Leedham-Green, McKay and others on p-groups of maximal class and the approach to p-groups in terms of their co-class. Naturally, this cannot be discussed without describing the earlier work of Blackburn, and the fundamental contributions of P. Hall in the 1930's. Next, we deal with some topics related to group rings, powers of the augmentation ideal, and dimension subgroups. Sjogren's paper on the integral dimension subgroups receives parti-cular attention. It gives the best results available so far on these

GROUP THEORY: essays for Philip Hall
ISBN 0-12-304880-X

subgroups, and introduces interesting new techniques. We go on to discuss some questions about residually nilpotent groups, dealing especially with wreath products and groups arising from free presentations $(F/R', F/\gamma_{c+1}(R)$, and so on). Šmel'kin has given an interesting approach to some of these questions using Lie algebras and the Mal'cev correspondence, and this is discussed at some length. Section 4 is devoted to the induced actions of automorphism groups of free groups on the lower central and related factors of the group.

A particularly notable omission is the recent work[‡] of Grunewald and Segal on the isomorphism problem for finitely generated nilpotent groups. They prove that this problem is algorithmically soluble as part of a study of more general algorithmic questions in algebra [23], by sophisticated machinery involving finiteness theorems for arithmetic groups, among other things. The introduction of this type of finiteness theorem into the theory of nilpotent groups is due to Pickel [79].

This essay was written while the author was visiting the Universities of Utah and of Singapore. It is a pleasure to thank both universities for their generous hospitality, and the United Kingdom–United States Educational Commission for a Fulbright Travel Grant.

1.   Finite p-groups

1.1   Introduction

There are many beautiful and deep theorems relating the structure of finite soluble groups to their nilpotent subgroups or sections. We have for example the p-length theory of Hall and Higman, the theory of Carter subgroups and system normalizers,[§] and the theory of fixed point free automorphism groups developed by Thompson, Dade, Berger and others. By contrast, a general finite p-group remains a rather mysterious object, and although there has been striking progress in the study of several more or less restricted classes of finite p-groups (regular p-groups and p-groups of maximal class, for instance) our understanding of properties common to all finite p-groups has probably not advanced greatly since the celebrated 1934 paper of P. Hall [28] and its successor [29], in which the foundations of the systematic study of finite p-groups were laid. He expresses the opinion there that 'the astonishing multiplicity of these groups is one of the main difficulties which beset the advance of finite-group theory', and we see that illustrated above. On the other hand, however, the classification of finite

‡ See the essay by Grunewald and Segal.
§ See Hawkes' essay, p. 16.

simple groups has been brought to its present state in spite of the unsatisfactory condition of finite p-group theory.

This 'astonishing multiplicity', which seems to rule out any fine
classification, was asymptotically determined by C. C. Sims [87].
Building on earlier work of G. Higman [47], he showed that the number
of (non-isomorphic) groups of order p^n is $p^{A(n)n^3}$, where

$$A(n) = \tfrac{2}{27} + O(n^{-1/3}).$$

In fact, Higman showed that the number of finite groups of order p^n
having a central elementary abelian Frattini subgroup is

$$\geqslant p^{2(n^3 - 6n^2)/27},$$

and the diversity of p-groups of nilpotency class two is frequently
emphasized by the emergence of new phenomena in these groups. For
example, Heineken [46] has constructed infinitely many such groups of
exponent p^2 ($p \geqslant 3$) all of whose normal subgroups are characteristic,
and Heineken and Liebeck [44] have shown that if p is odd and K is a
given finite group, then there exist finite p-groups G of nilpotency class
two such that $\operatorname{Aut} G/\operatorname{Aut}_c G \cong K$, where $\operatorname{Aut}_c G$ denotes the group of
central automorphisms of G. See also [45]. Torsion-free nilpotent
groups of class two are also remarkably complex, and the classification
of such groups G when G' is free abelian of rank two and G/G' has
a free abelian part of rank four is not a straightforward matter (see
Grunewald, Segal and Sterling [24]).

The basic tools introduced by Hall for the study of finite p-groups
were the various now standard types of commutator subgroup, and the
'commutator collection process', which provides the link between the
power and commutator structure of a finite p-group. These, together
with the Lie ring and, more recently, cohomology, remain the indispensable apparatus of the subject. The commutator collection process
of [28] is the process of rearranging the power product $(xy)^m$, formed
from two elements x, y of a group, as a product $x^m y^m \ldots$ in which $x^m y^m$
is followed by commutators in x and y of increasing complexity.
Nowadays, a number of other commutator collection processes, adapted
to other contexts, are in circulation, but we shall not be able to discuss
them here (cf. [103]). The important point in these processes is to keep
track of the form and powers of the commutators introduced. Let
$x = z_1, y = z_2, z_3, \ldots$ denote the formally distinct complex commutators
in the elements x and y of a group G, arranged in order of increasing
weight, but otherwise arbitrarily. Then the version of the commutator

collection process stated in [28] is that there exist integer valued polynomials $f_1(X), f_2(X), \ldots$, with rational coefficients, independent of G and with certain other properties, such that

$$(xy)^m = z_1^{f_1(m)} z_2^{f_2(m)} \ldots \ldots \qquad (1)$$

This infinite product is to be interpreted as a congruence modulo any desired term of the lower central series by omitting terms of higher weight. This formula, or one of its relatives such as the Hall–Petresco Formula, is the basis upon which the theory of regular p-groups is built.

HALL–PETRESCO FORMULA ([49], p. 317) *Let x and y be free generators of a free group F. Then there exist elements $c_i \in \gamma_i(F)$ $(i = 2, 3, \ldots)$ such that for all $m > 0$ we have*

$$x^m y^m = (xy)^m c_2^{\binom{m}{2}} \ldots c_{m-1}^{\binom{m}{m-1}} c_m.$$

The theory of regular p-groups is so well known that we shall not say much about it here (see [28] or [49]). We content ourselves with saying that the power structure of regular p-groups is very similar to that of finite abelian p-groups, even down to close analogues of the usual invariants by which the latter groups are classified (though they no longer provide a classification), and pointing out that if $\mho^1(G)$ denotes the group generated by the pth powers of the elements of G, then the finite p-group G is regular provided $|G/\mho^1(G)| < p^p$ ([29], p. 472).

Although the applications of (1) to the power structure of regular p-groups are well known, its applications to the power structure of general finite p-groups may be less so. In [29], Theorem 2.5, Hall proves

THEOREM 1.1.1 *Let G be any finite p-group and H any normal subgroup of G. Then there exists in H a sequence*

$$1 = L_0 \leqslant L_1 \leqslant \ldots \leqslant H \qquad (2)$$

of normal subgroups of G with the properties

(i) *L_i/L_{i-1} is of order at most p^{p-1} and exponent dividing p, and*
(ii) *either $|L_i| = p^{i(p-1)}$ exactly, or L_i is a characteristic subgroup of H, and consists of all the elements of H whose order divides p^i.*

Now the polynomial $f_j(X)$ occurring in (1) actually has degree $\leqslant w$, the weight of z_j in x and y, is integer valued, and takes the value 0 when $X = 0$. Hence $f_j(X)$ is a $\mathbb{Z}$-linear combination of binomial coefficients $\binom{X}{k}$ with $k \leqslant w$. Thus if $p^\alpha \geqslant p^\beta > w > 0$, then

$$p^{\alpha - \beta + 1} \mid f_j(p^\alpha).$$

Let $x \in G$ and $y \in L_i$. Then from (1),

$$(xy)^{p^i} = x^{p^i} y^{p^i} z_3^{f_3(p^i)} \ldots . \tag{3}$$

We have $y^{p^i} = 1$ by Theorem 1.1.1(i). The series (2) can be refined to a series with factors of order p, central in G. Hence if z_j has weight $w \geqslant 2$ and $p^{\beta - 1} \leqslant w < p^\beta \leqslant p^i$, then $z_j \in L_{i - \beta + 1}$. The exponent of this group divides $p^{i - \beta + 1}$, which in turn divides the exponent of z_j in (3). Commutators of weight $\geqslant p^i$ in (3) are obviously trivial. We obtain the following, in which only property (i) of the L_i is used.

COROLLARY 1.1.2   *If $x \in G$ and $y \in L_i$, then x and xy have the same p^ith power.*

This, when G equals H, is one of the ingredients which enables information to be obtained in [29] about the number of solutions of the equation

$$x^{p^k} = a \tag{4}$$

and much more general equations and systems of equations, in a finite group. For by Corollary 1.1.2, the solutions of (4) in a finite p-group constitute the union of a set of cosets of L_k, and so their number is divisible by $p^{k(p-1)}$, unless the set of solutions of $x^{p^k} = 1$ in G is a subgroup $L_k = \Omega_k(G)$ of order $< p^{k(p-1)}$. In that case, the number is divisible by $|\Omega_k|$.

It is also easy to deduce by induction on m, using Corollary 1.1.2, that if $L_{k,1}, L_{k,2}, \ldots, L_{k,m}$ are m subgroups, each of which enters as the kth term in a sequence (2) of $H = G$, then their product has exponent dividing p^k. Thus we obtain, in an arbitrary finite p-group G, a characteristic subgroup $\bar{\Omega}_k(G)$ of exponent dividing p^k, generated by all possible L_k. Further, if $x \in G$ and $y \in L_k$, the equation $x^{p^k} = (xy)^{p^k}$, which is the statement of Corollary 1.1.2, shows that x^{p^k} commutes with xy and hence with y. Thus

$$[\mho^k(G), \bar{\Omega}_k(G)] = 1,$$

where $\mho^k(G) = \langle x^{p^k} : x \in G \rangle$, as usual. Note that $|\bar{\Omega}_k(G)| \geqslant p^{k(p-1)}$ unless $\bar{\Omega}_k(G)$ consists of all elements in G of order dividing p^k. Thus, at least some of the theory of regular p-groups has an analogue in arbitrary finite p-groups; in the regular case, of course, $\bar{\Omega}_k(G) = \Omega_k(G)$, the set of elements of order dividing p^k.

1.2   Groups of maximal class

A p-group G of order p^n can have nilpotency class at most $n-1$, since either G is cyclic or p^2 divides the index of the Frattini subgroup of G. Groups of order p^n with $n > 2$ and nilpotency class exactly $n-1$ are known as *p-groups of maximal class*. Their structure was considerably elucidated in the fundamental paper of Blackburn [7] (see also Huppert [49], p. 361 *et seq.* for an account of these groups) and thanks to work of Shepherd [86], Miech [72, 73], and Leedham-Green and McKay [59, 60, 61, 62], they are now reasonably well understood. Although these may appear to be a rather restricted class of p-groups, their structure is by no means transparent, and developing our understanding of them to its present level must be regarded as one of the achievements of the subject. As we have pointed out, it seems necessary to study finite p-groups under quite heavy restrictions if significant progress is to be made; furthermore, p-groups of maximal class arise quite often in other contexts.

The 2-groups of maximal class are the dihedral, semi-dihedral and generalized quaternion groups ([49], III.11.9), and groups of order p^4 are also well understood. For these reasons, in considering a p-group G of order p^n and maximal class, it is customary to assume $p \geqslant 3$ and $n \geqslant 5$. The subgroup $G_1 = C_G(\gamma_2(G)/\gamma_4(G))$ (where $X = \gamma_1(X) \geqslant \gamma_2(X) \geqslant \ldots$ denotes the lower central series of a group X) plays a fundamental role in the study of these groups. It is a characteristic subgroup of index p in G. Write $G_i = \gamma_i(G)$ for $i \geqslant 2$. The *degree of commutativity* of G, introduced by Blackburn, is the largest integer l such that

$$[G_i, G_j] \leqslant G_{i+j+l}$$

for all $i, j \geqslant 1$, provided G_1 is not abelian. When G_1 is abelian, the degree of commutativity is taken as $n-3$, because $[G_1, G_1] = [G_1, G_2]$. One of the main achievements in Blackburn's paper is to show that the degree of commutativity is non-zero in most cases. Before making this precise, we note that

> if G is a finite p-group of maximal class and G_1 is not abelian, then the degree of commutativity of G is $\geqslant l$ if and only if $[G_1, G_i] \leqslant G_{i+1+l}$ for all $i \geqslant 1$.

For if $[G_1, G_i] \leqslant G_{i+1+l}$ for all $i \geqslant 1$, a simple induction on j gives $[G_i, G_j] \leqslant G_{i+j+l}$ for all $j \geqslant 1$. Namely, we have $G_{j+1} = [G_j, G]$, even if $j = 1$, so $[G_i, G_{j+1}] = [G_i, [G_j, G]] \leqslant [G_i, G, G_j][G_i, G_j, G] \leqslant G_{i+j+1+l}$. For this reason, we define a p-group G of maximal class to be *exceptional*, if and only if $G_1 \neq C_G(G_i/G_{i+2})$ for some i, which amounts to the same thing as $[G_1, G_i] \not\leqslant G_{i+2}$. Thus, the exceptional groups are those whose degree of commutativity is zero.

THEOREM 1.2.1 (Blackburn [7], see [49], III.14.6) *Let G be a finite p-group of maximal class with $|G| = p^n$ and $n \geqslant 5$. Then*

(i) *G/G_{n-1} is not exceptional. Thus $G_1 = C_G(G_i/G_{i+2})$ for $i \leqslant n-3$.*
(ii) *If G is exceptional, then $p > 3$ and $6 \leqslant n \leqslant p+1$; also n is even.*

Thus, for each prime p, there are only finitely many p-groups of maximal class whose degree of commutativity is zero. Examples of such groups, for each even n in the range $4 \leqslant n \leqslant p+1$, were constructed in [7]; see also [49], III.14.24. They may be described as follows.

LEMMA 1.2.2 *For each $r \geqslant 1$, let $X = X(r)$ be the $2r \times 2r$ matrix*

$$\begin{pmatrix} 1 & 0 & 0 & 0 & \cdots & & & \\ 1 & 1 & 0 & 0 & \cdots & & & \\ 0 & 1 & 1 & 0 & \cdots & & & \\ 0 & 0 & 1 & 1 & \cdots & & & \\ \vdots & & & & & 1 & 1 & 0 \\ 0 & 0 & 0 & 0 & \cdots & 0 & 1 & 1 \end{pmatrix}$$

Then there exists a $2r \times 2r$ skew-symmetric matrix A over $\mathbb{Z}$ such that $\det A = 1$ and $XAX^T = A$, where X^T denotes the transpose of X.

To see this, let the skew-symmetric matrix sought be (a_{ij}). The condition $XAX^T = A$ is expressed by

$$a_{i-1,j-1} + a_{i,j-1} + a_{i-1,j} = 0 \quad (1 < i, j \leqslant 2r)$$

together with $a_{1,j-1} = 0$ $(1 < j \leqslant 2r)$ and $a_{i-1,1} = 0$ $(1 < i \leqslant 2r)$. Thus the first row and column of A are zero except perhaps for their last entries, and if

$$\begin{pmatrix} u & v \\ w & t \end{pmatrix}$$

is any 2×2 submatrix formed from the intersection of two successive rows and columns of A, then $u+v+w = 0$. We use this to fill in the matrix in successive strips $i+j = n$ for $n = 2, 3, \dots$. For $i+j = 2, 3, \dots 2r$, the entries in the corresponding strips are zero. For $i+j = 2r+1$ we put $a_{r,r+1} = 1$, $a_{r+1,r} = -1$ and complete the strip with $+1$ and -1 alternately working from the diagonal towards the corners. The next strip is uniquely determined by the conditions on the 2×2 submatrices, since the mid-point entry is zero, but in the strip $i+j = 2r+2$ we can choose one of the entries next to the diagonal arbitrarily and then work outwards as before. In this way, we can complete the entire matrix.

The matrix A can be interpreted as the matrix of a non-singular skew-symmetric bilinear form over the field of p elements, and so if H is an extraspecial group of order p^{2r+1} and exponent p, where $p > 2$, then H admits an automorphism α whose effect on H/H' has matrix X with respect to a suitable basis. The lower central series of the semi-direct product $G = \langle \alpha \rangle H$ is

$$G > [H, \alpha]H' > \dots > [H, {}_{2r-1}\alpha]H' > H' > 1$$

where $[H, {}_{n+1}\alpha] = [[H, {}_{n}\alpha], \alpha]$ for $n > 0$. If $2r \leqslant p$, then α can be chosen to have order p, the order of X, so G has order p^{2r+2} and maximal class. If $r \geqslant 2$, then $G_1 = H$, $G_{2r-1} = [H, {}_{2r-1}\alpha]H'$, and $[G_{2r-1}, H] \neq 1$, so the degree of commutativity of G is zero. Even if $2r \geqslant p$, the lower central factors of G other than the first have order p, so G belongs to $CF(n, 2r+1, p)$ for suitable n, in the notation of [7].

We digress briefly to mention some analogues of the above facts for torsion-free groups. Blackburn introduced the class $NCF(m)$ of all nilpotent groups G of class $m-1$ in which $\gamma_i(G)/\gamma_{i+1}(G)$ is infinite cyclic for $i = 2, 3, \dots, m-1$. The subgroup G_1 can be introduced as before and the degree of commutativity defined. Blackburn ([7], Theorem 2.11) shows that if m is odd and >4 and if $G \in NCF(m)$, then G has positive degree of commutativity. Examples to show that this is not true for even m can be constructed on the basis of Lemma 1.2.2. Let W be the set of all row vectors $\mathbf{w} = (w_1, \dots, w_{2r})$ ($r \geqslant 1$) with integer entries, and let H be the set of all pairs $(n, \mathbf{w})$ ($n \in \mathbb{Z}$, $\mathbf{w} \in W$), with multiplication given by

$$(n, \mathbf{w})(n', \mathbf{w}') = (n + n' + \mathbf{w} A \mathbf{w}'^T, \mathbf{w} + \mathbf{w}')$$

where A is any matrix given by Lemma 1.2.2. The group H has an automorphism α given by

$$(n, \mathbf{w})\alpha = (n + w_1, \mathbf{w}X)$$

and it is not hard to see that the semi-direct product $G = \langle \alpha \rangle H$ belongs to $NCF(2r+2)$. However, $G_1 = H$, which does not centralize G_{2r}.

Now we revert to the discussion of finite p-groups of maximal class. Let G be such a group, of order p^n and degree of commutativity l. Recall that $|G/G_1| = |G_1/G_2| = p$, and choose elements $s, s_1, \ldots, s_n$ such that

$$s \in G \setminus G_1$$

$$s_1 \in G_1 \setminus G_2$$

$$s_{i+1} = [s_i, s] \quad (1 \leqslant i \leqslant n-1).$$

These elements play a key role in the investigation of the degree of commutativity of G, and we shall continue to use this notation without further comment.

LEMMA 1.2.3   *If $l > 0$, then $G_i = \langle s_i, G_{i+1} \rangle$ for $1 \leqslant i \leqslant n-1$.*

Proof.   This follows easily by induction. If the assertion holds for some $i \leqslant n-2$, then as G_1 centralizes G_i/G_{i+2}, we have

$$G_{i+1} = [G_i, G] = [G_i, s]G_{i+2} = \langle [s_i, s], G_{i+2} \rangle.$$

Blackburn's proof that $l > 0$ when $n \geqslant p+2$, as stated in Theorem 1.2.1, proceeds by induction on n. Once the case $n = p+2$ has been dealt with, one can assume G/G_{n-1} has positive degree of commutativity, in which case the conclusion of Lemma 1.2.3 will hold for $1 \leqslant i \leqslant n-2$. The inductive step involves studying the interactions of the commutators and powers of these elements by rather intricate calculations based on commutator collection formulae.

Using Blackburn's Theorem, Shepherd [86] and Leedham–Green and McKay [59] have shown that the degree of commutativity of G is close to $\frac{1}{2}n$, and in fact $\frac{1}{2}n - l$ is bounded above by a simple function of p. In particular, it turns out that the class of G_1 is usually not more than three, and, in a certain sense, G_1 almost has class $\leqslant 2$. Some of the detailed results are as follows.

THEOREM 1.2.4 [59], [86]   *With the notation above, if $n \geqslant 4$ then*

$$2l \geqslant n - 3p + 6.$$

Notice that if $n \leqslant p+1$ this says nothing, as is to be expected, since l may be zero for such values of n. Thus, in proving the theorem, Theorem 1.2.1 allows us to assume that $l > 0$, in which case Lemma 1.2.3 holds for the elements $s, s_1, \ldots, s_{n-1}$. We have

$$[s_i, s_j] \equiv s_{i+j+l}^{a(i,j)} \bmod G_{i+j+l+1}$$

provided $i+j+l < n$, where the integers $a(i,j)$ are uniquely determined modulo p. The power and commutator relations between s and the s_i give numerical relations between the $a(i,j)$, and these turn out to be inconsistent unless the desired inequality holds.

Now it is easy to see that $G_1' = [G_2, G_1]$; thus

$$\gamma_4(G_1) = [G_2, G_1, G_1, G_1] \leqslant G_{3l+5}.$$

From Theorem 1.2.4, $3l+5 \geqslant n$ if $n \geqslant 9p-28$, so we obtain

COROLLARY 1.2.5 G_1 *has class at most* 3 *if* $n \geqslant 9p-28$.

Shepherd [86] has proved that the class of G_1 is at most $\frac{1}{2}(p+1)$ in general. Theorem 1.2.1 and Corollary 1.2.5 illustrate the way p-groups of maximal class tend to become more civilized when n is large compared with p.

From the corollary, the derived length of G_1 is certainly at most three if n is large enough, and can be bounded explicitly in terms of p. The existence of such a bound was first proved (non-constructively) by Alperin [1]. In fact, no example is known in which the class of G_1 exceeds three.

In a similar way to the corollary, we have

$$\gamma_3(G_1) = [G_2, G_1, G_1] \leqslant G_{2l+4},$$

and $2l+4 \geqslant n-3p+10$. We deduce the following

COROLLARY 1.2.6 $|\gamma_3(G_1)| \leqslant p^{3p-10}$.

This can be interpreted as saying that in a p-group G of maximal class, there is a normal subgroup of order bounded in terms of p, modulo which the class of G_1 is at most two. When so viewed, it provides a spur to an attempt to construct all p-groups G of maximal class in which G_1 has class at most two. This is done successfully and in elegant fashion by Leedham-Green and McKay [59, 60, 61], and also by Miech [72]. Miech [73] classifies metabelian p-groups of maximal class for which $n \geqslant \max(p+1, 2m+3)$, where m is given by $[G_1, G_2] = G_{n-m}$.

It is not hard to see that this inequality implies that G_1 has class at most two (see the remarks after Lemma 1.2.8), and so Miech's metabelian groups are produced by the construction of Leedham-Green and McKay, though they apparently do not classify them. In fact, if G is a metabelian p-group of maximal class, then G_1 has class at most two provided that $n \geqslant 2(p-1)$. For Blackburn [7], Theorem 3.10, has shown that $l \geqslant n-p-1$ for such groups, whence

$$\gamma_3(G_1) = [G_1, G_2, G_1] \leqslant G_{2(n-p-1)+4} = 1$$

if $n \geqslant 2(p-1)$. On the other hand, metabelian p-groups G of maximal class can be constructed in which the class of G_1 exceeds two. We outline a construction.

Let a, b be integers such that $5 \leqslant b \leqslant 2p-5$ and $2 \leqslant a < b$. Thus, we need $p \geqslant 5$. Let θ be a primitive complex pth root of 1, $\mathcal{O}$ the ring of integers in $\mathbb{Q}[\theta]$, and $\not{p}$ the prime ideal of $\mathcal{O}$ generated by $\kappa = \theta - 1$. We shall repeatedly use the fact that $p\mathcal{O} = \not{p}^{p-1}$. Let H be a semi-direct product $M\langle s \rangle$, where $M \cong \mathcal{O}/\not{p}^b$, s has order p, and s acts on M as multiplication by θ. Thus we have an isomorphism $\lambda: M \to \mathcal{O}/\not{p}^b$ such that $(m^s)\lambda = (m\lambda)\theta$ $(m \in M)$. Then $[M, {}_rH]\lambda = \not{p}^r/\not{p}^b$ $(1 \leqslant r \leqslant b)$, and H has maximal class and order p^{b+1}. Let $m_0 = (1 + \not{p}^b)\lambda^{-1}$. Then H has an automorphism τ sending s to $m_0 s$ and acting on M like multiplication by $1 + \xi$, where ξ is an element chosen from $\not{p}^a \setminus \not{p}^{a+1}$. We see that if

$$a \geqslant b - (p-2) \tag{1}$$

then τ^p is an inner automorphism induced by an element w such that $w\lambda \in \not{p}^{p-2}/\not{p}^b$; thus also $w^\tau = w$. In this case we can form a group $G = H\langle s_1 \rangle$, in which $H \lhd G$, s_1 acts on H like τ, and $s_1^p = w$. It is not hard to check that this is a metabelian p-group of maximal class and order p^{b+2}. We have $G_1 = \langle M, s_1 \rangle$, and taking $a = \max(2, b-(p-2))$, we find that the class of G_1 exceeds two. In fact, for $p \geqslant 7$, we can take $b = p-2$ and $a = 2$, and the class of G_1 is unbounded as p increases.

Now let us outline the construction given by Leedham-Green and McKay. The following will be useful.

LEMMA 1.2.7 *If $l > 0$ then*

(i) $C_{G_1}(s) = G_{n-1}$,
(ii) $(sy)^p \in G_{n-1}$ *if* $y \in G_1$,
(iii) sG_2 *is the set of G-conjugates of s*,
(iv) $s^p = (sy)^p$ *if* $y \in G_2$.

Proof. If (i) is false, then s commutes with an element $w \in G_i \setminus G_{i+1}$, where $1 \leqslant i \leqslant n-2$ and so centralizes G_i/G_{i+2}. Since G_1 centralizes G_i/G_{i+2}, it follows that $G = \langle G_1, s \rangle$ does. This contradiction shows that (i) holds. Therefore $s^p \in C_G(s) = G_{n-1}$; since s is just an arbitrary element of $G \setminus G_1$, we have (ii). By (i) again, $|G: C_G(s)| = p^{n-2}$, so s has p^{n-2} conjugates in G, and since these all lie in sG_2, they constitute that coset. Thus, if $y \in G_2$, then $(sy)^p$ is conjugate to s^p, and so $(sy)^p = s^p$ by (ii).

As a matter of fact, $s^p \in G_{n-1}$ even when $l = 0$; see Blackburn [7], Theorem 3.2.

Now let $\mathcal{O}$, $\mathcal{p}$, θ, κ be as before. Suppose $l > 0$ and let $A = G_u/G_v$ be abelian of order p^k, where $1 \leqslant k \leqslant n-2$. If $a \in G_u$, then

$$(sa)^p = s^p a^{s^{p-1}} a^{s^{p-2}} \ldots a^s a.$$

If $u > 1$, then $(sa)^p = s^p$ by Lemma 1.2.7(iv), so

$$a^{s^{p-1}} a^{s^{p-2}} \ldots a^s a = 1,$$

while if $u = 1$, then $G_v \geqslant G_{n-1}$, and the same holds by Lemma 1.2.7(ii), according to which $(sa)^p$ and s^p both lie in G_{n-1}. Thus

$$1 + s + \ldots + s^{p-1} = 0$$

in the endomorphism ring of A, and we can make A into an $\mathcal{O}$-module by allowing θ to act as conjugation by s.

LEMMA 1.2.8 $A \cong \mathcal{O}/\mathcal{p}^k$ *as $\mathcal{O}$-modules.*

By Lemma 1.2.3, the G_v-cosets of the elements $s_u, s_{u+1}, \ldots, s_{v-1}$ generate A as an abelian group. Since $(s_i G_v)(\theta - 1) = s_{i+1} G_v$ $(u \leqslant i < v-1)$, it follows that A is a cyclic $\mathcal{O}$-module, generated by $s_u G_v$. The map $1 \to s_u G_v$ extends to an epimorphism from $\mathcal{O}$ to A. The kernel contains $\mathcal{p}^k$, and must be exactly $\mathcal{p}^k$, by order considerations.

If G is metabelian, we can take $A = G_2$ and $k = n-2$. Since $[s, s_1]$ $\in G_2, s_1$ acts on A as an element of $\mathrm{Hom}_{\mathcal{O}}(A, A)$, and since it has order p, s_1 acts as multiplication by $1+\xi$ say, where $\xi \in \not{p}^a \setminus \not{p}^{a+1}$ for some $a > 0$. Defining m by $[G_1, G_2] = G_{n-m}$, as in the remarks following Corollary 1.2.6, we see that G_{n-m} corresponds to $\not{p}^a / \not{p}^{n-2}$ under the isomorphism of Lemma 1.2.8, in other words, $a = n - m - 2$. If $n \geqslant 2m+3$, then $2a > n-2$, so $\xi^2 \equiv \mathcal{O} \bmod \not{p}^{n-2}$. This amounts to the assertion that G_1 has class at most two when $n \geqslant 2m+3$.

Now assume that G_1 has class at most two, and choose m (unrelated to the m used just above) so that $G_1' \leqslant G_m \leqslant Z(G_1)$, the centre of G_1. Commutation induces a group homomorphism δ of the exterior square (over $\mathbb{Z}$) $G_1/G_m \wedge G_1/G_m$ to G_m. This is a θ-homomorphism if we allow θ to act diagonally on the exterior square, as in Lemma 1.2.8. However, the exterior square is not an $\mathcal{O}$-module, as it is not annihilated by $1+\theta+\ldots+\theta^{p-1}$. We can fill in the top row uniquely to make a commutative square of θ-homomorphisms

$$
\begin{array}{ccc}
\mathcal{O}/\not{p}^{m-1} \wedge \mathcal{O}/\not{p}^{m-1} & \xrightarrow{\;\alpha\;} & \mathcal{O}/\not{p}^{n-m} \\
{\scriptstyle f \wedge f}\Big\downarrow & & \Big\downarrow{\scriptstyle g} \\
G_1/G_m \wedge G_1/G_m & \xrightarrow[\;\delta\;]{} & G_m
\end{array}
\qquad (*)
$$

in which the vertical maps come from Lemma 1.2.8, and are isomorphisms. The degree of commutativity of G can be expressed in terms of α. A non-zero element $\beta \in \mathrm{Hom}_{\langle\theta\rangle}(\mathcal{O}/\not{p}^x \wedge \mathcal{O}/\not{p}^x, \mathcal{O}/\not{p}^y)$ is defined to have *degree d*, if, for all $i, j \geqslant 0$, we have

$$
\beta(\not{p}^i/\not{p}^x \wedge \not{p}^j/\not{p}^x) \leqslant \not{p}^{i+j+d} + \not{p}^y/\not{p}^y,
$$

and d is as large as possible subject to this. The zero map is assigned the degree $n - m - 1$. One verifies quite easily that in the square $(*)$, if G has degree of commutativity l and α has degree d, then

$$
l = m + d - 2. \qquad (2)
$$

The theorem showing how to construct all finite p-groups G of maximal class $(p \neq 2)$ such that $l > 0$, $|G| \geqslant p^4$, and G_1 has class at most two, is the following.

THEOREM 1.2.9 *Let p be odd, let m, n, u, v be integers such that $n \geqslant m+2 \geqslant 4$, and let $\alpha \in \mathrm{Hom}_{\langle \theta \rangle}(\mathcal{O}/\not{p}^{m-1} \wedge \mathcal{O}/\not{p}^{m-1}, \mathcal{O}/\not{p}^{n-m})$. Then there exists a group G of order p^n and maximal class, with $l > 0$ and $G_1' \leqslant G_m \leqslant Z(G_1)$, such that $s^p = s_{n-1}^u, (ss_1)^p = s_{n-1}^v$, and the diagram (*) commutes, where the various maps are constructed as described above. Furthermore, G is uniquely determined by m, n, u, v and α, and every finite p-group H of maximal class, positive degree of commutativity, and order $\geqslant p^4$, with H_1 of class at most two, arises in this way (possibly several times).*

This is the version stated in [60] as 4.3, and differs slightly from the version in [59]. That every finite p-group H of the type described arises from the construction follows from the discussion just given, in particular, Lemmas 1.2.7 and 1.2.8. Uniqueness cannot be expected, since there is a great deal of flexibility in the choices for s and s_1. However, the given parameters determine multiplication completely. For G is generated by $s, s_1, \ldots, s_{n-1}$. The commutation relations among these elements are determined by the relations $s_{i+1} = [s_i, s]$ ($1 \leqslant i \leqslant n-2$) and α, so we just have to see that s_i^p is determined as a product of powers of $s_{i+1}, \ldots, s_{n-1}$. For $i = 1$, this is done by rearranging the relation $(ss_1)^p = s_{n-1}^v$ by means of a suitable commutator collection formula, and larger values of i can be dealt with via the relation $[s_{i-1}, s] = s_i$. So the main point at issue is the construction of G from the given data.

Let $A = \mathcal{O}/\not{p}^{n-m}$, $B = \mathcal{O}/\not{p}^{m-1}$ and $C_1 = \mathcal{O}/\not{p}^{n-1}$. We have two extensions

$$0 \longrightarrow A \overset{\varepsilon_1}{\longrightarrow} C_1 \overset{\pi_1}{\longrightarrow} B \longrightarrow 0,$$

$$0 \longrightarrow A \overset{\varepsilon_2}{\longrightarrow} C_2 \overset{\pi_2}{\longrightarrow} B \longrightarrow 0,$$

in which some confusion of additive and multiplicative notation seems inevitable. The first is the natural one. In the second,

$$C_2 = \{(a, b); a \in A, b \in B\}$$

with multiplication given by

$$(a_1, b_1)(a_2, b_2) = (a_1 + a_2 + \tfrac{1}{2}\alpha(b_1 \wedge b_2), b_1 + b_2).$$

Thus C_2 is a group of nilpotency class two if $\alpha \neq 0$, and $\varepsilon_2(a) = (a, 0)$; $\pi_2(a, b) = b$ ($a \in A, b \in B$). Let $\langle s \rangle$ be a cyclic group of order p. Then $\langle s \rangle$

operates on both C_1 and C_2; it acts as multiplication by θ on A, B and C_1, and as multiplication by θ on the coordinates of the elements of C_2. The maps in our two sequences are s-homomorphisms. We now form the Baer sum of these two extensions (a homological discussion of this construction is given in [60]). This is an extension

$$0 \longrightarrow A \overset{\varepsilon}{\longrightarrow} G_1 \overset{\pi}{\longrightarrow} B \longrightarrow 0$$

obtained as follows. Let

$$C = \{(c_1, c_2) \in C_1 \times C_2; \, \pi_1(c_1) = \pi_2(c_2)\},$$

$$D = \{(\varepsilon_1(a), \varepsilon_2(a)); \, a \in A\},$$

notice that D is normal, and even central, in C, and put $G_1 = C/D$. The maps are the obvious ones:

$$\varepsilon(a) = D(\varepsilon_1(a), 0) = D(0, \varepsilon_2(-a));$$

$$\pi(D(c_1, c_2)) = \pi_1(c_1) = \pi_2(c_2).$$

Because the π_i and ε_i are s-maps, s preserves C and D and acts on $C/D = G_1$. Putting $G_i = [G_{i-1}, s]$ ($i \geq 2$) we see easily that $G_m = (\varepsilon_1(A) \times \varepsilon_2(A))/D$, which has order p^{n-m}, and deduce easily that $|G_j| = p^{n-j}$ for $j = 1, 2, \ldots, n$. Thus, the semi-direct product $G = G_1 \langle s \rangle$ is a group of order p^n and class $n-1$. Its lower central series is $G > G_2 > G_3 > \ldots > G_n = 1$, and because $n - m \geq 2$, $G_1 = C_G(G_2/G_4)$, as it should be. Again, $G_{n-2} \leq G_m \leq Z(G_1)$, so by Theorem 1.2.1(i), G has positive degree of commutativity. It is amusing to note that by (2) we now know that α has degree at least $3-m$, which is not obvious *a priori*. Now, choosing $s_1, s_2, \ldots$ in the usual way, we see that G satisfies all the desired conditions, with the possible exception of $s^p = s_{n-1}^u$ and $(ss_1)^p = s_{n-1}^v$. In fact, we have $s^p = 1$ by construction, and $(ss_1)^p = s_{n-1}^w$ by Lemma 1.2.7. A slight modification yields the other conditions. Form $G \times A$, where $A = \langle a \rangle \times \langle b \rangle$ and a and b have order p^2. Put $t = sa$, $t_1 = s_1 b$, and $t_i = s_i$ ($2 \leq i \leq n-1$). Then $[t_i, t] = t_{i+1}$ ($1 \leq i \leq n-2$), and if $L = \langle t, t_1 \rangle$, then the lower central series of L is

$$L > G_2 > G_3 > \ldots > G_{n-1} > G_n = 1.$$

Factoring out the central subgroup $\langle t^p t_{n-1}^{-u}, (tt_1)^p t_{n-1}^{-v} \rangle$ from L gives a group having all the properties required of G.

The problems remaining after Theorem 1.2.9 are the calculation of $\mathrm{Hom}_{\langle\theta\rangle}(\mathcal{O}/\not{h}^{m-1} \wedge \mathcal{O}/\not{h}^{m-1}, \mathcal{O}/\not{h}^{n-m})$, and the solution of the isomorphism problem for the groups constructed. The former problem is dealt with in [59], [60], [61], and as it is rather technical, we shall not go into it. The isomorphism problem is discussed in [62] in a special case, the so-called *groups of type III*, in which, among other things, $u = v = 0$ and $2l \geqslant n-4$. It turns out that for these groups, the homomorphisms α that arise all lift to elements of $\mathrm{Hom}_{\langle\theta\rangle}(\mathcal{O} \wedge \mathcal{O}, \mathcal{O})$. Let K be the field of fractions of $\mathcal{O}$. Then it is not too hard to write down elements S_a ($2 \leqslant a \leqslant \frac{1}{2}(p-1)$) which form a K-basis of $\mathrm{Hom}_{\langle\theta\rangle}(\mathcal{O} \wedge \mathcal{O}, K)$ (see subsection 1.3 below). Given a group of type III, we make the choice $m = l+3$, and then a choice for s and s_1 leads to an element $\alpha \in \mathrm{Hom}_{\langle\theta\rangle}(\mathcal{O} \wedge \mathcal{O}, \mathcal{O})$, which determines the structure of the group. Then α in turn leads to $\frac{1}{2}(p-3)$ parameters $c_a \in K$, the coefficients obtained when α is expressed in terms of the S_a. It is in terms of these parameters that the isomorphism problem is discussed in [62], where it is transformed to a number theoretic question and discussed in detail for groups in which only one of the parameters c_a is non-zero.

We conclude this subsection by noting that the degree of commutativity of the groups constructed as in Theorem 1.2.9 can be read off from the degree of α and (2), and in this way, it can be seen that the bound for l given in Theorem 1.2.4 is close to best possible.

1.3   The coclass of a finite p-group

The detailed results just discussed for p-groups of maximal class can be viewed as the first stage of a programme to study finite p-groups in terms of their *coclass*, which is defined to be $n-c$ for a group of order p^n and class c. This programme was proposed by Leedham–Green and Newman [63], who put forward a number of interesting conjectures suggesting the direction such a programme might take. As p-groups of maximal class have coclass 1, these conjectures are largely based on information about that case. For a p-group G of maximal class, Theorem 1.2.4 shows immediately that G_{p-2} has class at most two. This suggests the following, which, like the other conjectures in this section, was put forward by Leedham-Green and Newman.

CONJECTURE 1.3.1   *An arbitrary finite p-group has a normal subgroup of class at most two and index bounded in terms of p and the coclass.*

The index of a normal abelian subgroup cannot be so bounded, except perhaps when $p = 2$. For $p \geqslant 5$, appropriate examples of maximal class can be obtained by choosing α suitably in Theorem 1.2.9. For an integer a such that $(a,p) = 1$, let σ_a be the automorphism of $\mathcal{O}$ mapping θ to θ^a. Assuming also $(a-1,p) = 1$, we have elements $S_a \in \mathrm{Hom}_{\langle\theta\rangle}(\mathcal{O} \wedge \mathcal{O}, \mathcal{O})$ given by

$$S_a(x \wedge y) = \sigma_a(x)\sigma_{1-a}(y) - \sigma_a(y)\sigma_{1-a}(x) \quad (x,y \in \mathcal{O}).$$

If K is the field of fractions of $\mathcal{O}$, then the S_a $(2 \leqslant a \leqslant \frac{1}{2}(p-1))$ form a K-basis for $\mathrm{Hom}_{\langle\theta\rangle}(\mathcal{O} \wedge \mathcal{O}, K)$ ([59], Theorem 3.7). Now, with $\kappa = \theta - 1$, we have

$$S_a(p^i \wedge p^j\kappa) = p^{i+j}(\sigma_{1-a}(\kappa) - \sigma_a(\kappa)) = p^{i+j}(\theta^{1-a} - \theta^a)$$

$$= \theta^{1-a}p^{i+j}(1 - \theta^{2a-1})$$

which belongs to $\not{p}^{(i+j)(p-1)+1}$ but to no higher power of $\not{p}$ (recall that $p\mathcal{O} = \not{p}^{p-1}$). Thus, if m and n are correctly chosen, then $S_a(p^i \wedge p^j\kappa) \neq 0 \bmod \not{p}^{n-m}$, and S_a determines a homomorphism α of $\mathcal{O}/\not{p}^{m-1} \wedge \mathcal{O}/\not{p}^{m-1}$ to $\mathcal{O}/\not{p}^{n-m}$. Constructing G as in Theorem 1.2.9, with this α and $u = v = 0$, we have $[G_{1+i(p-1)}, G_{1+j(p-1)}] \neq 1$. Taking $i = j = 1, 2, \ldots$, we obtain a sequence of p-groups of maximal class in which the smallest index of an abelian normal subgroup is unbounded.

Any 3-group G of maximal class has degree of commutativity $\geqslant n - 4$ ([7], Theorem 3.13) and hence G_2 is abelian. Thus the above cannot work for $p = 3$. Instead, we begin with the ring of integers in the p^rth cyclotomic field and use the same construction to produce groups of order p^{n+r-1} and coclass r. In particular, we can produce 3-groups of coclass two in this way, and taking $\alpha = S_a$ as above, with $a = 2$, we see that these do not possess abelian normal subgroups of bounded index.

Conjecture 1.3.1 is obviously stronger than

CONJECTURE 1.3.2 *The derived length of a finite p-group is bounded in terms of p and the coclass of the group.*

Other conjectures arise from considering a directed graph $\mathscr{C}(p,r)$ whose vertices are the (isomorphism types of) finite p-groups of coclass r. An edge runs from G to H if and only if H is an epimorphic image of G and $\mathrm{cl}\,G = 1 + \mathrm{cl}\,H$, where $\mathrm{cl}\,X$ denotes the nilpotency class of a nilpotent group X. In particular, if an edge runs from G to H, then $|G| = p|H|$. Consider a group G in $\mathscr{C}(p,1)$. If $|H| \geqslant p^4$ and H is an epimorphic image of G, then $H \cong G/G_d$ for some d, and $H_1 \cong G_1/G_d$. If $|G|$ is large compared with $|H|$, then H_1 will be abelian (Theorem 1.2.4) and H will split over it (Lemma 1.2.7). Then we find from Lemma 1.2.8 that

H is isomorphic to the semi-direct product of some $\mathcal{O}/\not{p}^k$ by an element of order p acting as multiplication by θ. Thus, these semi-direct products constitute the unique infinite maximal chain in $\mathscr{C}(p, 1)$.

If we have an infinite chain in $\mathscr{C}(p, r)$, we may choose epimorphisms from each group to the next and so obtain an inverse system

$$\longrightarrow G_n \xrightarrow{\phi_n} G_{n-1} \xrightarrow{\phi_{n-1}} \ldots \xrightarrow{\phi_2} G_1, \tag{1}$$

in which each G_i is a finite p-group of coclass r, the maps are epimorphisms, and cl $G_n = 1 + $ cl G_{n-1}. The kernel of ϕ_n must be the last non-trivial term of the lower central series of G_n. If we have another inverse system formed from the same chain, say

$$\longrightarrow H_n \xrightarrow{\psi_n} H_{n-1} \xrightarrow{\psi_{n-1}} \ldots \xrightarrow{\psi_2} H_1, \tag{2}$$

so that $H_n \cong G_n$ for all n, it follows that any isomorphism from G_n to H_n induces isomorphisms from G_i to H_i for all $i \leqslant n$, making the various squares

$$\begin{array}{ccc} G_i & \longrightarrow & G_{i-1} \\ \downarrow & & \downarrow \\ H_i & \longrightarrow & H_{i-1} \end{array}$$

commute. A simple argument with inverse limits of finite sets then allows us to choose isomorphisms from G_i to H_i, one for each i, making all the above squares commute. Thus, (1) and (2) have isomorphic inverse limits, and the inverse limit L of the groups in an infinite chain in $\mathscr{C}(p, r)$ depends only on the chain and not on the maps. Furthermore, it is easy to see that all the groups in (1) can be generated by a fixed number, d say, of elements. It can be deduced that

$$L/\gamma_c(L) \cong \varprojlim G_i/\gamma_c(G_i) \tag{3}$$

for all $c \geqslant 1$. The relevance of the bound on the number of generators here is that in a nilpotent group X generated by elements $g_1, \ldots, g_d$, every element of $\gamma_m(X)$ can be expressed as $[v_1, g_1] \ldots [v_d, g_d]$, with $v_i \in \gamma_{m-1}(X)$ (see [36] for instance). Of course, the maps in (3) are all isomorphisms after a certain point. If G_1 has class c_1 and order p^{c_1+r}, then G_i has class $c_1 + (i-1)$ and order $p^{c_1+(i-1)+r}$, and so if $c \geqslant c_1 + 1$, $G_i/\gamma_c(G_i)$ has class $c - 1$ and order p^{r+c-1} once $c_1 + (i-1) \geqslant c - 1$. It follows that $L = \varprojlim G_i$ is a pro-p group of finite coclass, in the sense of the following definition.

Definition 1.3.3. A pro-p group H has *finite coclass* r, if there exists an integer u such that $|H/\gamma_u(H)| = p^{u-1+r}$ and $|\gamma_j(H)/\gamma_{j+1}(H)| = p$ for $j \geqslant u$.

The above considerations show that there is a $(1, 1)$ correspondence between infinite maximal chains in $\mathscr{C}(p, r)$ and isomorphism types of pro-p-groups of coclass r.

CONJECTURE 1.3.4   *There are only finitely many of each of the objects just described, and pro-p groups of finite coclass are soluble.*

The last assertion amounts to saying that the derived lengths of the groups occurring in any infinite chain in $\mathscr{C}(p, r)$ are bounded—cf. Conjecture 1.3.2.

The so-called *p-adic uniserial space groups* provide an important source of pro-p groups of finite coclass. By definition, a p-adic uniserial space group is an extension P of a direct sum A of finitely many copies of $\mathbb{Z}_p$, the p-adic integers, by a finite p-group F that operates on A in such a way that $|A : [A, {}_iF]| = p^i$ for $i = 1, 2, \ldots$. As long as A has rank at least two, the last assertion is equivalent to requiring that the F-invariant subgroups of finite index of A (which necessarily have p-power index) are totally ordered under inclusion, in which case the subgroups $[A, {}_iF]$ are the only ones. If P is such a p-adic uniserial space group and we choose u so that $\gamma_u(P) \leqslant A$, then clearly $|\gamma_i(P)/\gamma_{i+1}(P)| = p$ for $i \geqslant u$, and so P is a pro-p group of finite coclass. It is not too hard to see that if H is any soluble pro-p group of finite coclass, then the hypercentre L of H is finite and H/L is a p-adic uniserial space group [63].

We see now that a first step in proving Conjecture 1.3.4 must be to show that there are only finitely many p-adic uniserial space groups of any given coclass r.$\ddagger$ If P is such a group, A is its 'translation subgroup' as above, and $F = P/A$, the uniseriality shows that F operates irreducibly on $V = A \otimes_{\mathbb{Z}_p} \mathbb{Q}_p$, where $\mathbb{Q}_p$ is the field of fractions of $\mathbb{Z}_p$, and so the dimension of V has the form $p^s - p^{s-1} = (\mathbb{Q}_p(\zeta) : \mathbb{Q}_p)$, where ζ is a primitive p^sth root of 1. For V will be induced from a proper subgroup of F unless every abelian normal subgroup of F is cyclic, in which case F is cyclic or generalized quaternion and the claim is quite easy to verify. The problem now is to try to show that s is bounded in terms of r.

$\ddagger$ This has now been done for odd p by Leedham-Green, McKay and Plesken [104].

One may also consider *p-uniserial space groups*—these are space groups G in the usual sense, in which the point group $F = G/T$ is a p-group operating on the translation subgroup T in such a way that $|T:[T, {}_iF]| = p^i$ for all i. The subgroups $[T, {}_iF]$ are then the only F-invariant subgroups of T of finite p-power index. These space groups also give rise to chains of finite p-groups of constant coclass, the groups $G/\gamma_i(G)$ for large enough i. Again, the rank of T necessarily has the form $p^s - p^{s-1}$. The simplest example of a p-uniserial space group is obtained by forming the semi-direct product of the ring $\mathcal{O}$ of integers in the pth cyclotomic field by an element of order p operating as multiplication by a primitive pth root of 1. Calling this $P(1)$ and defining $P(s+1)$ to be the wreath product of $P(s)$ by a cyclic group of order $p (s \geqslant 1)$, we find that $P(s)$ is a p-uniserial space group whose translation subgroup has rank $p^s - p^{s-1}$ and whose point group $W(s)$ is the wreath product of s cyclic groups of order p, that is, a Sylow p-subgroup of the symmetric group of degree p^s. Results of Vol'vačev [97] show that if P is any p-uniserial space group with point group F and translation subgroup T of rank $p^s - p^{s-1}$, then F can be identified with a subgroup of $W(s)$ (thought of as a subgroup of $GL(p^s - p^{s-1}, \mathbb{Z})$ as above) and T with a sublattice of the 'natural' $W(s)$-lattice just constructed. The question of which subgroups of $W(s)$ can occur is discussed in some detail in [63] and [15]; we shall not pursue it here, except to say that much seems to be unknown. Further discussion of these conjectures and related matters can be found in [63] and [15].

The p-uniserial space groups are receiving quite a bit of attention at the present time. In particular, they have been used to construct finite 2-groups whose class exceeds their 'breadth' [14]. The breadth of a finite p-group is the largest integer b such that G has a conjugacy class of size p^b, and it had been conjectured that the class of a finite p-group cannot exceed its breadth. For information about this 'class–breadth conjecture', see [8].

1.4   Automorphisms

Many authors have studied finite p-groups from the point of view of their automorphisms. In particular, we must mention the celebrated result of Gaschütz that every finite p-group of order $> p$ has an automorphism of p-power order that is not inner ([18], [49]; see [99] for a recent non-homological proof); variations on this in [85]; and the work of G. Higman, Shult, Shaw and Gross on finite p-groups G admitting a group of automorphisms of possibly restricted type that permutes the subgroups of order p transitively. A detailed account of the latter can be found in [8] for $p = 2$. See also [10], [50] for some recent results related to the case $p = 2$.

However, we wish here to discuss a different topic, namely, the restrictions on the structure of a finite p-group admitting an automorphism of prime order q with boundedly many fixed points. Probably the best known result here is that due to G. Higman [48], in which necessarily $q \neq p$:

> *If a finite p-group G admits a fixed point free automorphism of prime order q, then the class of G is bounded by a function $k(q)$ of q alone.*

Higman's work did not produce an explicit function $k(q)$, but one has subsequently been given by Kostrikin and Kreknin [55]; see [8]. These results are proved by considering the Lie ring of G, which is by definition

$$\mathscr{L}(G) = \bigoplus_{i \geqslant 1} \gamma_i(G)/\gamma_{i+1}(G),$$

the direct sum of the finite abelian groups $\gamma_i(G)/\gamma_{i+1}(G)$, with Lie multiplication $(\ , \ )$ given by

$$(x\gamma_{i+1}(G), y\gamma_{j+1}(G)) = [x,y]\gamma_{i+j+1}(G) \quad (x \in \gamma_i(G), y \in \gamma_j(G)).$$

The automorphism α acts as a fixed point free automorphism of $\mathscr{L}(G)$, but the advantage of the Lie ring here is that we can pass to $\mathscr{L}(G) \otimes_{\mathbb{Z}} \mathbb{Z}[\omega]$, where ω is a primitive qth root of 1, and this extended algebra then breaks up as a direct sum of α-eigenspaces which multiply together in the same way as the powers of ω. The fact that the 1-eigenspace is trivial can then be exploited.

There is a certain amount of evidence, though by no means an overwhelming amount, pointing towards an extension of Higman's Theorem along the following lines.

CONJECTURE 1.4.1   *If the finite p-group G admits an automorphism α of prime order q with at most p^m fixed points, then G has a subgroup H whose index divides $p^{f(m,q)}$, where $f(m,q)$ is a function of m and q alone, and whose class is bounded in terms of q.*

We can of course take H to be α-invariant and normal by modifying $f(m,q)$. We allow the possibility $q = p$ here, though the cases $q = p$ and $q \neq p$ can be expected to present quite different features.

Let us first consider the case $q \neq p$. Higman's Theorem shows that if G has class $> k(q)$, then α has a non-trivial fixed point on $G/\gamma_{k(q)+2}(G)$, and in this way one can at any rate bound the derived length of G. A similar result holds for arbitrary finite soluble G of order prime to q.

When $q = 2$, the conjecture holds with $f(m, 2) = 1^2 + 2^2 + \ldots + m^2$ and H of class two (Hartley and Meixner [37]). The proof of this is direct and does not use the Lie ring. Meixner [71], using the Lie ring, has proved the conjecture for metabelian G; in this case H may be taken to have class at most $2q + 1$. An obvious difficulty in using the Lie ring here is that subrings of $\mathscr{L}(G)$ need not correspond to subgroups of G, and so the Lie ring has to be used in a somewhat indirect way.

It may be worth remarking that there are finite p-groups ($p \neq 2$) of unbounded derived lengths that admit a fixed point free 4-group of automorphisms. To see this, let G_n be the group of all 2×2 matrices congruent to the identity modulo p, over the integers modulo p^n. Let α be the automorphism of G_n induced by conjugation with $\left(\begin{smallmatrix} -1 & 0 \\ 0 & 1 \end{smallmatrix} \right)$, and β the automorphism of G_n sending each matrix to its transposed inverse. Then $A = \langle \alpha, \beta \rangle$ is a 4-group, a trivial calculation reveals that $C_{G_n}(A) = 1$, and the groups G_n have unbounded derived lengths (Hall [32]).

Now let us pass to the case $q = p$; thus we have a finite p-group G admitting an automorphism α of order p with at most p^m fixed points. Alperin [1] has proved that this bounds the derived length of G in terms of m. His argument, which uses the Lie ring, runs as follows. First, α has at most p^m fixed points on any α-invariant section of G, and so the fixed points of α on $\mathscr{L}(G)$ lie in the ideal of elements annihilated by p^m. Then it is observed that the proof of Higman's Theorem (the one just quoted) yields the following.

If the Lie ring L admits an automorphism of prime order q whose fixed points all lie in an ideal I of L, then $(qL)^{k(q)} \leqslant I$.

Here $k(q)$ has the same meaning as above.

This implies that $p^{k(p)+m}$ annihilates $L^{k(p)}$, whence the lower central factors of G beyond the $k(p)$th have bounded exponent. Their ranks are easily seen to be bounded by mp and hence their orders are also bounded. Letting p^s be the bound obtained ($s = s(m, p)$) let $H = \gamma_{s+1}(G)$ and apply the same considerations to H. Now all the lower central factors of H, with the possible exception of the last, have order at least p^{s+1} (P. Hall [28], Theorem 2.56). At the same time, those beyond the $k(p)$th have order at most p^s. Thus H has class at most $k(p)$, and we see that G is actually an extension of a group of class at most $k(p)$ by one of class at most s.

When $q = p = 2$, the group G of the conjecture actually has an abelian subgroup of bounded index [37]. When $q = p$ in general, but $m = 1$, it is quite easy to see that the semi-direct product $T = G\langle\alpha\rangle$ has maximal class, as $[G,_i\alpha]/[G,_{i+1}\alpha]$, which is a section on which α operates trivially, must have order dividing p, and so the lower central series of T is $T \geqslant [G,\alpha] \geqslant [G,\alpha,\alpha] \geqslant \ldots$. In this case, we know that G has a nilpotent subgroup of class at most two and index dividing p^{p-2}, namely T_{p-2} (see the remark before Conjecture 1.3.1).

In general, however, we cannot hope for a subgroup H of bounded index and class less than $p-1$. Let $\mathcal{O}$ as usual be the ring of integers in the cyclotomic field generated by θ, a primitive pth root of 1, and let $R = \mathcal{O}/p^n\mathcal{O}$. Let U be the group of upper unitriangular $(p-1) \times (p-1)$ matrices over R, V the free R-module consisting of all row vectors of length $p-1$ over R, and $G = G_n$ the semi-direct product VU. Identifying θ with its image in R, we see that the diagonal matrix with entries $\theta, \theta^2, \ldots, \theta^{p-1}$ induces an automorphism α of G in the obvious way, and $|C_G(\alpha)| = p^m$, where $m = p-1+\frac{1}{2}(p-1)(p-2)$. But for any $k \geqslant 1$, we can choose n large enough to ensure that G has no normal subgroup of index dividing k and class $\leqslant p-2$.

We observe now that the conjecture is somewhat ambiguous as to whether m is supposed to be independent of p, and perhaps the above example indicates that it should be taken so.

2. Augmentation Ideals and the Lower Central Series

2.1 Formal power series

Linear methods of one kind or another have proved extremely powerful in both finite and infinite group theory. One of the fundamental papers in this area is that of Magnus [66], who studied free groups by embedding them in formal power series rings. Let $\{y_\lambda : \lambda \in \Lambda\}$ be a set of non-commuting indeterminates, and let A be the free ring they generate. Thus

$$A = A_0 \oplus A_1 \oplus \ldots,$$

where $A_0 = \mathbb{Z}1$ and A_t, the homogeneous component of degree t, has a $\mathbb{Z}$-basis consisting of the monomials

$$\xi = y_{\lambda_1} \ldots y_{\lambda_t} \quad (\lambda_i \in \Lambda)$$

of degree t. The formal power series ring $\hat{A}$ in the y_λ consists of all formal infinite expressions

$$a = a_0 + a_1 + \ldots + a_t + \ldots \quad (a_t \in A_t) \qquad (1)$$

multiplied and added in the usual way. Magnus [66] proved that in $\hat{A}$, *the elements* $x_\lambda = 1 + y_\lambda$ *freely generate a free group* F (see Magnus, Karrass and Solitar [68], Chapter 5, for an account of this material). (In fact, the same is true when we replace $\hat{A}$ by a formal power series ring with coefficients coming from an arbitrary commutative ring R with identity [58]; since we work in the subring of R generated by 1, the case $R = \mathbb{Z}/n\mathbb{Z}$ is all that need be considered.) This embedding makes certain properties of free groups particularly transparent. For example, writing J_n for the set of elements (1) with $a_0 = a_1 = \ldots = a_{n-1} = 0$, we see easily that $[1 + J_n, 1 + J_m] \leqslant 1 + J_{n+m}$, whence $\gamma_n(F) \leqslant 1 + J_n$, and

$$\bigcap_{n=1}^{\infty} \gamma_n(F) = 1.$$

The subgroup

$$D_n(F) = F \cap (1 + J_n)$$

of F, which seems to have been first considered by Magnus, is called the *nth dimension subgroup of* F (over $\mathbb{Z}$). It is the kernel of the natural homomorphism of F into the group of units of $\hat{A}/J_n$. When Λ is finite, $\hat{A}/J_n$ is additively a finitely generated free abelian group, and we obtain a faithful representation of $F/D_n(F)$ by matrices over $\mathbb{Z}$ by allowing it to act by right multiplication on $\hat{A}/J_n$. It was conjectured by Magnus [66], and subsequently proved by Grün [22] (see also Magnus [67], Witt [100]), that

$$D_n(F) = \gamma_n(F).$$

Dimension subgroups $D_n(R, F)$ of free groups over other commutative rings R with identity can be introduced similarly by considering power series over R. The group-theoretical description of these is more complicated.

THEOREM 2.1.1 (Lazard [58])

$$D_n(\mathbb{Z}/p^r\mathbb{Z}, F) = \Pi \gamma_i(F)^{p^j}$$

over all i and j such that $ip^{(j-r+1)} \geqslant n$.

Here, if H is a group, then $H^n = \langle h^n : h \in H \rangle$. The case $r = 1$ of this is due to Zassenhaus [101]. Lazard's theorem appears as part of a general study of embeddings of free groups in formal power series rings and subgroups arising from filtrations of the ring. The dimension subgroups of a free group over an arbitrary coefficient ring (commutative with identity, of course) follow easily ([58], p. 139) from the theorem.

In the above context, it is natural to consider the related *Lie rings*. Thus, in A, we can consider the Lie subring

$$L = L_1 \oplus L_2 \oplus \ldots \tag{2}$$

(under the Lie multiplication given by $(a, b) = ab - ba$) generated by the y_λ. Witt showed that this Lie ring is freely generated by the y_λ (strictly speaking, he proved this when $\mathbb{Z}$ is replaced by a field). A recent account of this can be found in ([8], II, p. 366 *et seq.*). We can also form the Lie ring

$$\mathscr{L}(F) = \bigoplus_{n=1}^{\infty} \gamma_n(F)/\gamma_{n+1}(F) \tag{3}$$

as on p. 81. If $u \in \gamma_n(F)$, then in the power series representation, $u = 1 + u_n + u_{n+1} + \ldots$ with $u_j \in A_j$, and one easily sees that the map

$$u\gamma_{n+1}(F) \mapsto u_n \tag{4}$$

extends by linearity to a Lie homomorphism $\mathscr{L}(F) \to L$. It is surjective, and since the elements $x_\lambda F'$, which generate $\mathscr{L}(F)$, map to free generators of L, the map (4) is an isomorphism.

THEOREM 2.1.2 $\mathscr{L}(F)$ *is freely generated by the elements $x_\lambda F'$ ($\lambda \in \Lambda$), and with respect to the map* (4), A *is the universal enveloping algebra of $\mathscr{L}(F)$.*

A basis for L_n, and hence for $\gamma_n(F)/\gamma_{n+1}(F)$, can be obtained from Hall's theory of basic commutators. The approach we outline is taken from P. Hall [30, 31]. Given a set $X_1 = \{z_\lambda : \lambda \in \Lambda\}$ of symbols, which we take finite for simplicity, and a formal operation of commutation [,], allowing us to form complex commutators in the z_λ, we define inductively a sequence $b_1, b_2, \ldots$ of commutators in the z_λ and a sequence $X_1, X_2, \ldots$ of sets of commutators in the z_λ as follows. We have X_1 as given. Having formed X_i, we choose b_i to be any element of smallest weight in X_i, and put

$$X_{i+1} = \{[x, _r b_i] : r \geqslant 0, x \in X_i \setminus \{b_i\}\}.$$

The sequence $b_1, b_2, \ldots$ is called a *basic sequence in the* z_λ, or a *sequence of basic commutators*. It is shown in [30] that the basic Lie commutators of weight n in the y_λ form a basis of L_n, and so the basic group commutators in the x_λ represent the elements of a basis of the free abelian group $\gamma_n(F)/\gamma_{n+1}(F)$. These basic commutators are an important computational tool. They have analogues in certain types of relatively free groups, which we shall not explore here; see [103].

2.2 Augmentation ideals over a field

It is natural to try to extend the above machinery to other types of group, where the power series representation may not be available. The key to this is the fact that in $\hat{A}$, *the elements of F are linearly independent* [19], and so span a copy of $\mathbb{Z}F$, the integral group ring of F. We find that $\mathbb{Z}F \cap J$ is the augmentation ideal $\mathfrak{f}$ of $\mathbb{Z}F$, and $\mathbb{Z}F \cap J_n = \mathfrak{f}^n$. Thus

$$D_n(F) = (1 + \mathfrak{f}^n) \cap F.$$

For an arbitrary group G and commutative ring R with 1, we define

$$D_n(R, G) = (1 + R\mathfrak{g}^n) \cap G,$$

the *nth dimension subgroup of G over R*.

These dimension subgroups are by no means well understood, except when R is a field. For fields of characteristic $p > 0$ they have been determined by Jennings [52], and turn out to be obtainable from a free presentation $F \to G \to 1$ as the images of the corresponding subgroups of F.

THEOREM 2.2.1 *If k is a field of characteristic $p > 0$, and G is any group, then*

$$D_n(k, G) = \Pi \gamma_i(G)^{p^j}$$

over all i, j such that $ip^j \geqslant n$.

For fields of characteristic zero they seem to have been determined first by Mal'cev [70] in another guise, though his argument is a little sketchy. See also Jennings [53] and P. Hall [30].

THEOREM 2.2.2 *If k is a field of characteristic zero and G is any group, then*

$$D_n(k, G) = \bar{\gamma}_n(G) = \{x \in G : x^m \in \gamma_n(G) \text{ for some } m \geqslant 1\},$$

the isolator of $\gamma_n(G)$ in G.

An expanded account of Mal'cev's argument is in [40], and proofs of Theorems 2.2.1 and 2.2.2, together with related matters, can be found in [77] and [78].

Now let k be any field. Analogously to $\mathscr{L}(G)$, we may form the Lie ring

$$\mathscr{D}_k(G) = \bigoplus_{n=1}^{\infty} D_n(k, G)/D_{n+1}(k, G) \tag{1}$$

in which, since $[D_n(k, G), D_m(k, G)] \leqslant D_{n+m}(k, G)$, multiplication is given by the same type of formula as before (cf. p. 81). As a substitute for the free associative ring A used when G is free, we may form the graded ring $\mathscr{G}_k(G)$ corresponding to the filtration of kG by the powers of its augmentation ideal $k\mathfrak{g}$; thus

$$\mathscr{G}_k(G) = \bigoplus_{n=0}^{\infty} k\mathfrak{g}^n/k\mathfrak{g}^{n+1}, \tag{2}$$

with the associative multiplication

$$(x + k\mathfrak{g}^{m+1})(y + k\mathfrak{g}^{n+1}) = xy + k\mathfrak{g}^{m+n+1} \quad (x \in k\mathfrak{g}^m, y \in k\mathfrak{g}^n).$$

The map

$$xD_{n+1}(k, G) \mapsto (x-1) + k\mathfrak{g}^{n+1} \quad (x \in D_n(k, G)) \tag{3}$$

extends to a Lie homomorphism of $\mathscr{D}_k(G)$ into $\mathscr{G}_k(G)$ and hence to a Lie homomorphism

$$k \otimes_{\mathbb{Z}} \mathscr{D}_k(G) \to \mathscr{G}_k(G). \tag{4}$$

A beautiful result of Quillen extends the classical work of Magnus and Witt to this situation. It has two versions, depending on the characteristic of k.

THEOREM 2.2.3 [81]   If k has characteristic zero, then with respect to the map (4), $\mathscr{G}_k(G)$ is the universal enveloping algebra of $k \otimes_{\mathbb{Z}} \mathscr{D}_k(G)$.

Quillen's proof uses Hopf algebra techniques. A more direct proof, based on Chapter 7 of Hall's Edmonton Notes [30], has been given by

Passi [77]. Here we indicate the proof of a slightly stronger result. We have $D_n(k, G) = \bar{\gamma}_n(G)$. Analogously, we introduce

$$\overline{\mathfrak{g}^n} = \{\alpha \in \mathbb{Z}G : m\alpha \in \mathfrak{g}^n \text{ for some } m \geqslant 1\}$$

and form

$$\mathscr{G}_{\mathbb{Z}}(G) = \bigoplus_{n=0}^{\infty} \overline{\mathfrak{g}^n}/\overline{\mathfrak{g}^{n+1}},$$

the graded ring corresponding to the filtration $\{\overline{\mathfrak{g}^n}\}$ of $\mathbb{Z}G$. The map

$$x\bar{\gamma}_{n+1}(G) \mapsto x - 1 + \overline{\mathfrak{g}^{n+1}} \quad (x \in \bar{\gamma}_n(G))$$

extends to a Lie homomorphism from $\mathscr{D}_{\mathbb{Q}}(G)$ to $\mathscr{G}_{\mathbb{Z}}(G)$ and hence gives a homomorphism

$$\theta : \mathscr{U}(\mathscr{D}_{\mathbb{Q}}(G)) \to \mathscr{G}_{\mathbb{Z}}(G), \tag{5}$$

where $\mathscr{U}(L)$ denotes the universal envelope of a Lie ring L; note that $\mathscr{D}_{\mathbb{Q}}(G)$ is not in general a Lie algebra over $\mathbb{Q}$.

THEOREM 2.2.3′ θ is an isomorphism.

This can be proved as follows. First, routine arguments (see [77], pp. 112–114) allow us to assume that G is finitely generated torsion-free nilpotent, of class c say. Then each $\bar{\gamma}_i(G)/\bar{\gamma}_{i+1}(G)$ is free abelian of finite rank. We can choose elements

$$x_m, x_{m-1}, \ldots, x_1$$

of G and integers

$$m = t_{c+1} > t_c > \ldots > t_1 = 0$$

such that the cosets $x_\alpha \bar{\gamma}_{i+1}(G)$ $(t_{i+1} \geqslant \alpha > t_i)$ form a basis of $\bar{\gamma}_i(G)/\bar{\gamma}_{i+1}(G)$. For $1 \leqslant \alpha \leqslant m$, define μ_α by

$$x_\alpha \in \bar{\gamma}_{\mu_\alpha}(G) \setminus \bar{\gamma}_{\mu_\alpha+1}(G).$$

Let

$$u_r^{(\alpha)} = 1 \quad \text{if } r = 0,$$
$$= x_\alpha^{-i}(x_\alpha - 1)^{2i} \quad \text{if } r = 2i,$$
$$= x_\alpha^{-i}(x_\alpha - 1)^{2i-1} \quad \text{if } r = 2i-1,$$

and for each vector $\mathbf{j} = (j_m, \ldots, j_1)$ of non-negative integers, define

$$u(\mathbf{j}) = u_{j_m}^{(m)} u_{j_{m-1}}^{(m-1)} \ldots u_{j_1}^{(1)}.$$

Assign to this element the 'weight'

$$\mu(\mathbf{j}) = \sum_{\alpha=1}^{m} j_\alpha \mu_\alpha.$$

THEOREM 2.2.4 [35, 41]   *If G is finitely generated torsion-free nilpotent, then*

$$\{u(\mathbf{j}) : \mu(\mathbf{j}) \geqslant a\}$$

is a $\mathbb{Z}$-basis of $\overline{\mathfrak{g}^a}$ ($a \geqslant 0$).

Now $x_\alpha^{\pm 1} \equiv 1 \bmod \mathfrak{g}$, and we see that

$$u(\mathbf{j}) \equiv v(\mathbf{j}) = (x_m - 1)^{j_m} \ldots (x_1 - 1)^{j_1} \bmod \overline{\mathfrak{g}^{a+1}}$$

if $\mu(\mathbf{j}) = a$. Consequently

LEMMA 2.2.5   *The elements $v(\mathbf{j}) + \mathfrak{g}^{a+1}$ such that $\mu(\mathbf{j}) = a$ form a $\mathbb{Z}$-basis of $\overline{\mathfrak{g}^a}/\overline{\mathfrak{g}^{a+1}}$.*

Now the elements $x_\alpha \bar{\gamma}_{\mu_\alpha + 1}(G)$ ($\alpha = m, m-1, \ldots, 1$) form a $\mathbb{Z}$-basis of $\mathscr{D}_\mathbb{Q}(G)$, and by the Poincaré–Birkhoff–Witt Theorem, their ordered products form a $\mathbb{Z}$-basis of $\mathscr{U}(\mathscr{D}_\mathbb{Q}(G))$. The images of these ordered products under θ are precisely the basis elements referred to in Lemma 2.2.5. Thus we have Theorem 2.2.3′.

The situation is particularly simple when the groups $G/\gamma_a(G)$ are all torsion free. Then

$$x_\alpha \in \gamma_{\mu_\alpha}(G), \quad u(\mathbf{j}) \in \mathfrak{g}^a \quad \text{if } \mu(\mathbf{j}) \geqslant a,$$

and so $\mathfrak{g}^a = \overline{\mathfrak{g}^a}$. This argument is only valid in the finitely generated torsion-free nilpotent case, but is easily adapted to the general situation. Of course, $\mathscr{D}_\mathbb{Q}(G) = \mathscr{L}(G)$ now, and $\mathscr{G}_\mathbb{Z}(G)$ becomes

$$\mathscr{G}_\mathbb{Z}(G) = \bigoplus_{n=0}^{\infty} \mathfrak{g}^n/\mathfrak{g}^{n+1}.$$

We obtain the following.

THEOREM 2.2.6   *If $\gamma_n(G)/\gamma_{n+1}(G)$ is torsion-free for all $n \geqslant 1$, then*

$$\mathscr{U}(\mathscr{L}(G)) \cong \mathscr{G}_{\mathbb{Z}}(G),$$

via the map (5). *Hence $D_n(G) = \gamma_n(G)$ for all $n \geqslant 1$.*

(Cf. Sandling and Tahara [84].) This is exactly analogous to the free case.

When k is a field of characteristic $p > 0$, $k \otimes_{\mathbb{Z}} \mathscr{D}_k(G)$ becomes a restricted Lie algebra over k. The analogue of Theorem 2.2.3, also due to Quillen, is

THEOREM 2.2.7 [81]   *If k has characteristic $p > 0$, then $\mathscr{G}_k(G)$ is the restricted universal envelope of $k \otimes_{\mathbb{Z}} \mathscr{D}_k(G)$.*

A proof of this can be given in the above style (cf. [77]). The case when G is free is due to Lazard [58]. The Poincaré–Birkhoff–Witt Theorem allows us to obtain relations between the ranks of the factors $D_n(k, G)/D_{n+1}(k, G)$ and the dimensions of the $k\mathfrak{g}^n/k\mathfrak{g}^{n+1}$, as was done by Witt in the free case (cf. [78], Chapter III).

2.3 Integral dimension subgroups

The situation is much less well understood once we depart from fields as coefficients. The dimension subgroups over an arbitrary commutative ring with 1 can be determined from those over $\mathbb{Z}$ and $\mathbb{Z}/p^r\mathbb{Z}$, where p ranges over the primes and $r \geqslant 1$ (Sandling [83]; see also [77], Theorem 2.1). Unfortunately, however, these are not understood group theoretically. For free groups everything is known by the results already discussed. We also know that if $F \to G \to 1$ is a presentation of a group G, then the dimension subgroups of G over a field of characteristic p are the images of those of F. But Moran [75] has observed that this need no longer be true with $\mathbb{Z}/p^r\mathbb{Z}$ as coefficients. The corresponding question with integral coefficients is just whether the dimension subgroups $D_n(G) = D_n(\mathbb{Z}, G)$ coincide with the corresponding terms of the lower central series. They were suspected to do so for quite a long time (the so-called Dimension Subgroup Conjecture), until the suspicion was eventually dispelled by Rips [82]. He gave a finite 2-group G of nilpotency class 3 such that $|D_4(G)| = 2$. More such examples have since been obtained by Tahara [95]. A number of results in favour of the Dimension Subgroup Conjecture had previously been discovered; most of these follow now from work of Sjogren [88], and we shall therefore give quite a detailed account of this. To set it in context, note that $D_n(G) \leqslant D_n(\mathbb{Q}, G) = \bar{\gamma}_n(G)$ (Theorem 2.2.2), so $D_n(G)/\gamma_n(G)$ is periodic.

THEOREM 2.3.1 [88] *Let b_m be the least common multiple of $\{1, 2, \ldots, m\}$ and let $c_1 = c_2 = 1$ and*

$$c_n = \prod_{i=1}^{n-2} b_i^{\binom{n-2}{i}}.$$

Then for any group G, the exponent of $D_n(G)/\gamma_n(G)$ divides c_n.[‡]

(This bound is the same as that given in Sjogren's paper.)

We have $c_3 = b_1 = 1$; $c_4 = b_1^2 b_2 = 2$; $c_5 = b_1^3 b_2^3 b_3 = 48$, and so on. Thus $D_3(G) = \gamma_3(G)$ and $D_4(G)/\gamma_4(G)$ has exponent dividing 2; these facts were previously known (see [77]) and are best possible. The bound of 48 for the exponent of $D_5(G)/\gamma_5(G)$ has been improved by Tahara to 6 [96]; it is not known whether this is best possible, and in fact the following problem is open.

Problem 2.3.2. *Is there a finite p-group G of odd order such that $D_n(G) \neq \gamma_n(G)$ for some n?*

We also see that $p \nmid c_n$ if $n \leqslant p+1$, and so we have, from Theorem 2.3.1, the following.

COROLLARY 2.3.3 *If G is a p-group, then $D_n(G) = \gamma_n(G)$ for $n \leqslant p+1$.*

For $n \leqslant p$ this was proved by Moran [75].

Sjogren's approach to the dimension subgroups of G is to write $G = F/R$, where F is free, and starting from the relation $D_n(F) = \gamma_n(F)$, to introduce relations in a controlled manner which allows one to observe how the D's and γ's diverge. This is done by using spectral sequences. On the one hand, Stallings [94] has shown how to write down a spectral sequence $\{\tilde{E}_{n,m}^r\}$ (actually $\tilde{E}_{n,m}^r = 0$ unless $n+m$ is 0 or 1) that converges on a diagonal to the homogeneous components of the graded ring $\mathscr{G}_{\mathbb{Z}}(G)$; more precisely, $\tilde{E}_{-n,n}^n \cong \mathfrak{g}^n/\mathfrak{g}^{n+1}$. Sjogren gives a similar spectral sequence $\{E_{p,q}^r\}$ converging on the same diagonal to

---

[‡] N. D. Gupta [105] has recently shown that if G is metabelian then the exponent of $D_n(G)/\gamma_n(G)$ divides $2b_1 \ldots b_{n-2}$ ($n \geqslant 3$). For related results see [106, 107]. C. K. Gupta and F. Levin [108] have shown that $D_n(G) = \gamma_n(G)$ if G is free centre-by-metabelian.

the homogeneous components of $\mathscr{L}(G)$; thus $E^n_{-n,n} \cong \gamma_n(G)/\gamma_{n+1}(G)$. The map $\phi: x \mapsto x-1$ induces a map from E to $\tilde{E}$ which is natural in a certain sense which we need not go into, and which, after convergence, is the usual map $x\gamma_{n+1}(G) \mapsto (x-1)+\mathfrak{g}^{n+1}$ of $\mathscr{L}(G)$ into $\mathscr{G}_{\mathbb{Z}}(G)$. The kernel of the map from $\gamma_n(G)/\gamma_{n+1}(G)$ is $D_{n+1}(G) \cap \gamma_n(G)/\gamma_{n+1}(G)$, and the spectral sequence machinery builds it up in an inductive manner from various previous kernels. However, although the spectral sequence machinery is probably the natural setting for this approach, it is possible to describe it in a more intrinsic way, as we shall see. The central part of the argument rests on the study of Lie elements in a free associative ring and a certain map ψ, reminiscent of the Dynkin–Specht–Wever operator, which Sjogren introduces. For more details of what follows, see [43].

Let $G = F/R$, where F is free and $R \lhd F$, and consider the subgroups

$$R(k) = \Pi[Y_1, Y_2, \ldots, Y_k]$$

over all choices in which each Y_i is either F or R and at least one is R. Thus $R = R(1) \geqslant [F, R] = R(2) \geqslant [F, R, F][F, F, R] = R(3) \geqslant \ldots$, and we have the sequence of natural homomorphisms

$$F/\gamma_n(F) \to F/\gamma_n(F)R(n-1) \to \ldots \to F/\gamma_n(F)R(1) = F/\gamma_n(F)R \qquad (1)$$

representing the gradual introduction of relations. Correspondingly, in $\mathbb{Z}G$ we form the sequence of ideals

$$J(k) = \Sigma \mathfrak{y}_1 \mathfrak{y}_2 \ldots \mathfrak{y}_k . \mathbb{Z}F$$

where the summation runs over all the same choices of $Y_1, \ldots, Y_k$ as before. For each k, we have $J(k) \geqslant \mathfrak{r}(k)$. Consider the two series

$$F > F' > \ldots > \gamma_k(F) = \gamma_k(F)R(k) \geqslant \gamma_{k+1}(F)R(k) \geqslant \ldots \geqslant R(k)$$

and

$$\mathbb{Z}F > \mathfrak{f}^2 > \ldots > \mathfrak{f}^k = \mathfrak{f}^k + J(k) \geqslant \mathfrak{f}^{k+1} + J(k) \geqslant \ldots \geqslant J(k).$$

For each n, we have a well defined homomorphism of

$$\gamma_{n-1}(F)R(k)/\gamma_n(F)R(k)$$

into $\mathfrak{f}^{n-1} + J(k)/\mathfrak{f}^n + J(k)$, under which an element $x\gamma_n(F)R(k)$ $(x \in \gamma_{n-1}(F))$

maps to the coset of $x-1$. Let $D_{n,k}/\gamma_n(F)R(k)$ be the kernel of this homomorphism. Then

$$D_{n,k} = \{\gamma_{n-1}(F) \cap (1 + \mathfrak{f}^n + J(k))\}R(k)$$

and

$$D_{n,k}/\gamma_n(F)R(k) \geqslant D_n(F/\gamma_n(F)R(k)) \cap (\gamma_{n-1}(F)R(k))/\gamma_n(F)R(k).$$

We have

$$\gamma_n(F) = D_{n,n} \leqslant D_{n,n-1} \leqslant \cdots \leqslant D_{n,1}$$

(cf. (1)), and

$$D_{n,1}/R = D_n(F/R) \cap \gamma_{n-1}(F)R/R.$$

By induction on $n-k$, integers $a_{n,k}$ are now calculated such that *the exponent of* $D_{n,k}/\gamma_n(F)R(k)$ *divides* $a_{n,k}$. Then it is not hard to check that $D_n(F/\gamma_n(F)R)$ has exponent dividing $\prod_{j=1}^n a_{j,1}$. The fact that $D_n(F) = \gamma_n(F)$ gives

$$a_{n,n} = 1, \tag{2}$$

and a reasonably short argument can be given to show that

$$a_{n,n-1} = 1. \tag{3}$$

The following lemma (cf. Sjogren [88], 2.3) is the key to the inductive step.

LEMMA 2.3.4 *If* $x \in D_{n,k} \cap \gamma_{n-1}(F)$, *then there exists* $f \in R(k)$ *and* $w \in J(k+1)$ *such that*

$$f - 1 + w \equiv b_k(x-1) \bmod \mathfrak{f}^n.$$

Here b_k *(as in Theorem 2.3.1) is the l.c.m. of* $\{1, 2, \ldots, k\}$.

It is a reasonably straightforward, if slightly technical, matter to deduce from this and the various sequences above that $D_{n,k}/\gamma_n(F)R(k)$ has exponent dividing $a_{k+2,k+1} \cdots a_{n,k+1}b_k$, if $k < n-1$, so that we can use this as the definition of $a_{n,k}$. It can then be rearranged to give

$$a_{n,k} = \prod_{i=k}^{n-2} b_i^{\binom{n-k-2}{i-k}}$$

and the theorem.

The proof of Lemma 2.3.4 involves a certain map ψ defined on a free associative algebra over $\mathbb{Q}$ as follows. First, let Γ be an arbitrary non-empty set and $S(\Gamma)$ the set of all finite sequences of elements of Γ. An *infiltration* I of a sequence $(I_1, \ldots, I_v)$ of elements of $S(\Gamma)$ is an indexing of each of $I_1, \ldots, I_v$ as subsequences of another sequence $J \in S(\Gamma)$, such that these subsequences cover J. We write $J = I^*$, the *infiltration sequence* determined by I. The infiltration is called a *shuffle*, if the subsequences of J corresponding to $I_1, \ldots, I_v$ are pairwise disjoint. Now let A be the free associative algebra over $\mathbb{Q}$ generated by elements y_i $(i \in \Gamma)$, and for $J = (j_1, \ldots, j_r) \in S(\Gamma)$, write

$$y_J = y_{j_1} \ldots y_{j_r}.$$

For $v \geq 1$, we define a $\mathbb{Q}$-linear map $\psi_v : A \to A$ by

$$\psi_v(y_J) = ((-1)^{v+1}/v) \sum_{(I_1, \ldots, I_v), I} y_{I_1} \ldots y_{I_v} \qquad (4)$$

where the sum ranges over all sequences $(I_1, \ldots, I_v)$ of elements of $S(\Gamma)$ and over all infiltrations I of $(I_1, \ldots, I_v)$ with $I^* = J$. A more instructive way to think of this is to write out the formal power series

$$\log((y_{j_1}+1) \ldots (y_{j_r}+1)) = \sum_{\tau=1}^{\infty} ((-1)^{\tau+1}/\tau)((y_{j_1}+1) \ldots (y_{j_r}+1)-1)^{\tau}$$

and to take from the term $\tau = v$ all the monomials which formally involve all the variables $y_{j_1}, \ldots, y_{j_r}$.

We obtain $\hat{\psi}_v(y_J)$ by taking the terms of degree r from (4); in other words, restricting the summation over I to shuffles. Then we fix an integer $l \geq 1$ and write

$$\psi = \sum_{v=1}^{l} \psi_v; \quad \hat{\psi} = \sum_{v=1}^{l} \hat{\psi}_v.$$

Thus there are really many different maps, depending on the value chosen for l. It is important to observe that although ψ and $\hat{\psi}$ do not

map $\mathbb{Z}[y_i : i \in \Gamma]$ into itself, *all the denominators introduced divide b_l.*
Note that

$$\psi(y_J) = \hat{\psi}(y_J) + terms\ of\ degree\ at\ least\ r+1. \tag{5}$$

The connection between $\hat{\psi}$ and the Lie elements of A is given by

LEMMA 2.3.5

(i) *If $r \leqslant l$ and $a \in A_r$, then $\hat{\psi}(a)$ is a Lie element of A.*

(ii) *If a is any Lie element of A, then $\hat{\psi}(a) = a$.*

Now we revert to our free group F, which we take to have basis $\{x_i : i \in \Gamma\}$. There is an identification θ of $\mathbb{Z}F/\mathfrak{f}^n$ with the ring

$$B = B_0 \oplus B_1 \oplus \ldots \oplus B_{n-1} \tag{6}$$

obtained from the free ring generated by non-commuting indeterminates $\tilde{x}_i$ ($i \in \Gamma$) by equating monomials of degree $\geqslant n$ to zero. Under θ, $x_i + \mathfrak{f}^n$ corresponds to $1 + \tilde{x}_i$. An element α of $\mathbb{Z}F$ corresponds in B to the element

$$\sum_{|J|<n} D(J)(\alpha)\tilde{x}_J,$$

where the summation runs over all sequences $J \in S(\Gamma)$ of length $\leqslant n-1$, and if $J = (j_1, \ldots, j_r)$, then $D(J)$ is the Fox derivative $\partial^r / \partial x_{j_1} \ldots \partial x_{j_r}$ followed by the augmentation [16]. What seems to play a central role here is that the $D(J)$ satisfy the following:

LEMMA 2.3.6 (Chen, Fox and Lyndon [12]) *If $I_1, \ldots, I_r \in S(\Gamma)$, then $D(I_1)D(I_2) \ldots D(I_r) = \Sigma_I D(I^*)$, where the sum runs over all infiltrations I of $(I_1, \ldots, I_r)$ and on the left-hand side the product is pointwise.*

Let R be the normal closure of a set of elements r_l ($l \in \Delta$), let S be the group freely generated by $\{s_l : l \in \Delta\}$, and let $F^* = F * S$. Let B^* be constructed analogously to (6) from indeterminates $\tilde{s}_l$ ($l \in \Delta$) and $\tilde{x}_i$ ($i \in \Gamma$). Then we have a commutative diagram

$$
\begin{array}{ccc}
\mathbb{Z}F^*/\mathfrak{f}^{*n} & \xrightarrow{\ \beta\ } & \mathbb{Z}F/\mathfrak{f}^n \\
{\scriptstyle \theta^*}\big\downarrow & & \big\downarrow{\scriptstyle \theta} \\
B^* & \xrightarrow{\ \delta\ } & B
\end{array}
$$

in which β and δ are ring homomorphisms,

$$\beta(s_l+\mathfrak{f}^{*n}) = r_l+\mathfrak{f}^n; \; \beta(x_i+\mathfrak{f}^{*n}) = x_i+\mathfrak{f}^n$$

and

$$\delta(\tilde{x}_i) = \tilde{x}_i; \quad \delta(\tilde{s}_l) = \sum_{J\neq\emptyset} D(J)(r_l)\tilde{x}_J.$$

The proof of Lemma 2.3.4 now runs as follows. As $x\in\gamma_{n-1}(F)$, $\theta(x)-1$ is a Lie element of B_{n-1}. Because $x\in D_{n,k}$, $x-1$ is congruent modulo $\mathfrak{f}^n$ to a $\mathbb{Z}$-linear combination of monomials in the elements x_i-1 and r_l-1, where each monomial has degree at least k and involves at least one r_l-1. In other words,

$$\theta(x)-1 = \delta(a),$$

where a is the element of B^* obtained by replacing in these monomials each r_l-1 by $\tilde{s}_l$ and each x_i-1 by $\tilde{x}_i$. Define

$$\psi = \sum_{v=1}^{k} \psi_v; \quad \hat{\psi} = \sum_{v=1}^{k} \hat{\psi}_v$$

which we view as acting on both $\mathbb{Q}\otimes_{\mathbb{Z}} B^* = \mathbb{Q}B^*$ and $\mathbb{Q}B$.

Now $\psi(b_k a)$ and $\hat{\psi}(b_k a)$ belong to B^*; also $\psi(b_k a) = \hat{\psi}(b_k a_k)+w'$, where a_k denotes the kth homogeneous component of a, w' has degree $\geq k+1$, and $\hat{\psi}(b_k a_k)$ is a Lie element of B_k^* (see (5) and Lemma 2.3.5). Therefore there is an element $f^*\in F^*$ such that f^* is a product of commutators of weight k in the x_i and s_l, each commutator involves at least one s_l, and

$$\theta^*(f^*)-1 = \hat{\psi}(b_k a_k)+v,$$

where v is a $\mathbb{Z}$-linear combination of monomials of degree at least $k+1$ in the $\tilde{x}_i$ and $\tilde{s}_l$, each involving at least one $\tilde{s}_l$. Thus $\delta(v)\in\theta(J(k+1)+\mathfrak{f}^n/\mathfrak{f}^n)$. Hence

$$\psi(b_k a) = \theta^*(f^*)-1+w'-v.$$

We apply δ to this. Everything turns on the relation between δ and ψ, which is as follows.

LEMMA 2.3.7 $\psi\delta = \delta\psi$.

So we obtain

$$\psi\delta(b_k a) = \delta(\theta^*(f^*) - 1) + \delta(w' - v),$$

or

$$\psi(b_k(\theta(x) - 1)) = \theta\beta(f^* - 1) + \delta(w' - v).$$

Now $\theta(x) - 1$ is a Lie element of B_{n-1}, so $\psi(b_k(\theta(x)-1)) = \hat{\psi}(b_k(\theta(x)-1)) = b_k(\theta(x) - 1)$, by Lemma 2.3.5. Putting $f = \beta(f^*)$, we now have

$$\theta(b_k(x - 1)) = \theta(f - 1) + \theta(w)$$

say, where $w \in J(k+1) + \mathfrak{f}^n/\mathfrak{f}^n$, and since θ is an isomorphism, this gives Lemma 2.3.4 and Sjogren's Theorem.

Two interesting questions remaining in this area are the following, which have been mentioned in several places.

Problem 2.3.8. Does there exist a function $f(c)$ such that $D_{f(c)}(G) = 1$ for every nilpotent group G of class c?

Problem 2.3.9. Is it true that $\bigcap_{n=1}^{\infty} D_n(G) = 1$ for every residually nilpotent group G?[‡]

The first problem reduces rapidly to one about a finite p-group, and a positive answer to it would yield a positive answer to the second. Very little is known (to me) about Problem 2.3.8. We can take $f(2) = 3$ but it does not even seem to be known whether $f(3)$ can be defined—a problem about finite 2-groups. Before making some observations about Problem 2.3.9, we remark again that dimension subgroups over $\mathbb{Z}/p^r\mathbb{Z}$ still seem to be shrouded in mystery.

2.4 Transfinite dimension subgroups

The transfinite powers of $\mathfrak{g}$ are defined by $\mathfrak{g}^0 = \mathbb{Z}G$; $\mathfrak{g}^{\alpha+1} = \mathfrak{g}^\alpha\mathfrak{g}$ and $\mathfrak{g}^\mu = \bigcap_{\alpha<\mu}\mathfrak{g}^\alpha$, for ordinals α and non-zero limit ordinals μ. (It should be noted that this definition is not left–right symmetric; for example $\mathfrak{g}^\omega\mathfrak{g} \neq \mathfrak{g}\mathfrak{g}^\omega$ in general.) The corresponding dimension subgroups are defined by $D_\alpha(G) = G \cap (1 + \mathfrak{g}^\alpha)$. Thus Problem 2.3.9 asks whether $D_\omega(G) = 1$ for every nilpotent group G. The following result of mine [41] (see also Kuškuleĭ [57]) 'reduces' it to the periodic case.

[‡] This is true if G is finitely generated metabelian [106].

THEOREM 2.4.1 *Let G be a nilpotent group of class $\leqslant c$ with torsion subgroup T. Then*

(i) $D_{\phi(b,\,c)}(G) \cap T \leqslant D_b(T)$ *for all* $b \geqslant 1$, *where* $\phi(b, c) = (b-1)c(c+1)+1$;
(ii) $D_\omega(G) = D_\omega(T)$.

These results are deduced from intersection theorems for powers of $\mathfrak{g}$; for example we have

$$\mathfrak{g}^{\phi(b,\,c)} \cap \mathbb{Z}T \leqslant \mathfrak{t}^b.$$

Some related results can be found in [42].

Some information about the structure of $D_\omega(G)$ can be obtained as follows. For a nilpotent group G, we define an element $g \in G$ to have *infinite p-height*, if the equation

$$x^{p^n} = g$$

has a solution in G for all $n \geqslant 1$. The set of elements of infinite p-height is a subgroup $G(p)$ of G. This follows from the following useful result of Mal'cev [69], which deserves to be better known.

LEMMA 2.4.2 *Let G be a nilpotent group of class c. If $m \geqslant 1$, then every product of m^cth powers in G is an mth power.*

Proof. This is by induction on c. Let $x_1, \ldots, x_t \in G$. We want to show that the equation

$$x_1^{m^c} \ldots x_t^{m^c} = u^m$$

can be solved with $u \in G$. We may assume that $G = \langle x_1, \ldots, x_t \rangle$ and $c > 1$. Let $H = \langle x_1^m, \ldots, x_t^m \rangle$. Applying the inductive hypothesis to the elements $x_1^m \gamma_c(H), \ldots, x_t^m \gamma_c(H)$, we find that

$$x_1^{m^c} \ldots x_t^{m^c} = y^m z, \quad \text{for some } y \in H,$$

and

$$z \in \gamma_c(H) = \langle [x_{i_1}^m, \ldots, x_{i_c}^m] : 1 \leqslant i_r \leqslant t \rangle.$$

Thus $z = w^m$ for some w in the centre of G, and $y^m z = (yw)^m$.

Now $G(p) = \bigcap_{n=1}^{\infty} G^{p^n}$, so $G/G(p)$ is residually of finite p-power exponent. If H is any nilpotent p-group of finite exponent or any torsion-free nilpotent group, then $\bigcap_{n=1}^{\infty} \mathfrak{h}^n = 0$ (see [33], or the next section). Hence, writing $L = T \cap (\cap_p G(p)) = \prod_p T_p(p)$, where T_p denotes the Sylow p-subgroup of T, we have $D_\omega(G) \leqslant L$. Now in a nilpotent group, every p-element commutes with every element of infinite p-height. This can be proved by an easy induction on the class. Hence we obtain the first part of the next result.

THEOREM 2.4.3

(i) *If G is nilpotent with torsion subgroup T, then $D_\omega(G)$ lies in the centre of T.*

(ii) *[98, p. 69] If G is residually nilpotent, then $D_{\omega+1}(G) = 1$.*

In proving the second part, we may assume that G is nilpotent. Defining L as above, we have $\mathfrak{g}^\omega \leqslant IG$, and hence $\mathfrak{g}^{\omega+1} \leqslant I\mathfrak{g}$. But since L is abelian, a well known isomorphism (see [20]) shows that $L \cap (1 + I\mathfrak{g}) = 1$, as required.

It is known that $D_\omega(G) = 1$ if G is residually nilpotent and $D_\omega(H) = 1$ for some subgroup H of finite index in G (Kuškuleĭ [57]; see also [41]). It follows easily from this and Theorem 2.4.2(ii) that $D_\omega(G) = 1$ *if G is a residually nilpotent group of finite (Mal'cev special, or Prüfer) rank.*

Some information about higher transfinite dimension subgroups can be found in [20] and [34].

Questions about dimension subgroups and powers of the augmentation ideal are often discussed in the language of *stable representations.*[‡] Let G be a group and M a right G-module. Then M is said to be *stable*, if it contains a series of submodules, on the factors of which G operates trivially. If there is a finite series of this type, then M is *finitely stable*. We see that a finitely stable module is one annihilated by some finite power of the augmentation ideal of G, and hence that G admits a faithful finitely stable module if and only if $D_n(G) = 1$ for some finite n. This point of view leads to an elaborate theory which we cannot explore here (see [98] and [80] for instance).

3. Some Residual Nilpotence Questions

3.1 Wreath products

If $\mathfrak{X}$ is some class of groups, a group G is said to be *residually-$\mathfrak{X}$* (often written $G \in \mathrm{R}\mathfrak{X}$) if given $1 \neq x \in G$, there exists a homomorphism ϕ

‡ For the related topic of stability groups, see Robinson's essay.

of G onto an $\mathfrak{X}$-group such that $\phi(x) \neq 1$. In particular, writing $\mathfrak{N}$ for the class of nilpotent groups,

$$G \in \mathrm{R}\mathfrak{N} \Leftrightarrow \bigcap_{n=1}^{\infty} \gamma_n(G) = 1.$$

The problem of when various group-theoretical constructions lead to residually nilpotent groups has received quite a bit of attention in the literature. Several of these questions are related to the residual nilpotence of wreath products. Conditions for the (restricted) wreath product

$$W = A \wr G$$

of two non-trivial groups A and G to be residually nilpotent were discussed at some length in [33], but although necessary and sufficient conditions on A and G are known when G is nilpotent, the general situation remains unclear. Of course, if W is residually nilpotent, so are A and G. To state the results of [33], let

$\mathfrak{A}$ denote the class of all abelian groups,
$\mathfrak{F}_p$ the class of finite p-groups,
$\mathfrak{N}_0$ the class of torsion-free nilpotent groups,
$\bar{\mathfrak{N}}_p$ the class of nilpotent groups of finite p-power exponent.

We shall be interested in the classes $\mathrm{R}\mathfrak{N}_0$ and $\mathrm{R}\bar{\mathfrak{N}}_p$, which have alternative descriptions as follows. For any group G, define an element $x \in G$ to be a *generalized periodic element*, if $x\gamma_n(G)$ is periodic for all $n \geqslant 1$. The set of all generalized periodic elements of G is a subgroup $T(G)$, which coincides with the torsion subgroup if G is nilpotent, and clearly

$$G \in \mathrm{R}\mathfrak{N}_0 \Leftrightarrow T(G) = 1.$$

The notion of a *generalized π-element*, where π is a set of primes, is defined similarly. We say that x has *generalized infinite p-height* in G, if $x\gamma_n(G)$ has infinite p-height in $G/\gamma_n(G)$ for all $n \geqslant 1$. By Lemma 2.4.2, the set of all elements of generalized infinite p-height is a subgroup $G(p)$, and we see easily that

$$G \in \mathrm{R}\bar{\mathfrak{N}}_p \Leftrightarrow G(p) = 1 \Leftrightarrow \bigcap_{n=1}^{\infty} G^{p^n}\gamma_n(G) = 1.$$

These concepts were introduced by Mal'cev [70].

THEOREM 3.1.1 *Let* $W = A \setminus G$, *where* $A \neq 1$ *and* $G \neq 1$. *If* W *is residually nilpotent, then one of the following holds:*

(i) *For some prime* p, $G \in \mathfrak{F}_p$ *and* $A \in \mathrm{R}\bar{\mathfrak{N}}_p$;

(ii) *G is infinite but not torsion-free,* $A \in \mathfrak{A}$, *and for some prime* p, $G \in \mathrm{R}\bar{\mathfrak{N}}_p$ *and* $A \in \mathrm{R}\bar{\mathfrak{N}}_p$;

(iii) *G is torsion free,* $A \in \mathfrak{A}$, *and* $G \in \mathrm{R}\bar{\mathfrak{N}}_p$ *whenever A contains an element of order* p. *Further, either* $G \in \mathrm{R}\mathfrak{N}_0$, *or G contains a non-trivial generalized* π-*element, in which case G is discriminated by* $\bigcup_{p \in \pi} \bar{\mathfrak{N}}_p$.

This is obtained from [33] Theorem A, together with an argument of Lichtman [64] which yields the last part. A group X is said to be *discriminated* by a family $\{K_\lambda : \lambda \in \Lambda\}$ of groups, if for any given finite set $x_1, \ldots, x_t$ of non-identity elements of X, there exists $\lambda \in \Lambda$ and a homomorphism $\phi : X \to K_\lambda$ such that $\phi(x_i) \neq 1$ $(i = 1, \ldots, t)$. Lichtman [65] has given an example of a group that is discriminated by $\bigcup_p \bar{\mathfrak{N}}_p$, but is not in $\mathrm{R}\bar{\mathfrak{N}}_p$ for any single prime p; this group contains generalized periodic elements but no generalized p-elements for any p.

In the converse direction to Theorem 3.1.1, we have from [33] the following results.

THEOREM 3.1.2 *If either* 3.1.1(i) *or* (ii) *holds, then* $W \in \mathrm{R}\bar{\mathfrak{N}}_p$.

THEOREM 3.1.3 *Suppose that* 3.1.1(iii) *holds, and that G is nilpotent. Then* $W \in \mathrm{R}\mathfrak{N}$, *and if in addition A is torsion free, then* $W \in \mathrm{R}\mathfrak{N}_0$.

The remaining question is thus

Problem 3.1.4. *Does* 3.1.1(iii) *without further restriction imply that* $W \in \mathrm{R}\mathfrak{N}$?

Assuming 3.1.1(iii) to hold, one might try to reduce the problem to the case when A is either a p-group or torsion free. This reduction can certainly be made if A splits over its torsion subgroup, but may not be easy in general. If A is a p-group, the above theorems show that $W \in \mathrm{R}\bar{\mathfrak{N}}_p$ unless A contains a non-trivial element of infinite p-height, while if A is torsion free, a straightforward argument based on the above results shows that $W \in \mathrm{R}\mathfrak{N}$ unless G contains a non-trivial generalized π-element and A contains a non-trivial element of infinite p-height for some $p \in \pi$. In particular this deals with the case when A is infinite cyclic, when the base group of W is isomorphic to $\mathbb{Z}G$.

3.2 Augmentation ideals

One outcome of the above discussion is the following.

THEOREM 3.2.1 (Lichtman [64]) $\mathfrak{g}^\omega = 0$ *if and only if either*

(i) $G \in \mathbb{R}\mathfrak{N}_0$, *or*
(ii) G *is discriminated by* $\bigcup_p \bar{\mathfrak{N}}_p$.

One of the main steps in proving these results is to show that $\mathfrak{g}^\omega = 0$ if G is torsion-free nilpotent (but not finitely generated). The author's proof of this involved bases of group rings like those described in Theorem 2.2.4, which itself derives from P. Hall's Edmonton Notes [30]. An alternative and rather elegant treatment, using Lie theory and the Campbell–Hausdorff Formula, has more recently been given by Šmel'kin [91, 92]. The main intermediate result, which is interesting in itself and has other applications, is

THEOREM 3.2.2   *Let L be a nilpotent Lie algebra, of possibly infinite dimension, over a field K of characteristic zero. Let U be the universal enveloping algebra of L, let J be the (associative) ideal of U generated by L, and let $\bar{U}$ be the completion of U in the J-adic topology. Then the elements $\exp x$ $(x \in L)$ of $\bar{U}$ are linearly independent over K.*

Note that $\bigcap_{n=1}^\infty J^n = 0$; the proof of this fact is not difficult and can be found in [40] for instance. Now let G be a torsion-free nilpotent group. In showing that $\mathfrak{g}^\omega = 0$, we may pass first to the Mal'cev completion, and so assume that roots can be extracted in G. Let L be the Lie algebra over $\mathbb{Q}$ associated to G by the Mal'cev correspondence, or inversion of the Campbell–Hausdorff Formula (see [2] or [58]), and let U, J be as in Theorem 3.2.2. Then $\{\exp x : x \in L\}$ is a multiplicative group isomorphic to G, and Theorem 3.2.2 tells us that we can identify $\mathbb{Q}G$ with a subalgebra of $\bar{U}$ in such a way that the augmentation ideal $\mathbb{Q}\mathfrak{g}$ lies in $\bar{J}$. Since $\bigcap_{n=1}^\infty J^n = 0$, we have $(\mathbb{Q}\mathfrak{g})^\omega = 0$.

The proof of Theorem 3.2.2 given in [92] does not seem to be quite correct, as applying induction does not appear to yield the conclusion claimed. Here is an alternative argument which I believe to be correct, but which uses the fact that $(K\mathfrak{g})^\omega = 0$ if G is *finitely generated* torsion-free nilpotent.

Proof of Theorem 3.2.2. Let $x_1, \ldots, x_t$ be distinct elements of L, let M be the (finite-dimensional) K-subalgebra of L they generate, let $\mathcal{U}(M)$ be its universal enveloping algebra, and $J(M)$ the ideal of $\mathcal{U}(M)$ generated by M. One can show quite easily, arguing along the lines of the proof that $\bigcap_{n=1}^{\infty} J^n = 0$, that $J^{nc} \cap \mathcal{U}(M) \leqslant J(M)^n$ for all $n \geqslant 1$. This implies that the $J(M)$-adic completion of $\mathcal{U}(M)$ is embedded in a natural way in $\bar{U}$. In this way, we can reduce to the case when L is finite dimensional and $x_1, \ldots, x_t$ generate it.

Let $g_i = \exp x_i$ $(i = 1, 2, \ldots, t)$ and $G = \langle g_1, \ldots, g_t \rangle$. Then G is finitely generated torsion-free nilpotent, and so $(K\mathfrak{g})^\omega = 0$. This is immediate from Theorem 2.2.4 for instance. Therefore there exists a finite-dimensional K-algebra A (namely $KG/(K\mathfrak{g})^m$ for suitably large m) and a homomorphism ϕ of G into $1 + E$, where $E = K\mathfrak{g}/(K\mathfrak{g})^m$ is a nilpotent ideal of A, such that $\phi(g_1), \ldots, \phi(g_t)$ are linearly independent over K. Now $\exp L$ is the Mal'cev completion of G and $1 + E$ admits unique extraction of roots, so ϕ extends uniquely to a homomorphism of $\exp L$ into $1 + E$. Then ϕ determines a Lie homomorphism of $L = \log \exp L$ into E, and hence an (associative) K-algebra homomorphism $U \to A$, which finally extends to $\bar{U}$. Since the elements $\phi(g_1), \ldots, \phi(g_t)$ are linearly independent over K, so are $g_1, \ldots, g_t$.

3.3 Free presentations

Now we pass to a question related to the matters just discussed. Let G be a group, F a free group, and

$$1 \to R \to F \to G \to 1 \tag{1}$$

a presentation of G with $R \neq 1$. When is F/R' residually nilpotent? More generally, when is F/S residually nilpotent, where S is a fully invariant, or perhaps merely characteristic, subgroup of R? Sufficient conditions for F/S to be residually nilpotent can be obtained from the following theorem of G. Baumslag.

THEOREM 3.3.1 [6] *Let F be free, $R \lhd F$, and S a fully invariant subgroup of R. If $\{H_i : i \in I\}$ and $\{K_j : j \in J\}$ are families of groups that discriminate R/S and F/R respectively, then $\{H_i \wr K_j : i \in I, \ j \in J\}$ discriminates F/S.*

Thus if, for example, F/R and R/S are residually finite p-groups, then so is F/S, whence one deduces easily that *free polynilpotent groups are residually finite p-groups*. This was first proved by Gruenberg [102].

When $S = R'$ we have that R/R' is free abelian, and taking the family

$\{K_j\}$ to consist simply of $G = F/R$, we find that every non-identity element of F/R' lies outside the kernel of some homomorphism of F/R' to $C_\infty \wr G$, where C_∞ is an infinite cyclic group. So if $C_\infty \wr G$ is residually nilpotent, so is F/R'. This also follows from the well known Magnus embedding.

On the other hand, we can view $\bar{R} = R/R'$ as a $\mathbb{Z}G$-module, the *relation module* for the presentation, in which G acts via F and F acts by conjugation. If F/R' is residually nilpotent, $\mathfrak{g}^\omega$ annihilates $\bar{R}$. When G is finite, a well known result of Gaschütz [17] shows that $\mathbb{Q} \otimes_\mathbb{Z} \bar{R} \cong (\mathbb{Q}G)^{d-1} \oplus \mathbb{Q}$, where d is the number of generators of F, so $\bar{R}$ is certainly a faithful $\mathbb{Z}G$-module if $d > 1$. Passi has extended this to the general case:

THEOREM 3.3.2 [76]   *If F is non-cyclic then $\bar{R}$ is a faithful $\mathbb{Z}G$-module.*

Therefore $\mathfrak{g}^\omega$ can annihilate $\bar{R}$ only if $\mathfrak{g}^\omega = 0$. Putting this information together with Theorem 3.2.1, we obtain the following rather satisfying theorem of Lichtman.

THEOREM 3.3.3 [64]   *For the free presentation* (1), *in which F is non-cyclic, the following are equivalent:*

(i) *F/R' is residually nilpotent;*
(ii) *$\mathfrak{g}^\omega = 0$;*
(iii) *either $G \in \mathrm{R}\mathfrak{N}_0$ or G is discriminated by $\bigcup_p \mathfrak{N}_p$.*

In fact, one sees that each of the individual conditions of (iii) passes from G to F/R', and hence to F/R'', F/R''', and so on.

Now we turn our attention to the more general problem of when F/S is residually nilpotent, where S is fully invariant in R. We shall assume that R/S is nilpotent, and begin by looking for necessary conditions. When $S = \gamma_{c+1}(R)$, one can use the fact [74] that the higher relation modules $\gamma_i(R)/\gamma_{i+1}(R)$ for G, like $\bar{R}$, are faithful $\mathbb{Z}G$-modules and argue as in the case of F/R' to show that if $F/\gamma_{c+1}(R)$ is residually nilpotent, then $\mathfrak{g}^\omega = 0$. A more general result is the following one.

LEMMA 3.3.4   *Let F, R, S be as just described, and suppose R/S is nilpotent and $F/S \in \mathrm{R}\mathfrak{X}$, where $\mathfrak{X}$ is some class of groups closed under taking homomorphic images.*

(i) *If R/S is not periodic, then $F/R' \in \mathrm{R}\mathfrak{X}$.*
(ii) *If R/S is periodic, then $F/R'R^p \in \mathrm{R}\mathfrak{X}$ for each prime p such that R/S has an element of order p.*

Proof. (i) We shall need the fact that R/R' and $R/R'R^p$ are faithful G-modules. This can be seen by reducing to the case when G is cyclic, in which case the structure of R/R' can be explicitly calculated. Write $H = F/S, K = R/S$. Then K is a relatively free and non-periodic nilpotent group, so its torsion subgroup T has finite exponent n and is contained in K'. Let

$$\mathbf{F} = \{N \lhd H: H/N \in \mathfrak{X}\}.$$

Thus $\bigcap_{N \in \mathbf{F}} N = 1$ and if $N \leqslant L \lhd H$ and $N \in \mathbf{F}$ then $L \in \mathbf{F}$. Put $T^* = \bigcap_{N \in \mathbf{F}} NT$. Then $(T^*)^n \leqslant \bigcap_{N \in \mathbf{F}} N = 1$. Hence $T^* \cap K = T \leqslant K'$, so $[T^*, K] \leqslant K'$. Therefore $T^* = T$, as $C_H(K/K') = K$.

Now we have shown that $H/T \in \mathfrak{R}\mathfrak{X}$, so we may now assume $T = 1$. Let K have class c and $1 < Z_1 < \ldots < Z_{c-1} < Z_c = K$ be its upper central series. It is not hard to see that $Z_{c-1} = K'$. Put $C_1 = C_H(K)$. Then if $N \in \mathbf{F}$, we have $[NC_1, K] \leqslant N$. Therefore $[\bigcap_{N \in \mathbf{F}} NC_1, K] = 1$, that is, $\bigcap_{N \in \mathbf{F}} NC_1 = C_1$, and $H/C_1 \in \mathfrak{R}\mathfrak{X}$. Similarly, defining

$$C_{i+1}/C_i = C_{H/C_i}(KC_i/C_i),$$

we have $H/C_i \in \mathfrak{R}\mathfrak{X}$ for all $i \geqslant 1$. But $C_i \cap K = Z_i$ and in particular $C_{c-1} \cap K = K'$. So $[C_{c-1}, K] \leqslant K'$, whence as before, $C_{c-1} \leqslant K$, and finally $C_{c-1} = K'$.

The proof of (ii) proceeds on similar lines.

Thus, if R/S is not periodic and F/S is residually nilpotent, then $\mathfrak{g}^\omega = 0$ and the conditions for this are known. However, as far as I am aware, little is known about the residual nilpotence of $F/R'R^p$.

Problem 3.3.5.   *When is $F/R'R^p$ residually nilpotent?*

Leaving this aside, suppose now that F, R, S are as before, with R/S torsion-free nilpotent for simplicity, and suppose $\mathfrak{g}^\omega = 0$. We would like to know whether F/S is residually nilpotent. According to Lichtman's Theorem (Theorem 3.2.1), two situations arise. The first, when $G \in \mathfrak{R}\mathfrak{N}_0$, has been handled by Šmel'kin by an ingenious indirect argument using Lie algebras. We will now describe Šmel'kin's work.

THEOREM 3.3.6 [91] *Let F be a free group, $R \lhd F$, and S be a fully invariant subgroup of R such that $R/S \in \mathfrak{R}\mathfrak{N}_0$. If $G = F/R \in \mathfrak{R}\mathfrak{N}_0$, then $F/S \in \mathfrak{R}\mathfrak{N}_0$.*

An essential ingredient in the proof of this theorem is the Andreev correspondence between varieties of Lie algebras over the rationals and varieties of groups in which the free groups are residually torsion-free nilpotent [3, 4]. (Note that [4] contains an incorrect result which is withdrawn in [5]). *All Lie algebras considered from this point will be Lie algebras over* $\mathbb{Q}$. Let $\mathfrak{V}$ be a variety of Lie algebras and L a free algebra in $\mathfrak{V}$. It is quite easy to see that $\mathfrak{V}$ can be defined by homogeneous, and even multilinear, identities, so L has a grading

$$L = L_1 \oplus L_2 \oplus \dots$$

with $(L_i, L_j) \leqslant L_{i+j}$. We can complete L to obtain an algebra $\tilde{L}$ of 'formal power series' $x = x_1 + x_2 + \dots$ $(x_i \in L_i)$ and introduce on $\tilde{L}$ an operation $\circ$ using the Campbell–Hausdorff Formula:

$$x \circ y = x + y + \tfrac{1}{2}(x, y) + \dots \tag{2}$$

where $(\ , \ )$ denotes Lie multiplication and the series on the right is the one occurring in the Campbell–Hausdorff Formula for $(\exp x)(\exp y)$. Then $\tilde{L}$ becomes a group under $\circ$. Actually, in this situation the universal enveloping algebra $\mathscr{U}(L)$ of L can be completed in the same way to give $\widetilde{\mathscr{U}(L)}$, the series for $\exp x$ converges in $\widetilde{\mathscr{U}(L)}$ if $x \in L$, and the group just described is isomorphic to the group of all $\exp x$ $(x \in L)$. Let $X = \{x_\lambda : \lambda \in \Lambda\}$ be a set of $\mathfrak{V}$-free generators for L. It is quite easy to see that any map of X into $\tilde{L}$ extends to a Lie endomorphism of $\tilde{L}$, and this Lie endomorphism will be a group endomorphism with respect to the operation $\circ$. From this, we see that $\langle x_\lambda : \lambda \in \Lambda \rangle$ is a relatively free group under $\circ$. Taking Λ to be countably infinite, this relatively free group determines a variety $\mathfrak{V}^\circ$ of groups in which the free groups are residually torsion-free nilpotent.

Conversely, given a relatively free group G belonging to $\mathrm{R}\mathfrak{N}_0$, we can obtain a (less obviously) relatively free Lie algebra

$$\bigoplus_{i \geqslant 1} \bar{\gamma}_i(G)/\bar{\gamma}_{i+1}(G) \otimes \mathbb{Q};$$

and so, given a variety $\mathfrak{U}$ of groups in which the relatively free groups are in $\mathrm{R}\mathfrak{N}_0$, we can construct the corresponding relatively free Lie algebras and form the variety $\mathfrak{U}^\dagger$ they generate. The correspondences $\mathfrak{V} \to \mathfrak{V}^\circ$ and $\mathfrak{U} \to \mathfrak{U}^\dagger$ are bijective on nilpotent varieties, but not in general. However, every variety of groups whose free groups belong to

R$\mathfrak{N}_0$ has the form $\mathfrak{V}^\circ$ for some $\mathfrak{V}$, and on these grounds, such varieties are called by Šmel'kin *varieties of Lie type*.

Another important notion in the proof of Theorem 3.3.6 is that of the *verbal wreath product* of groups or Lie algebras. Let $\mathfrak{U}$ be a variety of groups and A, B be groups. For simplicity, assume that $A \in \mathfrak{U}$. The verbal wreath product $A \wr_\mathfrak{U} B$ of A and B corresponding to $\mathfrak{U}$ is $W = A * B / U(A^F)$, where $F = A * B$ and $U(A^F)$ is the verbal subgroup of A^F (the normal closure of A in F) corresponding to $\mathfrak{U}$. Thus if A is $\mathfrak{U}$-freely generated by $\{x_\lambda : \lambda \in \Lambda\}$, then $W = \bar{A}B$, $\bar{A} \triangleleft W$, $\bar{A} \cap B = 1$, where $\bar{A}$ is $\mathfrak{U}$-free and $\mathfrak{U}$-freely generated by $\{x_\lambda^b : b \in B, \lambda \in \Lambda\}$. The verbal wreath product of Lie algebras is defined analogously; the normal closure of A is of course replaced by the ideal generated by A. When $\mathfrak{U}$ is the variety of abelian groups, we have the usual wreath product $A \wr B$, where, because of the unnecessary restrictions we have imposed, A must be taken abelian. We also have the classical Magnus embedding of F/R' in $A \wr G$, where $G = F/R$ and A is a free abelian group of the same rank as F. Analogously, in the situation of Theorem 3.3.6,

$$F/S \text{ can be embedded in } A \wr_\mathfrak{U} G, \qquad (3)$$

where $\mathfrak{U}$ is the variety generated by R/S and A is a free $\mathfrak{U}$-group on the same number of free generators as F [89, 90].

The proof of the theorem now turns on the properties of verbal wreath products of Lie algebras and their connections with verbal wreath products of groups.

Those readers familiar with relatively free groups may be surprised to learn that relatively free Lie algebras (over $\mathbb{Q}$) are always residually nilpotent. This is because a fully invariant ideal of a free Lie algebra is homogeneous, as can be seen by a simple linearization argument. Thus, no hypothesis on $\mathfrak{W}$ is necessary in the next result.

LEMMA 3.3.7 [91] *Let $\mathfrak{W}$ be a variety of Lie algebras, C a free $\mathfrak{W}$-algebra, and D a nilpotent Lie algebra. Then $W = C \wr_\mathfrak{W} D$ is residually nilpotent.*

This is valid for Lie algebras over any field of characteristic zero. Its proof is quite involved. Note, however, that when $\mathfrak{W}$ is the variety of abelian algebras, the 'base ideal' of $C \wr_\mathfrak{W} D$ is a free module over the universal enveloping algebra $\mathcal{U}(D)$ of D, and Lemma 3.3.7 amounts to the assertion that $\bigcap_{n=1}^\infty J(D)^n = 0$, where $J(D)$ is the ideal of $\mathcal{U}(D)$ generated by D.

We can now complete W with respect to the filter of ideals with nilpotent quotients. In the completion $\tilde{W}$, the series (2) converges and so we can introduce the circle operation to obtain a group $\tilde{W}^\circ$, which obviously belongs to $R\mathfrak{N}_0$. Actually it turns out that $\bigcap_{n=1}^{\infty} J(W)^n = 0$ as well, so $\mathscr{U}(W)$ can also be completed and the alternative interpretation of $\tilde{W}^\circ$ is available.

Let C be $\mathfrak{W}$-freely generated by elements x_λ, and C° the group generated by these elements under $\circ$. Thus C° is the free $\mathfrak{W}^\circ$-group generated by the x_λ. Further, let D° be the group consisting of the elements of D with the circle operation. Then we have the following beautiful and satisfying fact:

LEMMA 3.3.8 [91].   *The subgroup of $\tilde{W}^\circ$ generated by C° and D° is isomorphic to their $\mathfrak{W}^\circ$-wreath product.*

When $\mathfrak{W}$ is the variety of abelian algebras, $\mathfrak{W}^\circ$ is of course the variety of abelian groups, and the above lemma amounts to the same as Theorem 3.2.2. In fact, this is the crucial case as far as the proof is concerned.

Now we can prove Theorem 3.3.6 by assembling these facts and arguing as follows. First, we reduce by more or less routine arguments to the case when both F/R and R/S are torsion-free nilpotent, and R/S is relatively free in a variety of Lie type. By the Andreev correspondence, this variety has the form $\mathfrak{W}^\circ$ for some variety $\mathfrak{W}$ of Lie algebras. Thus there exists a $\mathfrak{W}$-free Lie algebra C such that $C^\circ \cong R/S$, and a nilpotent Lie algebra D such that D° is isomorphic to the Mal'cev completion of $G = F/R$. By Lemma 3.3.7, $W = C\wr_{\mathfrak{W}} D$ is residually nilpotent. The group $\tilde{W}^\circ$ belongs to $R\mathfrak{N}_0$. By Lemma 3.3.8, $\tilde{W}^\circ$ contains a copy of $C^\circ\wr_{\mathfrak{W}^\circ} D^\circ$ and hence, by (3), a copy of F/S. This concludes the proof.

The case $S = \gamma_{c+1}(R)$ of Theorem 3.3.6 was known before. For example, it is implicit in Gupta and Gupta [25]; see also Gupta and Passi [27] for related results.

Šmel'kin's paper contains a number of other interesting results proved along similar lines. G. Baumslag has introduced the term *Magnus variety* to describe a variety of groups whose free groups are residually nilpotent and have torsion-free lower central factors. Šmel'kin uses Lie-theoretical methods to prove

THEOREM 3.3.9 [91]   *The product of two Magnus varieties is a Magnus variety.*

This is the usual varietal product; $\mathfrak{U}\mathfrak{B}$ consists of all extensions of $\mathfrak{U}$-groups by $\mathfrak{B}$-groups. A particular case of this is an unpublished result obtained some time ago by P. Hall—*the lower central factors of free polynilpotent groups are torsion free* (cf. [102]). This, together with the fact that free polynilpotent groups are residually finite p-groups (see the remark after Theorem 3.3.1) shows that *free polynilpotent groups are residually torsion-free nilpotent.*

Reverting to the problem of when F/S is residually nilpotent, when S is fully invariant in a normal subgroup R of F and R/S is torsion-free nilpotent, we see now that the outstanding case is when F/R is discriminated by $\bigcup_p \mathfrak{N}_p$. Let $S = V(R)$, where V is a set of words, and let $\mathbf{X}$ be a system of normal subgroups of F containing R such that if $X \in \mathbf{X}$, then $F/X \in \mathfrak{N}_p$ for some p, and every finite set of non-identity elements of F/R is disjoint from some X/R with $X \in \mathbf{X}$. Then by the proof of a theorem of Dunwoody [13], $V(R) = \bigcap_{X \in \mathbf{X}} V(X)$. In this way, as was pointed out to me by Lichtman, the problem of showing that $F/S \in \mathfrak{R}\bar{\mathfrak{N}}_p$ can be reduced to the case $F/R \in \bar{\mathfrak{N}}_p$.

Problem 3.3.10. If F, R, S are as just described, and $F/R \in \bar{\mathfrak{N}}_p$, does it follow that $F/S \in \mathfrak{R}\bar{\mathfrak{N}}_p$?‡

A positive answer would mean that in Theorem 3.3.3, R' could be replaced by any fully invariant subgroup S of R with torsion-free nilpotent quotient.

The case when R/S is a nilpotent p-group is obviously of considerable interest, and is not, on the face of it, susceptible to the above methods.

3.4 Free products

Necessary and sufficient conditions for a free product to be residually nilpotent have been obtained by Lichtman, building on earlier work of Mal'cev.

THEOREM 3.4.1 [65] *Let $\{G_j : j \in J\}$ be a family of non-trivial groups with $|J| \geqslant 2$. Their free product is residually nilpotent if and only if one of the following conditions holds.*

(i) *All the groups G_j $(j \in J)$ belong to $\mathfrak{R}\mathfrak{N}_0$.*
(ii) *For any finite set of non-identity elements $g_1, \ldots, g_k \in \bigcup_{j \in J} G_j$, there exists a prime p such that none of these elements is of generalized infinite p-height in the corresponding G_j.*

In particular, in case (ii), each G_j will be discriminated by $\bigcup_p \bar{\mathfrak{N}}_p$

‡ This question has now been answered positively [108].

4. Actions on the Lower Central Factors of a Free Group

4.1 Higher relation modules

Let R be a free group, freely generated by a set of elements r_i $(i \in I)$ and let H be some group of automorphisms of R. Then H induces automorphisms of the lower central factors $\gamma_i(R)/\gamma_{i+1}(R)$. The study of these induced actions, and related actions such as those on the factors of the lower central p-series $\{R_{i,p}\}$ of R, defined by

$$R_{1,p} = R; \quad R_{i+1,p} = [R_{i,p}, R]R_{i,p}^p$$

is often important. Frequently $R \lhd F$, another free group. We take $H = F$, acting by conjugation, and obtain an induced action of $G = F/R$. For example, the *relation modules* $\bar{R} = R/R'$ for G arise in this way. These have been extensively studied, especially when G is finite (see [21]), and we recall the basic result of Gaschütz already mentioned.

THEOREM 4.1.1 [17]   *If F is free of finite rank d, $R \lhd F$ and $G = F/R$ is finite, then*

$$\bar{R} \otimes_{\mathbb{Z}} \mathbb{Q} \cong \mathbb{Q} \oplus (\mathbb{Q}G)^{d-1}.$$

To study the higher factors $\gamma_i(R)/\gamma_{i+1}(R)$ in terms of R/R', we can form the Lie ring $\mathscr{L}(R) = \oplus_{c=1}^{\infty} \gamma_c(R)/\gamma_{c+1}(R)$ and embed it in its universal envelope A as described at the beginning of section 2. The ring A is freely generated by $|I|$ elements, and

$$A = \bigoplus_{c=0}^{\infty} A_c,$$

where A_c is the homogeneous component of degree c. Let

$$L = \bigoplus_{c=1}^{\infty} L_c$$

be the set of Lie elements in L, that is, the Lie subring generated by A_1. We have an identification of R/R' with A_1 and of $\gamma_c(R)/\gamma_{c+1}(R)$ with L_c. Our automorphism group H acts $\mathbb{Z}$-linearly on A_1 and so acts as a group of ring automorphisms of A leaving L invariant. The induced action of H on L_c is compatible with the identification of $\gamma_c(R)/\gamma_{c+1}(R)$ with L_c, and so the group R itself can largely be ignored from this point.

When A is finitely generated and H is finite, character theory is available for the study of the modules $L_c \otimes_{\mathbb{Z}} \mathbb{Q}$. Let $E = \text{End}_{\mathbb{Z}} A_1$ be the ring of additive endomorphisms of A_1, let $\Gamma = \text{Aut}_{\mathbb{Z}} A_1$, and for $\varepsilon \in E$, let $\varepsilon^{(c)}$ and $\varepsilon^{[c]}$ denote the additive endomorphisms induced by ε on A_c and L_c respectively. The maps $\varepsilon \to \varepsilon^{(c)}$ and $\varepsilon \to \varepsilon^{[c]}$ are multiplicative but not additive. In the case under discussion, the trace of $\varepsilon^{[c]}$ is given by a formula of Brandt [9]:

$$\text{tr}\, \varepsilon^{[c]} = \frac{1}{c} \sum_{e \mid c} \mu(e)\, (\text{tr}(\varepsilon^e))^{c/e},$$

where μ is the Möbius function. This enables the character of H on L_c to be computed in terms of that of H on $A_1 = L_1$. If R is normal in a free group F of finite rank d, $G = F/R$ is finite, and G plays the role of H, the latter character can be read off from Theorem 4.1.1. This enabled Gupta, Laffey and Thomson to compute the characters χ_c of the *higher relation modules* $\gamma_c(R)/\gamma_{c+1}(R)$ of a finite group G.

THEOREM 4.1.2 [26]   *If $g \in G$, then*

$$\chi_c(g) = \frac{1}{c} \sum_{e \mid c} \mu(e)\, (\chi(g^e))^{c/e},$$

where for $x \in G$, $\chi(x) = \chi_1(x) = 1$ if $x \neq 1$,

$$= 1 + |G|\, (d-1)   \text{if } x = 1.$$

The nature of this character is influenced by the elements of G of order dividing c. When $(c, |G|) = 1$, the only such element is the identity, from which it follows that

$\gamma_c(R)/\gamma_{c+1}(R)$ is a free $\mathbb{Q}G$-module if $c \geq 2$ and $(c, |G|) = 1$.

When $(c, |G|) \neq 1$, a lower bound can be given for the multiplicities of the absolutely irreducible characters of G in χ_c, whence it follows that the above modules always contain a large free module (see [26]). The multiplicity of the trivial character in χ_c can be computed exactly, and this gives the size of the centre of $F/\gamma_{c+1}(R)$:

THEOREM 4.1.3 [26]   *If F is free of finite rank $d > 1$, $R \triangleleft F$ and $G = F/R$ is finite, then the centre of $F/\gamma_{c+1}(R)$ is a free abelian group of rank r_c*

given by

$$r_1 = d;$$

$$r_c = \frac{1}{c\,|G|} \sum_{e\,|\,c} \mu(e) v_e(G)\,(m^{c/e} - 1) \quad (c \geqslant 2),$$

where $v_e(G)$ is the number of elements of G of order dividing e, and $m = 1 + (d-1)\,|G|$ is the rank of R.

4.2 Relative higher relation modules and faithfulness questions

Continuing with the supposition that R is non-trivial and normal in a (now non-cyclic) free group F, and writing $G = F/R$, we drop the assumption that G is finite and allow F to have infinite rank. We have mentioned Passi's Theorem that $\bar{R} = R/R'$ is a faithful $\mathbb{Z}G$-module. The extension of this result to the higher relation modules $\gamma_c(R)/\gamma_{c+1}(R)$ is due to Mital and Passi [74]. A 'relative' version, corresponding to the replacement of $\gamma_c(R)/\gamma_{c+1}(R)$ by the last non-trivial term of the lower central series of $R/V(R)$, where $V(R)$ is a verbal, or merely characteristic, subgroup of R such that $R/V(R)$ is torsion-free nilpotent, was considered in [39]. More precisely, we define a *cth relative higher relation module* (r.h.r.m.) for a group G to be a $\mathbb{Z}G$-module of the form $\gamma_c(R)/N$, where F is a non-cyclic free group, $1 \to R \to F \to G \to 1$ is a presentation of G, and N is a characteristic subgroup of R such that $\gamma_{c+1}(R) \leqslant N < \gamma_c(R)$ and $\gamma_c(R)/N$ is additively torsion-free. Of course, $\gamma_c(R)/N$ is to be viewed as a G-module via conjugation in F. Extending Passi's work, we consider whether these modules are faithful for $\mathbb{Z}G$. Now one can see that $\bar{R}$ is $\mathbb{Z}G$-faithful by noting first that $\bar{R}$ can be embedded in a free $\mathbb{Z}G$-module, namely $\mathfrak{f}/\mathfrak{fr}$; this follows from the well known 'relation sequence' [21]

$$0 \to \bar{R} \to \mathfrak{f}/\mathfrak{fr} \to \mathfrak{g} \to 0.$$

Therefore the annihilator J of $\bar{R}$ in $\mathbb{Z}G$ is the annihilator of a subset of $\mathbb{Z}G$—an 'annihilator ideal'—and by a result of M. K. Smith ([93] or [78], Theorem 4.3.17) J is controlled by $\Delta^+(G)$, the product of the finite normal subgroups of G, in the sense that $J = (J \cap \mathbb{Z}\Delta^+(G)) . \mathbb{Z}G$.[‡] This enables the problem to be reduced easily to the case when G is finite, when Gaschütz's Theorem (Theorem 4.1.1) can be invoked. It is natural to try to imitate this strategy in the more general situation.

‡ See Passman's essay, p. 228.

Studying these relative higher relation modules from the Lie-theoretic point of view leads us to consider $\mathbb{Z}H$-modules, not necessarily derived from a free presentation, of the form L_c/N, where H is a subgroup of $\Gamma = \mathrm{Aut}_{\mathbb{Z}}A_1$ and N is a $\Gamma^{(c)}$-invariant subgroup of L_c with non-trivial torsion-free quotient. Here, of course, $\Gamma^{(c)}$ denotes the image of Γ under the map $\varepsilon \mapsto \varepsilon^{(c)}$. Similarly we have $E^{(c)}$, where $E = \mathrm{End}_{\mathbb{Z}}A_1$.

THEOREM 4.2.1 [39] *Let $H \leqslant \Gamma$ and let N be a $\Gamma^{(c)}$-submodule of L_c such that L_c/N is additively torsion-free. Then*

(i) *N is $E^{(c)}$-invariant.*
(ii) *If $A_1 = L_1$ can be embedded in a free $\mathbb{Z}H$-module, so can L_c/N.*

The first assertion implies that *if M is a characteristic subgroup of a free group R such that R/M is torsion free and M lies between two successive terms of the lower central series of R, then M is fully invariant in R.* And since every relation module for a group G can be embedded in a free $\mathbb{Z}G$-module as we have remarked, we obtain from (ii) the following.

COROLLARY 4.2.2 *Every r.h.r.m. for a group G can be embedded in a free $\mathbb{Z}G$-module.*

The proof of Theorem 4.2.1 involves studying the connection between the actions of $\Gamma^{(c)}$ and $E^{(c)}$ on A_c, which is isomorphic to the cth tensor power $A_1^{\otimes c}$ of A_1, and that of the symmetric group S_c, which acts by permuting the tensor components. Write $\mathbb{Q}A$ for the free associative $\mathbb{Q}$-algebra $\mathbb{Q} \otimes_{\mathbb{Z}} A$, which can also be viewed as the tensor algebra on $\mathbb{Q}A_1$, and let $\mathbb{Q}A_c$, $\mathbb{Q}L$, $\mathbb{Q}L_c$ have the obvious meanings. Let $E_{\mathbb{Q}} = \mathrm{End}_{\mathbb{Q}}(\mathbb{Q}A_1)$ and $\Gamma_{\mathbb{Q}} = \mathrm{Aut}_{\mathbb{Q}}(\mathbb{Q}A_1)$. When A is finitely generated, say by m elements, $\Gamma_{\mathbb{Q}} \cong GL(m, \mathbb{Q})$, and in looking at $\Gamma_{\mathbb{Q}}^{(c)}$, we are essentially considering the action of $GL(m, \mathbb{Q})$ on the space of tensors of weight c. Thus, classical theory assures us that in this case, each of the rings $\mathbb{Q}(\Gamma_{\mathbb{Q}}^{(c)})$ and $\mathbb{Q}S_c$, acting on $\mathbb{Q}A_c$, induces the full endomorphism ring of the other. In the general case, something similar is true. Let S and T be sets of linear transformations of a vector space U. We say that *S is dense with respect to T* if, given finitely many elements $u_1, \ldots, u_k \in U$ and $\tau \in T$, there exists $\sigma \in S$ such that $\sigma(u_i) = \tau(u_i)$ $(1 \leqslant i \leqslant k)$.

LEMMA 4.2.3 [39] *Each of the rings $\mathbb{Q}\Gamma^{(c)}$, $\mathbb{Q}E^{(c)}$, $\mathbb{Q}(\Gamma_{\mathbb{Q}}^{(c)})$, $\mathbb{Q}(E_{\mathbb{Q}}^{(c)})$ and $\mathrm{End}_{\mathbb{Q}S_c}(\mathbb{Q}A_c)$ of linear transformations of $\mathbb{Q}A_c$ is dense with respect to each of the others.*

Now mutually dense sets of linear transformations obviously have the same invariant subspaces, and so Theorem 4.2.1(i) follows easily from the density of $\mathbb{Q}\Gamma^{(c)}$ in $\mathbb{Q}E^{(c)}$. What is more, $\mathbb{Q}A_c$ is certainly completely reducible under $\mathbb{Q}S_c$ and so, by a ring-theoretic principle ([51], Theorem VI.1.1), it is completely reducible under $\mathrm{End}_{\mathbb{Q}S_c}(\mathbb{Q}A_c)$. By Lemma 4.2.3, $\mathbb{Q}A_c$ *is then completely reducible under* $\mathbb{Q}\Gamma^{(c)}$, $\mathbb{Q}E^{(c)}$, and so on. Now in proving Theorem 4.2.1(ii), we know that $\mathbb{Q}A_1$ can be embedded in a free $\mathbb{Q}H$-module, and hence so can its tensor power $\mathbb{Q}A_c$. Since we now know that $\mathbb{Q}N$ is complemented in $\mathbb{Q}A_c$, we see that $\mathbb{Q}\otimes_{\mathbb{Z}}(L_c/N)$ can be embedded in a free $\mathbb{Q}H$-module. A little more argument yields the result.

Using Corollary 4.2.2, we can reduce the question of whether an r.h.r.m. for G is necessarily $\mathbb{Z}G$-faithful to the case when G is finite, in the same way as before. Unfortunately they need not be faithful in this case, though they usually are. Suppose that A is finitely generated, say by m free generators. The mth exterior power of $\mathbb{Q}A_1$ gives us a one-dimensional $E_{\mathbb{Q}}^{(m)}$-submodule of $\mathbb{Q}A_m$ affording the determinant representation of $\Gamma_{\mathbb{Q}}$, and then we find that for any $t \geqslant 1$, $\mathbb{Q}A_{tm}$ will contain a one-dimensional $E_{\mathbb{Q}}^{(tm)}$-submodule affording the tth power of the determinant representation. Now work of Kljačko [54] shows that when $s \neq 4, 6$, every irreducible $E_{\mathbb{Q}}^{(s)}$-submodule of $\mathbb{Q}A_s$ is isomorphic to a submodule of $\mathbb{Q}L_s$, with the exception of the symmetric and exterior powers of $\mathbb{Q}A_1$. Thus, apart from these exceptions, $\mathbb{Q}L_s$ will contain one-dimensional $E_{\mathbb{Q}}^{(s)}$-submodules when $m \mid s$, and these will determine r.h.r.m.'s that cannot be faithful. Fortunately, this is the only obstacle, and the following result is obtained, giving quite a bit more than fidelity.

THEOREM 4.2.4 [39] *Let G be finite of order g and let V be a cth r.h.r.m. for G arising from a (non-cyclic) free group F. Then V contains a non-zero free $\mathbb{Z}G$-module unless all the following conditions are satisfied:*

(i) *F has finite rank, d say;*
(ii) *c is a non-trivial multiple of the rank $m = 1 + g(d-1)$ of R;*
(iii) *$|G/C_G(V)| \leqslant 2$.*

THEOREM 4.2.5 [39] *Let V be a cth r.h.r.m. for a group G. Then V is $\mathbb{Z}G$-faithful unless G is finite and (i)–(iii) of Theorem 4.2.6 are all satisfied.*

The proof of Theorem 4.2.4 again turns on the interplay between the representations of $GL(m, \mathbb{Q})$ and S_c on the space of tensors of weight c.

4.3 The modular case

Rather little seems to be known about the 'modular' versions of the questions discussed in this section, and it seems worth formulating the following rather general problem.

Problem 4.3.1. *Let $1 \to R \to F \to G \to 1$ be a free presentation of a group G. What can be said about G-modules of the form S/T, where S/T is central in R and of exponent p, and $T \geqslant \gamma_m(R)$ for some m?*

The reader intending to pursue this would be well advised to consult Kovács' informative article [56] for some Lie-theoretic background and references to other related literature.

We turn briefly to the lower central p-series of a free group R, defined at the beginning of this section. Write $R_i = R_{i,p}$. We can think of R/R_2 as a vector space over the field $\mathbb{F}_p$ of p elements, and form the free Lie algebra Λ over $\mathbb{F}_p$ generated by a basis of $R/R_2 = \Lambda_1$. We have

$$\Lambda = \bigoplus_{c=1}^{\infty} \Lambda_c,$$

where Λ_c is the homogeneous component of degree c of Λ. Any $\mathbb{F}_p$-linear automorphism α of Λ_1 extends to an algebra automorphism of Λ leaving the homogeneous components invariant. Also, α can be lifted to an automorphism of R, when it induces an automorphism of each R_c/R_{c+1} that is easily seen to be independent of the lifting. Thus each Λ_c and R_c/R_{c+1} can be viewed as a Σ-module, where $\Sigma = \operatorname{Aut} R/R_2$.

THEOREM 4.3.2 *There is an exact sequence of Σ-modules*

$$0 \to \Lambda_2 \oplus \ldots \oplus \Lambda_c \to R_c/R_{c+1} \to \Lambda_1 \to 0,$$

that splits if $p \neq 2$.

A proof of this is outlined in [11]. Bryant and Kovács go on to study the action of finite groups acting linearly on free Lie algebras, and prove the following useful result.

THEOREM 4.3.3 [11] *Let $L = \oplus_{c=1}^{\infty} L_c$ be a free Lie algebra over an arbitrary field K, where the L_c are the homogeneous components, and let G be a finite subgroup of $\operatorname{Aut}_K L_1$. Let m be the order of the subgroup of G consisting of the elements acting by scalar multiplication. Then if i is large enough, $L_i \oplus L_{i+1} \oplus \ldots \oplus L_{i+m-1}$ contains a non-zero free KG-submodule.*

Taking these results together, we obtain information about the actions of finite groups on the factors of the lower central p-series of free groups. Bryant and Kovács use this machinery to show that *given any linear group H of finite degree over* $\mathbb{F}_p$, *there exists a finite p-group P whose automorphism group induces on the Frattini factor group of P a group of automorphisms that is isomorphic, as a linear group, to H.* This has found applications in the theory of Fitting classes[‡] of finite soluble groups. In quite a different direction, Hartley and Robinson [38] have used these results to construct a family of finite complete groups as semi-direct products of p-groups by other groups. For example, finite complete supersoluble groups of odd order can be obtained in this way.

It seems likely that this method of constructing finite groups with specified properties (of a suitable kind) will find further applications.

References

[1] Alperin, J. L., Automorphisms of solvable groups, *Proc. Amer. Math. Soc.* **13** (1962), 175–180.

[2] Amayo, R. K., Stewart, I. N., 'Infinite-Dimensional Lie Algebras', Noordhoff, Leyden (1974).

[3] Andreev, K. K., Nilpotent groups and Lie algebras, *Algebra i Logika* **74** (1968), 4–14 (Russian); *Algebra and Logic* **7** (1968), 206–211.

[4] Andreev, K. K., Nilpotent groups and Lie algebras II, *Algebra i Logika* **8** (1969), 625–635 (Russian); *Algebra and Logic* **8** (1969), 353–358.

[5] Andreev, K. K., Letter to the editors, *Algebra i Logika* **10** (1971), 226–228 (Russian); *Algebra and Logic* **10** (1971), 143–144.

[6] Baumslag, G., Wreath products and extensions, *Math. Z.* **81** (1963), 286–299.

[7] Blackburn, N., On a special class of p-groups, *Acta Math.* **100** (1958), 45–92.

[8] Blackburn, N., Huppert, B., 'Finite Groups II, III', Springer-Verlag, Berlin (1982).

[9] Brandt, A., The free Lie ring and Lie representations of the full linear group, *Trans. Amer. Math. Soc.* **56** (1944), 528–536.

[10] Brjukhanova, E. G., Automorphism groups of 2-automorphic 2-groups, *Algebra i Logika* **20** (1981), 5–21 (Russian); *Algebra and Logic* **20** (1981), 1–12.

[11] Bryant, R. M., Kovács, L. G., Lie representations and groups of prime power order, *J. London Math. Soc.* (2) **17** (1978), 415–421.

[12] Chen, K. T., Fox, R. H., Lyndon, R., Free differential calculus IV: The quotient groups of the lower central series, *Ann. Math.* **68** (1958), 81–95.

[13] Dunwoody, M., On verbal subgroups of free groups, *Arch. Math.* **16** (1965), 153–157.

‡ See Hawkes' essay.

[14] Felsch, W., Neubüser, J., Plesken, W., Space groups and groups of prime power order IV: Counterexamples to the class–breadth conjecture, *J. London Math. Soc.* (2) **24** (1981), 113–122.
[15] Finken, H., Neubüser, J., Plesken, W., Space groups and groups of prime power order II, *Arch. Math.* **35** (1980), 203–209.
[16] Fox, R. H., Free differential calculus I, II, *Ann. Math.* **57** (1953), 547–560; *ibid.* **59** (1954), 196–210.
[17] Gaschütz, W., Über modulare Darstellungen endlicher Gruppen, die von freien Gruppen induziert werden, *Math. Z.* **60** (1954), 274–286.
[18] Gaschütz, W., Nichtabelsche p-Gruppen besitzen äussere p-Automorphismen, *J. Algebra* **4** (1966), 1–2.
[19] Gruenberg, K. W., Cohomological topics in group theory, *Springer Lecture Notes in Math.* **143** (1970).
[20] Gruenberg, K. W., Roseblade, J. E., The augmentation terminals of certain locally finite groups, *Canad. J. Math.* **24** (1972), 221–238.
[21] Gruenberg, K. W., Relation modules of finite groups, *CBMS Regional Conference Series in Mathematics* **25**, Amer. Math. Soc., Providence (1976).
[22] Grün, O., Über die Faktorgruppen freier Gruppen I, *Deutsche Math.*, Jahrgang 1, Heft 6 (1936), 772–782.
[23] Grunewald, F. J., Segal, D., Some general algorithms, I: Arithmetic groups, *Ann. Math.* **112** (1980), 531–583; II: Nilpotent groups, *ibid.*, 585–617.
[24] Grunewald, F. J., Segal, D., Sterling, L. S., Nilpotent groups of Hirsch length six, *Math. Z.* **179** (1982), 219–235.
[25] Gupta, C. K., Gupta, N. D., Generalized Magnus embeddings and some applications, *Math. Z.* **160** (1978), 75–87.
[26] Gupta, N. D., Laffey, T. J., Thomson, M. W., On the higher relation modules of a finite group, *J. Algebra* **59** (1979), 172–187.
[27] Gupta, N. D., Passi, I. B. S., Some properties of Fox subgroups of free groups, *J. Algebra* **43** (1976), 198–211.
[28] Hall, P., A contribution to the theory of groups of prime-power order, *Proc. London Math. Soc.* (2) **36** (1934), 29–95.
[29] Hall, P., On a theorem of Frobenius, *Proc. London Math. Soc.* (2) **40** (1935), 468–501.
[30] Hall, P., Nilpotent groups, Lectures given at Canadian Congress Summer Seminar, Univ. of Alberta (1957); republished in *Queen Mary College Lecture Notes in Math.*, Queen Mary College, London (1969).
[31] Hall, P., Some word-problems, *J. London Math. Soc.* **33** (1958), 482–496.
[32] Hall, P., A note on $\overline{SI}$-groups, *J. London Math. Soc.* **39** (1964), 338–344.
[33] Hartley, B., The residual nilpotence of wreath products, *Proc. London Math. Soc.* (3) **20** (1970), 365–392.
[34] Hartley, B., Augmentation powers of locally finite groups, *Proc. London Math. Soc.* (3) **32** (1976), 1–24.
[35] Hartley, B., On residually finite p-groups, *Symp. Math.* **17** (1976), 225–234.
[36] Hartley, B., Subgroups of finite index in profinite groups, *Math. Z.* **168** (1979), 71–76.
[37] Hartley, B., Meixner, Th., Periodic groups in which the centralizer of an involution has bounded order, *J. Algebra* **64** (1980), 285–291.
[38] Hartley, B., Robinson, D. J. S., On finite complete groups, *Arch. Math.* **35** (1980), 68–74.

[39] Hartley, B., Lichtman, A. I., Relative higher relation modules, *J. Pure Appl. Algebra* **22** (1981), 75–89.

[40] Hartley, B., Powers of the augmentation ideal, Proc. Int. Math. Conf., Singapore (1981), *North-Holland Math. Studies* **74**, 49–57, North-Holland, Amsterdam (1982).

[41] Hartley, B., Powers of the augmentation ideal in group rings of infinite nilpotent groups, *J. London Math. Soc.* (2) **25** (1982), 43–69.

[42] Hartley, B., An intersection theorem for powers of the augmentation ideal in group rings of certain nilpotent p-groups, *J. London Math. Soc.* (2) **25** (1982), 425–434.

[43] Hartley, B., Dimension and lower central subgroups—Sjogren's Theorem revisited, *Lecture Notes* **9**, Dept. of Math., Nat. Univ. of Singapore (1982).

[44] Heineken, H., Liebeck, H., The occurrence of finite groups in the automorphism groups of nilpotent groups of class 2, *Arch. Math.* **25** (1974), 8–16.

[45] Heineken, H., Gruppen mit kleinen abelschen Untergruppen, *Arch. Math.* **29** (1977), 20–31.

[46] Heineken, H., Nilpotente Gruppen, deren sämtliche Normalteiler charakteristisch sind, *Arch. Math.* **33** (1979), 497–503.

[47] Higman, G., Enumerating p-groups I. Inequalities, *Proc. London Math. Soc.* (3) **10** (1960), 24–30.

[48] Higman, G., Groups and rings having automorphisms without nontrivial fixed elements, *J. London Math. Soc.* **32** (1957), 321–334.

[49] Huppert, B., 'Endliche Gruppen I', Springer-Verlag, Berlin (1967).

[50] Ivanov, N. D., On automorphic algebras over $GF(2)$, *Vestnik Mosk. Univ.* 1982, No. 2, 69–72.

[51] Jacobson, N. A., 'Structure of rings', American Math. Soc., Providence (1956).

[52] Jennings, S. A., The structure of the group ring of a p-group over a modular field, *Trans. Amer. Math. Soc.* **50** (1941), 175–185.

[53] Jennings, S. A., The group ring of a class of infinite nilpotent groups, *Canad. J. Math.* **7** (1955), 169–187.

[54] Kljačko, A. A., Lie elements in a tensor algebra, *Sibirsk. Mat. Ž.* **15** (1974), 1296–1304 (Russian); *Siberian Math. J.* **15** (1974), 914–921.

[55] Kostrikin, A. I., Kreknin, V. A., Lie algebras with a regular automorphism, *Dokl. Akad. Nauk SSSR* **149** (1963), 249–251 (Russian); *Soviet Math. Dokl.* **4** (1963), 355–357.

[56] Kovács, L. G., Varieties of nilpotent groups of small class, *Springer Lecture Notes in Math.* **697** (1978), 205–229.

[57] Kuškuleĭ, A. H., On finitely stable representations of nilpotent groups, *Latv. Mat. Yežegodnik* **16** (1975), 39–45.

[58] Lazard, M., Sur les groupes nilpotents et les anneaux de Lie, *Ann. École Normale Supérieure* (3) **71** (1954), 101–190.

[59] Leedham-Green, C. R., McKay, S., On p-groups of maximal class I, *Quart. J. Math.* (2) **27** (1976), 297–311.

[60] Leedham-Green, C. R., McKay, S., On p-groups of maximal class II, *ibid.* **29** (1978), 175–186.

[61] Leedham-Green, C. R., McKay, S., On p-groups of maximal class III, *ibid.* 281–299.

[62] Leedham-Green, C. R., McKay, S., On the classification of p-groups of maximal class (preprint), Queen Mary College, London (1982).

[63] Leedham-Green, C. R., Newman, M. F., Space groups and groups of prime power order I, *Arch. Math.* **35** (1980), 193–202.

[64] Lichtman, A. I., The residual nilpotence of the augmentation ideal and the residual nilpotence of some classes of groups, *Israel J. Math.* **26** (1977), 276–293.

[65] Lichtman, A. I., Necessary and sufficient conditions for the residual nilpotence of free products of groups, *J. Pure Appl. Algebra* **12** (1978), 49–64.

[66] Magnus, W., Beziehungen zwischen Gruppen und Idealen in einem speziellen Ring, *Math. Ann.* **111** (1935), 259–280.

[67] Magnus, W., Über beziehungen zwischen höheren Kommutatoren, *J. reine angew. Math.* **177** (1937), 105–115.

[68] Magnus, W., Karrass, A., Solitar, D., 'Combinatorial Group Theory', New York (1966).

[69] Mal'cev, A. I., On homomorphisms onto finite groups, *Učen. Zap. Ivanovsk. Ped. Inst.* **18** (1958), 49–60.

[70] Mal'cev, A. I., Generalized nilpotent algebras and their adjoint groups, *Mat. Sb. N.S.* **25** (67) (1949), 347–366 (Russian); *Amer. Math. Soc. Translations* (2) **69** (1968), 1–21.

[71] Meixner, Th., Über endliche Gruppen mit Automorphismen, deren Fixpunktgruppe beschränkt sind, Dissertation, Naturwissenschaftliche Fachbereiche der Friedrich-Alexander-Universität Erlangen-Nürnberg (1979).

[72] Miech, R. J., Some *p*-groups of maximal class, *Trans. Amer. Math. Soc.* **189** (1974), 1–47.

[73] Miech, R. J., Counting commutators, *ibid.*, 49–61.

[74] Mital, J. N., Passi, I. B. S., Annihilators of relation modules, *J. Austral. Math. Soc.* **16** (1973), 228–233.

[75] Moran, S., Dimension subgroups modulo *n*, *Proc. Camb. Phil. Soc.* **68** (1970), 578–582.

[76] Passi, I. B. S., Annihilators of relation modules II, *J. Pure Appl. Algebra* **6** (1975), 235–237.

[77] Passi, I. B. S., Group rings and their augmentation ideals, *Springer Lecture Notes in Math.* **715** (1979).

[78] Passman, D. S., 'The Algebraic Structure of Group Rings', Wiley–Interscience, New York (1977).

[79] Pickel, P. F., Finitely generated nilpotent groups with isomorphic finite quotients, *Trans. Amer. Math. Soc.* **160** (1971), 327–341.

[80] Plotkin, B. I., Varieties of group representations, *Usp. Mat. Nauk* **32:5** (1977), 3–68 (Russian); *Russian Math. Surveys* **32:5** (1977), 1–72.

[81] Quillen, D., On the associated graded ring of a group ring, *J. Algebra* **10** (1968), 411–418.

[82] Rips, E., On the fourth integer dimension subgroup, *Israel J. Math.* **12** (1972), 342–346.

[83] Sandling, R., Dimension subgroups over arbitrary coefficient rings, *J. Algebra* **21** (1972), 250–265.

[84] Sandling, R., Tahara, K.-I., Augmentation quotients of group rings and symmetric powers, *Math. Proc. Camb. Phil. Soc.* **85** (1979), 175–194.

[85] Schmid, P., Normal *p*-subgroups in the outer automorphism group of a finite *p*-group, *Math. Z.* **147** (1976), 271–277.

[86] Shepherd, R., Ph.D. Thesis, University of Chicago (1971).

[87] Sims, C. C., Enumerating *p*-groups, *Proc. London Math. Soc.* (3) **15** (1965), 151–166.

[88] Sjogren, J. A., Dimension and lower central subgroups, *J. Pure Appl. Algebra* **14** (1979), 175–194.

[89] Šmel'kin, A. L., Wreath products and varieties of groups, *Izv. Akad. Nauk SSSR, Ser. Mat.* **29** (1965), 149–170 (Russian).

[90] Šmel'kin, A. L., A remark on my paper, 'Wreath products and varieties of groups', *Izv. Akad. Nauk SSSR, Ser. Mat.* **31** (1967), 443–444 (Russian).

[91] Šmel'kin, A. L., Wreath products of Lie algebras and their applications to group theory, *Trudy Mosk. Mat. Obšč.* **29** (1973), 247–260 (Russian); *Trans. Moscow Math. Soc.* **29** (1973), 239–252.

[92] Šmel'kin, A. L., A connection between Lie algebras and groups, *Usp. Mat. Nauk.* **33**: 3 (1978), 193–194 (Russian); *Russian Math. Surveys* **33**: 3 (1978), 177–178.

[93] Smith, M. K., On group algebras, *J. Algebra* **18** (1971), 477–499.

[94] Stallings, J. R., Quotients of the powers of the augmentation ideal in a group ring, in Knots, Groups and 3-Manifolds, *Ann. Math. Stud.* **84** (1975), 101–118.

[95] Tahara, Ken-ichi, On the structure of $Q_3(G)$ and the fourth integer dimension subgroup, *Japan J. Math. (N.S.)* **3** (1977), 381–394.

[96] Tahara, Ken-ichi, The augmentation quotients of group rings and the fifth dimension subgroups, *J. Algebra* **71** (1981), 141–173.

[97] Vol'vačev, R. T., Sylow p-subgroups of the general linear group, *Izv. Akad. Nauk SSSR, Ser. Mat.* **27** (1963), 1031–1054 (Russian); *Amer. Math. Soc. Translations* (2) **64** (1967), 216–243.

[98] Vovsi, S. M., 'Triangular Products of Group Representations and Their Applications', Birkhaüser, Boston (1981).

[99] Webb, U. H. M., An elementary proof of Gaschütz' Theorem, *Arch. Math.* **35** (1980), 23–26.

[100] Witt, E., Treue Darstellung Liescher Ringe, *J. reine angew. Math.* **177** (1937), 152–160.

[101] Zassenhaus, H., Ein verfahren, jeder endlichen p-Gruppe eine Lie-Ring mit der Charakteristik p zuzuordnen, *Abh. Mat. Sem. Hamburg* **13** (1940), 200–207.

Additional references

[102] Gruenberg, K. W., Residual properties of infinite soluble groups, *Proc. London Math. Soc.* (3) **7** (1957), 29–62.

[103] Viennot, G., Algèbres de Lie Libres et Monoïdes Libres, *Springer Lecture Notes in Math.* **691** (1978).

[104] Leedham-Green, C. R., McKay, S., Plesken, W., Space groups and groups of prime power order V: A bound to the dimension of space groups with fixed coclass, to appear.

[105] Gupta, N. D., Sjogren's Theorem—the metabelian case (to appear).

[106] Gupta, N. D., Hales, A. W., Passi, I. B. S., Dimension subgroups of metabelian groups, *J. reine angew. Math.* **346** (1984), 194–198.

[107] Gupta, N. D., Tahara, K.-I., Dimension and lower central subgroups of metabelian p-groups, to appear.

[108] Gupta, C. K., Levin, F., Dimension subgroups of free centre-by-metabelian groups, to appear.

[109] Hartley, B., A note on free presentations and residual nilpotence, *J. Pure Appl. Algebra*, to appear.

Reflections on the Classification of Torsion-free Nilpotent Groups

FRITZ GRUNEWALD and DAN SEGAL

1. General principles   121
2. Different equivalence relations   130
3. Some questions   135
4. Groups defined by central relations   138
5. Groups defined by ring extensions   142
6. $\mathscr{T}_2$-groups with small centre   148
7. Binary forms   153
References   157

Philip Hall's 'Edmonton Notes' [15] are just 25 years old as we write. In tribute to their stimulating influence over a quarter of a century, we would like to discuss what seem to us some interesting developments in the theory of finitely generated nilpotent groups, to raise questions and to present some modest contributions to their study.

The first two sections give a sketch of the general classification theory, as we see it. This is meant to motivate the problems, which appear throughout the article but mostly in section 3. The rest of the essay is devoted to the discussion of various classes of examples; those in section 4 form the basis of an important existence theorem (stated as 3.4 below).

The results from 2.6 on we believe to be new; we have been sparing with the proofs, however, preferring to emphasise the open questions which arise.

1. General Principles

By a $\mathscr{T}$-group we mean a torsion-free finitely generated nilpotent group. Hall's lectures contain a wealth of theorems about the structure

GROUP THEORY: essays for Philip Hall
ISBN 0-12-304880-X

of these groups, but little about their classification. This topic has seen some significant progress since 1957; many salient facts had been discovered, by A. I. Mal'cev, Hall and S. R. Jennings, but a fresh chapter was inaugurated when L. Auslander, G. Baumslag, A. Borel and P. F. Pickel started to look at these facts in a new way ([1], [21]). The excellent notes of Baumslag [4] give more details and proofs of some of the following material.

Let G be a $\mathscr{T}$-group. Then G has a *Mal'cev basis* $(x_1, \ldots, x_h)$: a generating set such that the subgroups $\langle x_1, \ldots, x_i \rangle$ form the successive terms in a central series of G with infinite cyclic factors. Each element of G is uniquely expressible as $x_1^{a_1} \ldots x_h^{a_h}$ with $a_1, \ldots, a_h$ in $\mathbb{Z}$. Representing such an element by the h-tuple $(a_1, \ldots, a_h)$ we identify G with $\mathbb{Z}^h$. The group operations in G are then given by polynomial mappings with rational coefficients. These mappings endow the h-dimensional affine space $\mathbf{A}_h$ with the structure of a unipotent affine algebraic group $\mathfrak{G}$, which is defined in fact not only over $\mathbb{Q}$ but over every *binomial ring* (a commutative ring R such that $r(r-1) \ldots (r-n+1)/n! \in R$ for every r in R and $n > 0$). (See [23], Appendix.)

For any ring R over which $\mathfrak{G}$ is defined, we write

$$G^R = \mathfrak{G}(R)$$

for the group of R-points of $\mathfrak{G}$ (the points with coordinates in R); statements involving the symbol G^R will always be meant to imply that $\mathfrak{G}$ is indeed defined over R. Then G^R is a nilpotent group containing $G^{\mathbb{Z}} = G$. The mapping $(n, g) \mapsto g^n$ from $\mathbb{Z} \times G$ to G is also given by a polynomial function $\mathbb{Z} \times \mathbb{Z}^h \to \mathbb{Z}^h$, which extends to a map $R \times G^R \to G^R$ and turns G^R into an *R-powered group* or *R-operator group*. In particular, $G^{\mathbb{Q}}$ is a radicable group, the *Mal'cev completion* or *radicable hull* of G, and G^R is torsion-free whenever the ring R has no $\mathbb{Z}$-torsion. Writing $\mathbb{Z}_p$ for the ring of p-adic integers, we can identify $G^{\mathbb{Z}_p}$ with the pro-p completion of G. Now for any family (R_λ) of rings, one has $G^{\Pi R_\lambda} \cong \Pi G^{R_\lambda}$. In particular, if $\mathbb{A}_\infty = \Pi_p \mathbb{Z}_p$ denotes the ring of finite integral adeles, we find

$$G^{\mathbb{A}_\infty} = \prod_p G^{\mathbb{Z}_p} = \hat{G}$$

where $\hat{G}$ denotes the *profinite completion* of G.

If R is a subring of the ring S then G^R is a subgroup of G^S, and the structure of G^R (as an operator group) determines that of G^S. This is so because G^R is dense in the Zariski topology of G^S and $R \times G^R$ is dense in $S \times G^S$. So the equivalence relation

$$G \underset{S}{\cong} H \Leftrightarrow G^S \cong H^S \quad \text{as } S\text{-operator groups}$$

of *S-isomorphism* is weaker than that of R-isomorphism. If classifying $\mathcal{T}$-groups up to R-isomorphism is too hard, it may be useful to attempt the classification up to S-isomorphism first. This approach has been very successful in other contexts, such as the theory of quadratic forms and the theory of semi-simple algebraic groups, and some powerful machinery is available for the purpose.

Let k be a field of characteristic zero. The unipotent affine group $\mathfrak{G}$ is defined over k, and the tangent space at the identity carries the structure of a nilpotent Lie algebra defined over k, the Lie algebra of $\mathfrak{G}$. Geometrically, $\mathfrak{G}$ is just an affine space, so $\mathfrak{G}$ is its own tangent space. It follows that $\mathfrak{G}(k)$ has the structure of a nilpotent Lie algebra over k, which we denote $\mathscr{L}_{\mathfrak{G}}(k)$ or $\mathscr{L}_G(k)$. (Note that, although we are identifying $\mathscr{L}_G(k)$ with G^k set-theoretically, and each has the structure of a vector space over k, these two structures are distinct unless G is abelian.) A concrete realisation, due to Jennings, is as follows (see [4], Chapter 4). One embeds G into the $\mathfrak{g}$-adic completion $\overline{\mathbb{Q}G}$ of the rational group algebra $\mathbb{Q}G$, where $\mathfrak{g}$ is the augmentation ideal. The power series $\log g$ converges in $\overline{\mathbb{Q}G}$ for each g in G, and the subset $\log G$ spans a Lie subalgebra $\mathbb{Q}(\log G)$ in the commutation Lie algebra on $\overline{\mathbb{Q}G}$. This subalgebra is $\mathscr{L}_G(\mathbb{Q})$. Using log and its inverse exp to identify G with $\log G$ inside $\mathscr{L}_G(\mathbb{Q})$, we obtain the following formulae for the Lie algebra operations on $G^{\mathbb{Q}}$:

$$g+h = \exp(\log g + \log h), \quad rg = g^r = \exp(r \log g),$$

$$(g,h) = \exp(\log g . \log h - \log h . \log g)$$

for $g, h \in G$ and $r \in \mathbb{Q}$. Everything then extends by linearity to $G^k = \mathscr{L}_G(k)$.

The group operations in G^k can be recovered from the Lie structure; since $g . h = \log(\exp g \exp h)$, we have $g . h = g * h$ where $g * h$ is given by the Baker–Campbell–Hausdorff formula

(1.1) $$g * h = g + h - \tfrac{1}{2}(g,h) + \tfrac{1}{12}(g,h,h) + \dots .$$

It follows that if G and H are $\mathcal{T}$-groups and

$$\theta: G^k \longrightarrow H^k$$
$$\| \qquad\qquad \|$$
$$\mathscr{L}_G(k) \qquad \mathscr{L}_H(k)$$

is a bijective mapping, then θ is a (k-operator) group isomorphism if and only if it is a Lie algebra isomorphism (over k). (For an elementary proof, see Segal [28], p. 116.)

Now let L be a finite-dimensional nilpotent Lie algebra over $\mathbb{Q}$. Then (1.1) defines a binary operation $*$ which makes L into a torsion-free radicable nilpotent group of finite rank. Let G be a finitely generated subgroup, of the same rank. Then G is a $\mathscr{T}$-group, $G^{\mathbb{Q}} = L$ and $L = \mathscr{L}_G(\mathbb{Q})$. So we have Mal'cev's correspondence which we state as

1.2 THEOREM   *The correspondence $G \mapsto \mathscr{L}_G(\mathbb{Q})$ sets up a bijection between the set of all $\mathbb{Q}$-isomorphism classes of $\mathscr{T}$-groups and the set of all isomorphism classes of finite-dimensional nilpotent Lie algebras over $\mathbb{Q}$. For every field k of characteristic zero, this induces a bijective correspondence between k-isomorphism classes of $\mathscr{T}$-groups and isomorphism classes of finite-dimensional nilpotent Lie algebras over k which are defined over $\mathbb{Q}$.*

The importance of 1.2 is that it enables us to study $\mathscr{T}$-groups by means of Lie algebras. Unlike groups, these lend themselves to investigation by the *theory of descent*, which we now discuss.

(i) Let k be a field and L, M be Lie algebras over k (all algebras will be finite-dimensional). *If $F \otimes_k L \cong F \otimes_k M$ for some field $F \supseteq k$, then $\bar{k} \otimes_k L \cong \bar{k} \otimes_k M$ where $\bar{k}$ is the algebraic closure of k, and indeed $K \otimes_k L \cong K \otimes_k M$ for some finite extension field K of k.* This follows from Hilbert's Nullstellensatz. It tells us that the first step in a classification programme should be

Problem 1.   *Classify the nilpotent Lie algebras over $\mathbb{Q}$ (or $\mathbb{C}$).*

The sort of answer one would hope for is discussed in section 3 below.

(ii) At the same time, we must classify the *k-forms* of a given algebra L; that is, the k-Lie algebras M such that $\bar{k} \otimes M \cong \bar{k} \otimes L$. Associated with L there is an algebraic group $\mathfrak{A}_L$, which we call the *algebraic automorphism group* of L: this is the subgroup of $GL(\bar{k} \otimes L)$ defined by the equations which stipulate that a linear automorphism also preserve the Lie operation. For any field $F \supseteq k$, $\mathfrak{A}_L(F) = \mathrm{Aut}(F \otimes L)$. Let M be a k-form of L and $\alpha: \bar{k} \otimes L \to \bar{k} \otimes M$ an isomorphism. If σ is any k-automorphism of $\bar{k}$, we can make σ act on $\bar{k} \otimes M$ and on $\bar{k} \otimes L$ by $r \otimes a \mapsto r^\sigma \otimes a$ ($r \in \bar{k}$, a in L or M). Then $\alpha^\sigma = \sigma^{-1}. \alpha. \sigma$ is another isomorphism $\bar{k} \otimes L \to \bar{k} \otimes M$, and

$$\Delta(\sigma) = \alpha^\sigma . \alpha^{-1} \in \text{Aut}(\bar{k} \otimes L) = \mathfrak{A}_L(\bar{k}).$$

The map $\Delta_M = \Delta \colon \text{Gal}(\bar{k}/k) \to \mathfrak{A}_L(\bar{k})$ is in fact a 1-cocycle. It depends not only on M but also on the choice of α; but defining equivalence of cocycles in the sense of non-abelian cohomology, one finds that the class $[\Delta]$ of Δ in $H^1(\text{Gal}(\bar{k}/k), \mathfrak{A}_L(\bar{k}))$—abbreviated henceforth to $H^1(k, \mathfrak{A}_L)$—depends only on the isomorphism type of M. Using the fact that $H^1(k, GL_h) = 0$, one can then verify

1.3 THEOREM   *Let L be a Lie algebra over k. The assignment $M \mapsto [\Delta_M]$ sets up a bijection between the set of all isomorphism types of k-forms of L and the Galois cohomology set $H^1(k, \mathfrak{A}_L)$.*

There is also a relative version:

1.4 THEOREM   *Let L be as in 1.3 and let F be a field containing k. There is a natural map $\omega_F \colon H^1(k, \mathfrak{A}_L) \to H^1(F, \mathfrak{A}_L)$, and for k-forms M and N of L we have*

$$F \otimes M \cong F \otimes N \iff [\Delta_M]\omega_F = [\Delta_N]\omega_F.$$

These are the salient features of the descent to $\mathbb{Q}$. For proofs and other applications, see J-P. Serre [29], Chapter 10, W. C. Waterhouse [35], Part V, or T. A. Springer [30]. The first conclusion to be drawn is that the algebraic automorphism groups $\mathfrak{A}_L$ are crucial to the classification programme. This prompts

Problem 2.   Which linear algebraic groups over $\mathbb{Q}$ can arise as $\mathfrak{A}_L$ for a nilpotent Lie algebra L over $\mathbb{Q}$?

The second conclusion is that results about Galois cohomology can give information about $\mathscr{T}$-groups. One good thing about Galois cohomology is that it is often computable, and we shall give examples of this in section 7. But there are also general theorems, which may be fruitfully interpreted in the present context. For example, the magnificent finiteness theorem of Borel and Serre [7], 7.1, says that for every linear algebraic group $\mathfrak{A}$ defined over $\mathbb{Q}$, the map

(1.5) $\omega = (\omega_{\mathbb{Q}_p}) \colon H^1(\mathbb{Q}, \mathfrak{A}) \to \prod_p H^1(\mathbb{Q}_p, \mathfrak{A})$

has finite kernel (here the Cartesian product may be taken over any set consisting of almost all the primes). This at once yields Borel's observation

1.6 THEOREM   *Let G be a $\mathscr{T}$-group. Then the $\mathscr{T}$-groups H which are $\mathbb{Q}_p$-isomorphic to G for all primes p lie in finitely many $\mathbb{Q}$-isomorphism classes.*

Two $\mathscr{T}$-groups H and K are $\mathbb{Q}$-isomorphic if and only if they are *commensurable*, i.e. there are subgroups H_1 and K_1, of finite index in H and K respectively, with $H_1 \cong K_1$. The group-theoretic meaning of the 'local' condition of $\mathbb{Q}_p$-isomorphism for all p will be elucidated in section 2; this condition is certainly implied by that of $\mathbb{Z}_p$-isomorphism for all p, and this in turn is equivalent to the two groups having *isomorphic finite quotients*. So the group-theoretic meaning of 1.6 is a little stronger than

1.7 COROLLARY   *A family of $\mathscr{T}$-groups having isomorphic finite quotients contains only finitely many mutually non-commensurable groups.*

Results like 1.6 and 1.7 of course raise the question of quantitative estimates: we discuss this in section 3.

One feature worth demanding of a respectable classification theory is that it be *effective*. Almost any reasonable question about a local field is effectively decidable; this is due to A. Tarski for $\mathbb{R}$ and to J. Ax and S. Kochen for p-adic fields ([32], [3]). The same holds for the algebraic closure of $\mathbb{Q}$, by classical elimination theory.

If $\mathfrak{A}$ is an algebraic group over $\mathbb{Q}$ such that the map in (1.5) is *injective* (when the Cartesian product goes over all primes including ∞, with $\mathbb{Q}_\infty = \mathbb{R}$), one says that the *Hasse principle* holds for $\mathfrak{A}$. If $\mathfrak{A} = \mathfrak{A}_L$ for a $\mathbb{Q}$-algebra L, we also say that the Hasse principle holds for L (or that L satisfies the principle). By 1.4, this is equivalent to the following property: another $\mathbb{Q}$-algebra M is isomorphic to L if and only if $\mathbb{Q}_p \otimes M \cong \mathbb{Q}_p \otimes L$ for all primes p (including $p = \infty$). One can then decide whether M is isomorphic to L or not, by the following procedure. First decide whether or not $\bar{\mathbb{Q}} \otimes M \cong \bar{\mathbb{Q}} \otimes L$; if the answer is 'yes', then $M \cong L$ if and only if $\mathbb{Q}_p \otimes M \cong \mathbb{Q}_p \otimes L$ for all primes p (including ∞), a local question which is effectively decidable (in [2] Ax shows how to deal with all the primes at once). Thus there is a good philosophical motivation for seeking a Hasse principle. Unfortunately, there are cases where the Hasse principle does not hold; but all is not lost: for Sarkisian ([25, 26]) has proved the important result

1.8 THEOREM   *Let $\mathfrak{A}$ be a linear algebraic group over $\mathbb{Q}$. Then there is an algorithm which decides whether any two explicitly given cocycles are equivalent in $H^1(\mathbb{Q}, \mathfrak{A})$.*

Together with methods from Grunewald and Segal [13] or [25], this is enough to establish

1.9 THEOREM   *The commensurability problem for $\mathscr{T}$-groups is effectively soluble.*

(This means that there is an effective procedure for deciding whether two $\mathscr{T}$-groups, given explicitly by generators and relations, are commensurable or not.) It is worth mentioning also that the general results of Ax and Kochen make it quite easy (in principle) to decide, for $\mathscr{T}$-groups G and H, whether or not

(a) $G \underset{\mathbb{Z}_p}{\cong} H$ for a given prime p;

(b) $\hat{G} \cong \hat{H}$;

(c) $G \underset{\mathbb{Q}_p}{\cong} H$ for a given p;

(d) $G \underset{\mathbb{A}}{\cong} H$, see p. 129.

(iii) Our descent is not quite finished: we are down to $\mathbb{Q}$-isomorphism, but we really want to know about $\mathbb{Z}$-isomorphism. We shall now consider $\mathscr{T}$-groups in a fixed $\mathbb{Q}$-isomorphism class: thus we fix a Lie algebra L over $\mathbb{Q}$ and study $\mathscr{T}$-groups G such that $\mathscr{L}_G(\mathbb{Q}) = L$. The theory of integral quadratic forms, for example, suggests that one should study the orbits of $\mathrm{Aut}(L)$ in the set of all full $\mathbb{Z}$-lattices in L. Unfortunately G, as a subset of L, is not necessarily a lattice; but fortunately G is contained as a subgroup of finite index in a $\mathscr{T}$-group $\bar{G}$ which is a lattice in L (see e.g. Segal [28], p. 108). We call such a group $\bar{G}$ a *lattice group*. The results we discuss next are proved by reducing to the case of lattice groups. This is not always a trivial step, but to keep things simple we gloss over it here.

Fix a basis for L, so that L may be identified with $\mathbb{Q}^h$, and $k \otimes L$ with k^h for any field k of characteristic zero. The set of all full $\mathbb{Z}$-lattices in k^h corresponds bijectively with the coset space $GL_h(\mathbb{Z}) \backslash GL_h(k)$. Suppose G and H are lattice groups such that $\mathscr{L}_G(k) = \mathscr{L}_H(k) = k \otimes L$; then $G = \mathbb{Z}^h . \alpha_G$ and $H = \mathbb{Z}^h . \alpha_H$ for some α_G, α_H in $GL_h(k)$. Now G is isomorphic to H as a group if and only if there exists a Lie algebra automorphism of $k \otimes L$ mapping G onto H, so we have

$$(1.10) \qquad G \cong H \iff GL_h(\mathbb{Z}) \alpha_G \mathfrak{A}(k) = GL_h(\mathbb{Z}) \alpha_H \mathfrak{A}(k),$$

where $\mathfrak{A}(k) = \mathrm{Aut}(k \otimes L)$. Thus the set of isomorphism classes of lattice groups in the k-isomorphism class associated to L gets mapped injectively into the double coset space

$$GL_h(\mathbb{Z}) \backslash GL_h(k) / \mathfrak{A}(k),$$

the class of a group G mapping to the double coset containing α_G.

This is the final stage of the descent. Like the first stage it forms the basis both for a decidability theorem and for a finiteness theorem.

Suppose we want an effective procedure to decide whether two lattice groups G and H are isomorphic. We first decide whether or not they are $\mathbb{R}$-isomorphic, using Tarski's decision procedure for questions about $\mathbb{R}$. Supposing the answer is 'yes', we take $k = \mathbb{R}$ in the above discussion, put $L = \mathscr{L}_G(\mathbb{Q})$, and use coordinates with respect to a $\mathbb{Z}$-basis of the lattice G. Then without loss of generality we may take $\alpha_G = 1$, and what has to be decided is whether or not $\alpha_H \in GL_h(\mathbb{Z}) . \mathfrak{A}(\mathbb{R})$. This kind of question is also effectively decidable: though expressed somewhat differently, the main result of Grunewald and Segal [12] is

1.11 THEOREM   *Let $\mathfrak{A}$ be an algebraic subgroup of GL_h defined over $\mathbb{Q}$. Then there is an algorithm which decides whether a given left coset of $\mathfrak{A}(\mathbb{R})$ in $GL_h(\mathbb{R})$ (defined explicitly as an algebraic set over $\mathbb{Q}$) meets $GL_h(\mathbb{Z})$.*

Thus one obtains

1.12 THEOREM   *The isomorphism problem for $\mathscr{T}$-groups is effectively soluble.*

For details of the proof, see Grunewald and Segal [13] or Segal [28], Chapter 6. (The accounts presented there differ both from each other and from the one given here. As is the case with most of the other results mentioned, the theorem extends to all finitely generated nilpotent groups; indeed, to all polycyclic-by-finite groups, but that is another story.)

Another application of the 'descent to $\mathbb{Z}$' is the celebrated finiteness theorem of Pickel [22]. This concerns groups which are $\mathbb{A}_\infty$-isomorphic. In view of 1.7, an $\mathbb{A}_\infty$-isomorphism class of $\mathscr{T}$-groups meets only finitely many $\mathbb{Q}$-isomorphism classes. We fix a lattice group E and let $\mathscr{C}$ be the set of all lattice groups G such that $G^{\mathbb{Q}} = E^{\mathbb{Q}}$ and $G \underset{\mathbb{A}_\infty}{\cong} E$.

Take $L = \mathscr{L}_E(\mathbb{Q})$ and use coordinates with respect to a $\mathbb{Z}$-basis of E. Then $E = \mathbb{Z}^h$, and for each G in $\mathscr{C}$ there exists α_G in $GL_h(\mathbb{Q})$ with $G = \mathbb{Z}^h . \alpha_G$. Also each $G \in \mathscr{C}$ is $\mathbb{Z}_p$-isomorphic to E, and the argument leading to 1.10 works as well with $\mathbb{Z}_p$ in place of $\mathbb{Z}$, if one takes $k = \mathbb{Q}_p$ and reads 'isomorphism' as '$\mathbb{Z}_p$-isomorphism'. Consequently, since $\alpha_E = 1$,

(1.13) $\alpha_G \in GL_h(\mathbb{Z}_p) . \mathfrak{A}(\mathbb{Q}_p)$

for each $G \in \mathscr{C}$ and each prime p. Now $\mathbb{A}_\infty$ is a subring of the *finite adele ring*

$$\mathbb{A} = \{x = (x_p) \in \prod_p \mathbb{Q}_p \mid x_p \in \mathbb{Z}_p \text{ for almost all } p\},$$

and so is $\mathbb{Q}$, via the diagonal embedding, so that $\mathbb{A}_\infty \cap \mathbb{Q} = \mathbb{Z}$. It follows that the natural map

$$j: GL_h(\mathbb{Z}) \backslash GL_h(\mathbb{Q})/\mathfrak{A}(\mathbb{Q}) \to GL_h(\mathbb{A}_\infty) \backslash GL_h(\mathbb{A})/\mathfrak{A}(\mathbb{Q}),$$

induced by the inclusion $GL_h(\mathbb{Q}) \to GL_h(\mathbb{A})$, is injective. Using square brackets to denote double cosets, (1.13) implies that

$$[\alpha_G]j \in GL_h(\mathbb{A}_\infty) \backslash GL_h(\mathbb{A}) . \mathfrak{A}(\mathbb{A})/\mathfrak{A}(\mathbb{Q})$$

for each G in $\mathscr{C}$. The set of double cosets on the right is in $1:1$ correspondence with $\mathfrak{A}(\mathbb{A}_\infty) \backslash \mathfrak{A}(\mathbb{A})/\mathfrak{A}(\mathbb{Q})$ and there is a very general theorem of Borel [5] which now applies:

1.14 THEOREM   *If $\mathfrak{A}$ is a linear algebraic group defined over $\mathbb{Q}$, then the double coset space $\mathfrak{A}(\mathbb{A}_\infty) \backslash \mathfrak{A}(\mathbb{A})/\mathfrak{A}(\mathbb{Q})$ is finite.*

Together with the injectivity of the map j and (1.10), it shows that the groups in $\mathscr{C}$ lie in finitely many isomorphism classes.

Thus we have outlined the proof of Pickel's Theorem [22] (cf. Grunewald, Pickel and Segal [10] or Segal [28], Chapter 10):

1.15 THEOREM   *The $\mathscr{T}$-groups having the same finite quotients as a given $\mathscr{T}$-group lie in finitely many isomorphism classes.*

This concludes our sketch of the background against which a distinctive theory of $\mathscr{T}$-groups might be developed. The results we have described may all be regarded as general facts about 'finite-dimensional algebraic structures'. It can be argued that any general fact about the classification of $\mathscr{T}$-groups is necessarily of this kind—we explain this more fully in section 3.

It seems to us now that further progress is most likely to come from studying the interplay between the specifically group-theoretic features of $\mathscr{T}$-groups and the general concepts we have been discussing.

2.   Different Equivalence Relations

We shall make some elementary observations regarding the group-theoretic substance of the arithmetically inspired definitions discussed in section 1. Throughout, let G and H be $\mathscr{T}$-groups, $\mathscr{F}(G)$ the set of isomorphism types of finite quotient groups of G and $\mathscr{F}_p(G)$ the set of isomorphism types of finite p-quotients.

The statements (2.1)–(2.5) are all well known or easy to verify; see Hall [15] Chapter 6, Baumslag [4] Chapters 2 and 4, [28] Chapters 6 and 10, or [10] §2.

(2.1) $G \underset{\mathbb{Q}}{\cong} H \Longleftrightarrow G^{\mathbb{Q}} \cong H^{\mathbb{Q}}$ as   abstract   groups $\Longleftrightarrow G$ and H are commensurable.

(2.2) $G \underset{\mathbb{Z}_p}{\cong} H \Longleftrightarrow G^{\mathbb{Z}_p} \cong H^{\mathbb{Z}_p}$ as abstract groups $\Longleftrightarrow \mathscr{F}_p(G) = \mathscr{F}_p(H)$.

(2.3) $G \underset{\mathbb{A}_\infty}{\cong} H \Longleftrightarrow G^{\mathbb{A}_\infty} \cong H^{\mathbb{A}_\infty}$ as abstract groups $\Longleftrightarrow \mathscr{F}(G) = \mathscr{F}(H)$.

(2.4) $G^{\mathbb{Q}_p}$ is the radicable hull of $G^{\mathbb{Z}_p}$.

(2.5) $G^{\mathbb{A}}$ is the radicable hull of $G^{\mathbb{A}_\infty} = \hat{G}$.

2.6 LEMMA   *The following are equivalent:*

(a) $G \underset{\mathbb{A}}{\cong} H$;

(b) *there is a finite set S of primes such that $G^{\mathbb{Z}_p}$ and $H^{\mathbb{Z}_p}$ are commensurable for each $p \in S$ and isomorphic for each prime $p \notin S$;*

(c) $\hat{G}$ and $\hat{H}$ are commensurable.

Proof. Suppose that (a) holds and assume without loss of generality that $G^{\mathbb{A}} = H^{\mathbb{A}}$. Since H is finitely generated, (2.5) then implies that $H_1 = H \cap \hat{G}$ has finite index in H. Then $\hat{H}_1 \leqslant \hat{G}$ and $\hat{H}_1$ has finite index in $\hat{H}$; by symmetry we infer that (c) holds (in the strong sense that $\hat{G} \cap \hat{H}$ has finite index in both $\hat{G}$ and $\hat{H}$). Now for a prime p, $G^{\mathbb{Z}_p}$ is the intersection of all subgroups of finite index coprime to p in $\hat{G}$ (and likewise with H in place of G); so if S is the set of prime factors of $|\hat{G} : \hat{G} \cap \hat{H}| \cdot |\hat{H} : \hat{G} \cap \hat{H}|$ and $p \notin S$ then $G^{\mathbb{Z}_p} = H^{\mathbb{Z}_p}$. For p in S, the same argument shows that $G^{\mathbb{Z}_p} \geqslant H_1^{\mathbb{Z}_p}$, and as above we may infer that $G^{\mathbb{Z}_p}$ and $H^{\mathbb{Z}_p}$ are commensurable. Thus (a) $\Rightarrow$ (c) $\Rightarrow$ (b). The remaining implications all follow easily.

The group-theoretic meaning of 2.6 is revealed in

2.7 PROPOSITION $G \underset{\mathbb{A}}{\cong} H$ *if and only if there exists a $\mathscr{T}$-group K commensurable with G such that $\mathscr{F}(K) = \mathscr{F}(H)$.*

The symmetry of $\mathbb{A}$-isomorphism now implies that if $G^{\mathbb{Q}} \cong F^{\mathbb{Q}}$ and $\hat{F} \cong \hat{H}$ for some $\mathscr{T}$-group F, then $\hat{G} \cong \hat{K}$ and $K^{\mathbb{Q}} \cong H^{\mathbb{Q}}$ for some other $\mathscr{T}$-group K, a fact which does not seem obvious. Before giving the proof, we want to mention another consequence of 2.7. Each $\mathbb{A}$-isomorphism class is partitioned in two ways, into $\mathbb{Q}$-classes and into $\mathbb{A}_\infty$-classes. We know from Pickel's theorem that each $\mathbb{A}_\infty$-class contains only finitely many isomorphism classes; so if we picture an $\mathbb{A}$-class as Fig. 1, with vertical strips representing $\mathbb{A}_\infty$-classes and horizontal

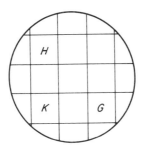

FIG. 1

strips representing $\mathbb{Q}$-classes, then each vertical strip meets only finitely many horizontal ones. What 2.7 says is that each vertical strip meets *every* horizontal strip. Hence there are only finitely many horizontal strips (this also follows directly from 1.6). The number $c'(G)$ of mutually incommensurable $\mathscr{T}$-groups K such that $\mathscr{F}(K) = \mathscr{F}(G)$ is called the *weak class number* of G. From 2.7 we deduce

2.8 COROLLARY　$c'(G)$ *is the number of commensurability classes of* $\mathscr{T}$*-groups* K *such that* $K \underset{A}{\cong} G$. *If* H *is commensurable with* G *then* $c'(H) = c'(G)$.

We shall mention an application of this in the next section.

　　The 'if' statement in 2.7 is clear. The 'only if' is a consequence of the following rather stronger result. Here we denote by $\mathscr{C}(G)$ the lattice of all subgroups of G^Q that are commensurable with G:

2.9 THEOREM　*If* $G \underset{A}{\cong} H$, *then there is an index-preserving lattice iso-morphism* $f: \mathscr{C}(G) \to \mathscr{C}(H)$ *such that* $\hat{K} \cong \widehat{f(K)}$ *for every* $K \in \mathscr{C}(G)$.

Proof.　Assume that $G^A = H^A$. For $K \in \mathscr{C}(G)$ define

$$f(K) = \hat{K} \cap H^Q.$$

The first thing to check is that $f(K) \in \mathscr{C}(H)$. From 2.6, there exists a natural number n such that $\hat{K}^n \leqslant \hat{H}$. Let c be the nilpotency class of H. Then the group

$$S = \langle x \in H^Q \mid x^{n^c} \in H \rangle$$

contains H as a subgroup of finite index, and $H \leqslant S^{n^c}$. Therefore

$$\hat{H} \leqslant \widehat{S^{n^c}} = \hat{S}^{n^c}$$

([28], p. 223) and it follows that every element of $\hat{H}$ has an nth root in $\hat{S}$ ([28], p. 113). Since extraction of roots is unique in the torsion-free nilpotent group $H^A = S^A$, we have $\hat{K} \leqslant \hat{S}$. Writing $L = f(K)$, we thus have $\hat{L} \leqslant \hat{K} \leqslant \hat{S}$, whence

(2.10)　　　　　$L = \hat{L} \cap L^Q \leqslant \hat{S} \cap H^Q = \hat{S} \cap S^Q = S.$

But

(2.11)　　　　　$|\hat{S} : \hat{L}| = |S : L| = |\hat{S} \cap H^Q : \hat{K} \cap H^Q| \leqslant |\hat{S} : \hat{K}|,$

and $|\hat{S}:\hat{K}|$ is finite by 2.6; so $|S:L|$ is finite and L is commensurable with H as required.

Since $\hat{L} \leqslant \hat{K} \leqslant \hat{S}$, (2.11) forces $\hat{L} = \hat{K}$, so we have also established

(2.12) $\widehat{f(K)} = \hat{K}$.

To show that f is bijective, define $g : \mathscr{C}(H) \to \mathscr{C}(G)$ by $g(F) = \hat{F} \cap G^{\mathbb{Q}}$. Then (2.12) (with the fact that $\hat{K} \cap K^{\mathbb{Q}} = K$ for $K \in \mathscr{C}(G)$) shows that gf is the identity map on $\mathscr{C}(G)$; by symmetry it follows that f is bijective. (2.12) also implies that f is index-preserving and intersection-preserving (see [28], p. 223, Ex. 4). Finally, it is an easy exercise to show that any intersection-preserving bijection $\mathscr{C}(G) \to \mathscr{C}(H)$ must be a lattice isomorphism.

We can now fulfil a promise made in section 1 by proving

2.13 PROPOSITION   *Two $\mathscr{T}$-groups G and H are $\mathbb{Q}_p$-isomorphic for all primes p if and only if $G \cong_{\mathbb{A}} H$.*

This is an application of a general fact about algebraic groups (see the final remark in Borel and Serre [7]),

2.14 PROPOSITION   *Let $\mathfrak{A}$ be an algebraic $\mathbb{Q}$-subgroup of GL_n and X a $\mathbb{Q}$-closed right coset of $\mathfrak{A}$ in GL_n. If $X \cap GL_n(\mathbb{Q}_p)$ is non-empty for all primes p then $X \cap GL_n(\mathbb{Z}_p)$ is non-empty for almost all p.*

Proof of 2.13.   Assume without loss of generality that G and H are lattice groups, put $L = \mathscr{L}_G(\mathbb{Q})$ and $M = \mathscr{L}_H(\mathbb{Q})$, and choosing suitable bases in L and M, identify G with $\mathbb{Z}^n \subset \mathbb{Q}^n = L$ and H with $\mathbb{Z}^n \subset \mathbb{Q}^n = M$. Here n is the Hirsch length, which we may clearly suppose is the same for both groups. We may also suppose that G and H are $\bar{\mathbb{Q}}$-isomorphic; let β in $GL_n(\bar{\mathbb{Q}})$ represent a suitable Lie algebra isomorphism $\bar{\mathbb{Q}} \otimes L \to \bar{\mathbb{Q}} \otimes M$, and let $\mathfrak{A} = \mathfrak{A}_L \leqslant GL_n$ denote the algebraic automorphism group of L. We now make the following observations.

(a)   The coset $\mathfrak{A}\beta$ is a $\mathbb{Q}$-closed subset of GL_n. For if $\sigma \in \mathrm{Gal}(\bar{\mathbb{Q}}/\mathbb{Q})$ then β^σ represents another isomorphism of $\bar{\mathbb{Q}} \otimes L$ onto $\bar{\mathbb{Q}} \otimes M$, so $\beta^\sigma \beta^{-1} \in \mathfrak{A}$, whence $(\mathfrak{A}\beta)^\sigma = \mathfrak{A}\beta$, and (a) follows by [28], p. 163, Prop. 7.

(b) $G \cong_{\mathbb{Q}_p} H$ if and only if $\mathfrak{A}\beta \cap GL_n(\mathbb{Q}_p) \neq \varnothing$. For $\mathbb{Q}_p \otimes L \cong \mathbb{Q}_p \otimes M$ if and only if there exists $\alpha_p \in GL_n(\mathbb{Q}_p)$ such that $\alpha_p \beta^{-1} \in \mathfrak{A}$.

(c) $G \underset{\mathbb{Z}_p}{\cong} H$ if and only if $\mathfrak{A}\beta \cap GL_n(\mathbb{Z}_p) \neq \emptyset$. For a matrix α_p as above represents an isomorphism sending $G^{\mathbb{Z}_p} = \mathbb{Z}_p^n$ onto $H^{\mathbb{Z}_p} = \mathbb{Z}_p^n$ if and only if $\alpha_p \in GL_n(\mathbb{Z}_p)$.

The 'if' in 2.13 follows from 2.6. To prove the 'only if' statement, we apply 2.14 with $X = \mathfrak{A}\beta$.

Remark. Thus we have yet another characterisation of $c'(G)$, as the number of commensurability classes of $\mathscr{T}$-groups which are $\mathbb{Q}_p$-isomorphic to G for all primes p. In particular, it shows that if $c'(G) = 1$ then $\mathscr{L}_G(\mathbb{Q})$ satisfies the Hasse principle (the converse does not hold, because the 'infinite prime' has been left out of account in the definition of $c'(G)$).

To conclude this section we discuss the weakest of the equivalence relations, $\mathbb{C}$-isomorphism.

2.15 PROPOSITION　*The following are equivalent for $\mathscr{T}$-groups G and H:*

(a) $G \underset{\mathbb{C}}{\cong} H$;

(b) $G \underset{\mathbb{Q}_p}{\cong} H$ *for some prime p;*

(c) $G \underset{\mathbb{Q}_p}{\cong} H$ *for infinitely many primes p;*

(d) $\mathscr{F}_p(G) = \mathscr{F}_p(H)$ *for some prime p;*

(e) $\mathscr{F}_p(G) = \mathscr{F}_p(H)$ *for infinitely many primes p.*

Proof. As we observed in section 1, $G \underset{\mathbb{C}}{\cong} H$ if and only if $G \underset{k}{\cong} H$ for some field k of characteristic zero, and if this holds at all then it holds with $k = \mathbb{Q}$. So if we can show that (a) implies (e), then the rest will follow, using (2.2) and (2.4). Suppose then that (a) holds, and use the notation of the above proof of 2.13. Part (c) of that proof, with (2.2), reduces the problem to showing that the matrix β lies in $GL_n(\mathbb{Z}_p)$ for infinitely many primes p. Now $\beta \in GL_n(k)$ for some algebraic number field k. Infinitely many primes p have a prime factor of residue-class degree one in k (see [8], Chapter 5, section 3), and for almost all of these (the unramified ones), k may be embedded in $\mathbb{Q}_p$. With finitely many exceptions, these embeddings send β into $GL_n(\mathbb{Z}_p)$ as required. This completes the proof.

3. Some Questions

We discuss Problem 1 first. Fix an algebraically closed field K of characteristic zero ($\mathbb{C}$ will do), and positive integers n and c. Let V be the space K^n, and put $G = GL(V) = GL_n(K)$ and $H = \operatorname{Hom}_K(V \otimes V, V)$. Each θ in H defines a (not necessarily associative) K-algebra structure on V by

$$a \cdot b = (a \otimes b)\theta \quad \text{for } a, b \in V,$$

and every K-algebra on V arises in this way. The group G has a natural action on H (via $(a \otimes b)\theta^g = ((ag^{-1} \otimes bg^{-1})\theta)g)$, and it is easy to see that the orbits of G correspond to isomorphism classes of K-algebras. Now the element θ of H defines a Lie algebra if and only if

$$(a \otimes b)\theta + (b \otimes a)\theta = 0$$

and

$$((a \otimes b)\theta \otimes c)\theta + ((b \otimes c)\theta \otimes a)\theta + ((c \otimes a)\theta \otimes b)\theta = 0$$

for all a, b and c in V. The set of all such θ forms an algebraic set W_n in H, and those θ in W_n which satisfy $(\ldots(a_0 \otimes a_1)\theta \otimes a_2)\theta \otimes \ldots)\theta \otimes a_c)\theta = 0$ for all $a_0, \ldots, a_c \in V$ form a closed subset $W_{n,c}$ corresponding to the Lie algebras which are nilpotent of class at most c. Evidently, each $W_{n,c}$ is invariant under the action of G. Problem 1 asks for an explicit description of the set of orbits.

As a first step, one should probably consider

3.1 *Describe the algebraic variety $W_{n,c}$.* What are its irreducible components? What are their dimensions? Are the different irreducible components definable by structural conditions on the corresponding Lie algebras?

Of course, $W_{n,n}$ 'is' the set of all nilpotent Lie algebra structures on V. M. Vergne [33] showed that $W_{n,n}$ is irreducible if $1 \leqslant n \leqslant 6$, and reducible if $n \geqslant 11$; beyond that, we know of no other results on this question.

Let us turn to the question of orbits. Since G is a reductive algebraic group, a general theorem of D. Mumford (see J. Fogarty [18], Chapter V) ensures the existence of a 'weak quotient' of $W_{n,c}$ by G: this is an affine variety $Y_{n,c}$, together with a morphism $\hat{\pi}: W_{n,c} \to Y_{n,c}$, such that

π is constant on G-orbits, and such that the whole thing is universal with these properties.

3.2 *Describe the variety $Y_{n,c}$ and the morphism $\pi\colon W_{n,c} \to Y_{n,c}$.*

An answer to 3.2 will only be a step towards the final solution of Problem 1; for in general each fibre of π (corresponding to a point of $Y_{n,c}$) will consist of several orbits of G. In particular, if X is an orbit of G then π is constant on the Zariski closure $\bar{X}$ of X. It is known that $\bar{X} \setminus X$ always consists of orbits of strictly smaller dimension than X: for example, $W_{3,3} = W_{3,2}$ has two G-orbits, say X_1 and X_0, with X_1 corresponding to the non-abelian Lie algebra on K^3 and X_0 to the abelian Lie algebra on K^3. So X_0 has one point and it is easy to verify that $\dim(X_1) = 4$ and $\bar{X}_1 = X_1 \cup X_0$. Thus even $Y_{3,3}$ is not quite good enough to classify 3-dimensional nilpotent Lie algebras. To complete the picture, we would like answers to

3.3 (i) *Given a Lie algebra L corresponding to θ in $W_{n,c}$, describe the set of all Lie algebras corresponding to points in the closure of the orbit θ^G. In particular, is $\overline{\theta^G} \subseteq \theta^G \cup W_{n,c-1}$? For which algebras L is θ^G closed?*

   (ii) *Does each fibre of π consist of the closure of a single G-orbit in $W_{n,c}$? If not, then how many?*

Let us turn to Problem 2. This is probably too hard to be a sensible problem, so let us weaken it a little. Every linear algebraic group $\mathfrak{A}$ defined over $\mathbb{Q}$ has a *unipotent radical* $u(\mathfrak{A})$, and $\mathfrak{A}$ is a semi-direct product $u(\mathfrak{A}) \rtimes r(\mathfrak{A})$ where $r(\mathfrak{A})$ is a reductive group defined over $\mathbb{Q}$. Now if $\mathfrak{N}$ is any unipotent normal subgroup of $\mathfrak{A}$ defined over $\mathbb{Q}$, and k is any field of characteristic zero, the natural map $H^1(k, \mathfrak{A}) \to H^1(k, \mathfrak{A}/\mathfrak{N})$ is a *bijection*. Thus $H^1(k, \mathfrak{A})$ may be identified with $H^1(k, r(\mathfrak{A}))$, and for the purposes of descent theory it suffices to know what the group $r(\mathfrak{A})$ can be, or indeed what $\mathfrak{A}/\mathfrak{N}$ can be for any suitably chosen $\mathfrak{N}$. Now let L be a nilpotent Lie algebra over $\mathbb{Q}$ and put $L' = (L, L)$.

There is a natural map

$$\pi\colon \mathfrak{A}_L \to \mathfrak{A}_{L/L'},$$

and we define

$$\mathfrak{B}_L = \pi(\mathfrak{A}_L)$$

to be the image of this map, so that $\mathfrak{B}_L$ is the algebraic subgroup of

$\mathfrak{A}_{L/L'}$ induced by automorphisms of L. Now the kernel of π is a unipotent algebraic group (this is essentially Hall's theorem [15], 3.7 bis). The fact that unipotent algebraic groups have trivial Galois cohomology (used implicitly above) implies that, for every field k of characteristic zero, $\mathfrak{B}_L(k)$ is exactly the group of automorphisms of '$k \otimes L$ abelianised' induced by automorphisms of $k \otimes L$. (Note that in general, a surjective morphism of algebraic groups need not be surjective on the k-rational points.)

As we have seen, $H^1(k, \mathfrak{A}_L) = H^1(k, \mathfrak{B}_L)$. The question 'which algebraic groups can arise as $\mathfrak{B}_L$ for a nilpotent Lie algebra L over $\mathbb{Q}$?' is answered by

3.4 THEOREM (Bryant and Groves [36])   *Let $n \geqslant 2$. Then every algebraic subgroup of GL_n defined over $\mathbb{Q}$ can be realised as $\mathfrak{B}_L$ for some nilpotent Lie algebra L over $\mathbb{Q}$ with $L/(L, L)$ of dimension n. Moreover, L may be chosen to have derived length at most 4.*

This is the substance of our remark at the end of section 1: the theory of descent for $\mathscr{T}$-groups is no different from descent theory in general, and it is no use hoping for specific results in this area. (Thus the failure of the Hasse Principle, for example, should occasion no surprise.) A rather general way to construct Lie algebras L with specified $\mathfrak{B}_L$ is discussed in the next section where we also sketch the application to the result of Bryant and Groves. Further examples are mentioned in section 5. We should also mention that J. L. Dyer [9] constructs a 2-generator nilpotent Lie algebra L over $\mathbb{Q}$ such that $\mathfrak{A}_L$ is unipotent; though superseded by 3.4, even this result is far from obvious.

As an open-ended project, we suggest

3.5 *Identify the algebraic group $\mathfrak{B}_L$, and if possible $\mathfrak{A}_L$, when $L = \mathscr{L}_G(\mathbb{Q})$ for interesting classes of $\mathscr{T}$-groups G.*

We end this section with some arithmetical questions. The finiteness theorems of section 1 each imply a challenge: to estimate the finite number whose existence is affirmed. Define the *class number* of a $\mathscr{T}$-group G to be the number $c(G)$ of isomorphism classes of $\mathscr{T}$-groups H with $\mathscr{F}(H) = \mathscr{F}(G)$; recall that the weak class number $c'(G)$ is the number of commensurability classes of such groups H. Thus $c'(G) \leqslant c(G)$. Remeslennikov asked (in lectures at Bielefeld in 1979) if every $\mathscr{T}$-group is commensurable with one of class number 1. That this is not so follows from the fact that c' is a commensurability invariant, by 2.8, together with the existence (established in unpublished work by G.

Higman some years ago (cf. Baumslag [4])) of $\mathscr{T}$-groups with weak class number greater than 1. We shall give several examples of such groups. There are also $\mathscr{T}$-groups G with $c'(G) = 1$ not commensurable with one of class number 1—this follows from the results of Grunewald, Segal and Sterling [14]; but we shall not go into that here. The first natural question is

3.6 *Characterise $\mathscr{T}$-groups with class number* 1.

This is certainly too hard, for an answer would imply by [11] or [14] a characterisation of the quadratic number fields with class number 1. However, it is perhaps not unreasonable to ask

3.7 *Characterise $\mathscr{T}$-groups G with $c'(G) = 1$.*

4. Groups Defined by Central Relations

Let F be a free nilpotent group of class $c \geqslant 2$ on n generators, and put $M = \gamma_c(F)$, the last non-trivial term of the lower central series of F. The classification of groups G of the form $G = F/K$ with $K \leqslant M$ can easily be reduced to certain questions which look reasonably tractable. We shall describe this reduction, which is probably well known, and formulate some relevant questions.

Let R be an integral domain of characteristic zero, with field of fractions k. For any group G, write $G^{ab} = G/G'$. We shall identify F^{ab} with $\mathbb{Z}^n$ and $(F^R)^{ab}$ with R^n, so that the group of R-automorphisms of $(F^R)^{ab}$ is $GL_n(R)$. For any $\mathscr{T}$-group G, we shall write

$$\mathfrak{A}_G = \mathfrak{A}_L, \quad \mathfrak{B}_G = \mathfrak{B}_L, \quad \text{where } L = \mathscr{L}_G(\mathbb{Q}),$$

$\mathfrak{A}_L$ and $\mathfrak{B}_L$ being the algebraic groups defined on pages 124 and 136, respectively. Recall that there is a natural morphism $\pi \colon \mathfrak{A}_L \to \mathfrak{B}_L$.

4.1 *Every R-automorphism of $(F^R)^{ab}$ lifts to an R-automorphism of F^R and so $\mathfrak{B}_F(R) = GL_n(R) = \pi(\mathrm{Aut}_R(F^R))$.*

Proof. It is not hard to verify that $(F^R)' = (F')^R$, whence $(F^R)^{ab}$ can be identified with $(F^{ab})^R$. Now let $\alpha \colon (F^R)^{ab} \to (F^R)^{ab}$ be an R-isomorphism. Then $\alpha|_{F^{ab}}$ induces $\alpha_1 \colon F \to (F^R)^{ab}$ and since F is relatively free α_1 lifts to a homomorphism $\theta \colon F \to F^R$. If $H = F\theta$, then

$$\frac{H^R \cdot (F^R)'}{(F^R)'} = (F^{ab} \alpha)^R = (F^R)^{ab} \alpha = \frac{F^R}{(F^R)'}.$$

Since F^R is nilpotent, $H^R = F^R$; therefore $H^k = F^k$, and comparing the k-dimensions of the upper central factors of H^k and F^k we infer that H and F have the same Hirsch length. As H is also a homomorphic image of F, this shows that θ is an isomorphism of F onto H. Hence θ induces an R-isomorphism $\bar{\theta}: F^R \to H^R = F^R$; and clearly $\pi(\bar{\theta}) = \alpha$.

Using the fact that $\ker \pi$ acts trivially on the lower central factors of F, it is easy to verify

4.2 *There is a morphism $\rho: \mathfrak{B}_F \to GL_m$, where m is the rank of M, such that the diagram*

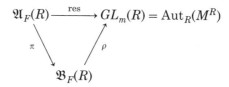

commutes for every allowed ring R (res denotes the restriction map).

For the next lemma, recall that a subgroup P of M is *isolated* if M/P is torsion-free.

4.3 *Let P and Q be isolated subgroups of M. Put $G = F/P$ and $H = F/Q$. Then for each R-isomorphism $\alpha: G^R \to H^R$ there exists $\alpha^* \in \mathfrak{B}_F(R)$ such that $P^R \rho(\alpha^*) = Q^R$. Conversely if $\beta \in \mathfrak{B}_F(R)$ is such that $P^R \rho(\beta) = Q^R$ then $\beta = \alpha^*$ for some such α.*

Proof. Since $(G^R)^{ab} = (H^R)^{ab} = (F^R)^{ab}$, α induces an automorphism $\alpha^* \in \mathfrak{B}_F(R)$. By 4.1 we have $\alpha^* = \pi(\gamma)$ for some $\gamma \in \mathrm{Aut}_R(F^R)$. Then γ acts like $\rho(\alpha^*)$ on M^R, hence γ acts like α on M^R/P^R, since the effect of α on the lower central factors of G^R is determined by α^*. It follows that $P^R \gamma \leqslant Q^R$, and by symmetry this is an equality. Thus

$$P^R \rho(\alpha^*) = P^R \rho\pi(\gamma) = P^R \gamma = Q^R.$$

The converse is proved similarly, also using 4.1.

These observations are enough to establish

4.4 PROPOSITION *Let $G = F/P$ and $H = F/Q$, where P and Q are isolated subgroups of M. Then $G \cong_R H$ if and only if P^R and Q^R lie in the same orbit of $GL_n(R)$, acting via ρ on M^R.*

4.5 PROPOSITION *Let $G = F/P$ where P is an isolated subgroup of M. Then $\mathfrak{B}_G(R)$ is the stabiliser of P^R in $GL_n(R)$, acting via ρ on M^R.*

To make use of these results, one has to know about the representation ρ. This is best studied in another form: let L be the free n-generator Lie algebra over $\mathbb{Q}$, with cth homogeneous component L_c. Then GL_n acts by linear substitutions on the generators of L, giving the standard representation of GL_n on $L_1 = \mathbb{Q}^n$, and the representation ρ on L_c.

4.6 THEOREM (Bryant and Groves [36]) *Let V be a finite-dimensional module for GL_n and assume that the representation of GL_n on V is a homogeneous polynomial representation of degree t say, defined over $\mathbb{Q}$. Let D be the 1-dimensional module for GL_n afforded by the determinant. Then there exists $r \geq 1$ such that $D^{\otimes r} \otimes V$ is a direct summand of L_c, where $c = nr + t$.*

Here $D^{\otimes r}$ denotes the rth tensor power of D.

To show how this sort of result is used, let us deduce a special case of 3.4. A subgroup $\mathfrak{H}$ of GL_n is called *homogeneous* if $\mathfrak{H}$ contains all the scalar matrices.

4.7 COROLLARY *Every homogeneous algebraic subgroup of GL_n defined over $\mathbb{Q}$ occurs as $\mathfrak{B}_G$ for some n-generator $\mathcal{T}$-group G (and G may be taken to be a one-relator group in its nilpotent variety).*

Now let $\mathfrak{H}$ be any $\mathbb{Q}$-closed subgroup of GL_n. Then there exist polynomials $f_1, \ldots, f_l$ in $\mathbb{Q}[T_{11}, T_{12}, \ldots, T_{nn}]$ such that

(4.8) $\mathfrak{H}(\mathbb{C}) = \{g \in GL_n(\mathbb{C}) \mid f_i^g \in \mathbb{C}f_i, \quad i = 1, \ldots, l\},$

where $GL_n(\mathbb{C})$ acts on the polynomial ring $S = \mathbb{C}[T_{11}, \ldots, T_{nn}]$ in the usual way, i.e.

$$f^g(\mathbf{T}) = f(\mathbf{T}g) = f\left(\sum_i T_{1i}g_{i1}, \ldots, \sum_i T_{ni}g_{in}\right).$$

Moreover, the f_i are semi-invariants of $\mathfrak{H}$ of the same weight, i.e. there is a character $\chi \colon \mathfrak{H}(\mathbb{C}) \to \mathbb{C}^\times$ such that

$$(4.9) \qquad f_i^g = \chi(g)f_i \quad (\text{all } g \in \mathfrak{H}(\mathbb{C}), \quad i = 1, \ldots, l).$$

(See Borel [6], §5.4.) Hence if $\mathfrak{H}(\mathbb{C}) \geqslant \mathbb{C}^\times 1_n$, then the polynomials f_i must all be homogeneous of the same degree, say t. Now let S_t denote the tth homogeneous component of S, and take V to be the direct sum of l copies of S_t. By 4.6, there exists $r \geqslant 1$ such that $D^{\otimes r} \otimes V$ is a GL_n-submodule of L_c, where $c = nr + t \geqslant 3$. Then

$$N = D^{\otimes r} \otimes (f_1, f_2, \ldots, f_l)$$

is a one-dimensional subspace of $D^{\otimes r} \otimes V$, and it is easy to see from (4.8) that its stabiliser in GL_n is exactly $\mathfrak{H}$.

Identify L_c with $M^{\mathbb{Q}} = \gamma_c(F^{\mathbb{Q}})$ and put $G = F/(M \cap N)$. Then 4.5 shows that $\mathfrak{B}_G = \mathfrak{H}$ as required.

To get explicit examples, let $\Phi(X, Y)$ be a binary form over $\mathbb{Q}$. The *projective automorphism group* of Φ is the algebraic subgroup of GL_2 consisting of all matrices g such that

$$\Phi^g(X, Y) = \Phi(g_{11}X + g_{12}Y, g_{21}X + g_{22}Y) = \lambda(g)\Phi(X, Y)$$

for some scalar $\lambda(g)$. It is easy to verify that this group satisfies the Hasse principle (see section 1) if and only if whenever a binary form over $\mathbb{Q}$ is projectively equivalent to Φ over $\mathbb{Q}_p$, for all primes p (including ∞ with $\mathbb{Q}_\infty = \mathbb{R}$), it is projectively equivalent to Φ over $\mathbb{Q}$. Here we call two forms Φ, Ψ over a ring R *projectively equivalent* if there is a g in $GL_2(R)$ and λ in $R^\times$ such that $\Phi^g = \lambda\Psi$. We shall show in section 7 that there exists a binary octic form Φ over $\mathbb{Q}$ which violates this principle. If $\mathfrak{P}$ is the projective automorphism group of Φ, we can represent $\mathfrak{P}$ as the stabiliser of a line in L_{10} (one can check directly that if V is the GL_2-module of all binary octic forms, then $D \otimes V$ occurs as a summand in L_{10}). Our general recipe then gives a $\mathcal{T}$-group G with $\mathfrak{B}_G = \mathfrak{P}$, and by general nonsense, or by 4.4, we see that G violates the Hasse Principle. This yields

4.10 COROLLARY *There exist 2-generator $\mathcal{T}$-groups G and H of class 10 which are $\mathbb{Q}_p$-isomorphic for all primes p (including ∞), but are not commensurable. In particular, $c'(G) \geqslant 2$.*

A construction like this, using quadratic rather than octic forms, was made by Remeslennikov [24], in order to obtain 2-generator $\mathscr{T}$-groups of class 4 having class number greater than one. Naturally, to obtain non-homogeneous algebraic groups as $\mathfrak{B}_G$ for suitable $\mathscr{T}$-groups G is harder. The main step of Bryant and Groves is

4.11 THEOREM *Let $n \geqslant 2$ and let L be the free n-generator Lie algebra over $\mathbb{Q}$. Let U be a homogeneous polynomial module for GL_n, defined over $\mathbb{Q}$. Then there exists r_0 such that for every $r \geqslant r_0$, $D^{\otimes r} \otimes U$ is a direct summand of the GL_n-module L''/L''' (where L'', L''' denote the second and third derived algebras of L respectively).*

To apply 4.5 and thereby deduce Theorem 3.4, one has to find fully invariant ideals $I \supset J$ in L such that L/J is nilpotent, I/J is abelian, $\ker \pi$ acts trivially on the factor I/J, and such that I/J contains an arbitrary pre-assigned finite direct sum of GL_n-modules like $D^{\otimes r} \otimes U$.

It would be interesting to find out the group-theoretic significance, for a $\mathscr{T}$-group G, of various natural conditions on the algebraic group $\mathfrak{B}_G$. What does it mean for G, for example, if $\mathfrak{B}_G$ is (a) connected, (b) homogeneous, (c) the whole of GL_n, (d) unipotent? In the case of 2-generator groups, one is dealing with algebraic subgroups of GL_2, and there are not many essentially different possibilities; it should be quite feasible to classify the $\mathscr{T}$-groups G for which $\mathfrak{B}_G$ is GL_2, SL_2, a torus, etc., and the results might be suggestive.

5. Groups Defined by Ring Extensions

In section 1 we associated with each $\mathscr{T}$-group G and binomial ring R the nilpotent R-operator group G^R. In certain cases the affine algebraic group $\mathfrak{G}$ associated to G can be defined over $\mathbb{Z}$. If G is such a group and R is a *finite $\mathbb{Z}$-algebra*, i.e. a ring which is free and finitely generated as a $\mathbb{Z}$-module, then $G^R = \mathfrak{G}(R)$ is again a $\mathscr{T}$-group. This suggests another classification project: (a) to describe all *irreducible $\mathscr{T}$-groups*, namely those which are not commensurable with any group of the form G^R with R containing $\mathbb{Z}$ strictly; (b) for a given G and for varying finite $\mathbb{Z}$-algebras R, to classify the groups G^R up to isomorphism as abstract groups.

As far as we know project (a) has hardly been begun. One relevant result, which emerges from [14], is that there are very few irreducible $\mathscr{T}$-groups of class 2 and Hirsch length 6. Here we discuss some issues connected with (b).

Definition. A $\mathcal{T}$-group G is *rigid* if $G^R \cong G^S$ implies $R \cong S$.

The main result of [11] implies that the group of 3×3 uni-triangular matrices over $\mathbb{Z}$ (which is the free nilpotent group of class two on 2-generators) is rigid. We do not know whether this extends to all free nilpotent groups, but we shall generalise the result in two other directions. We first remark that obviously no infinite abelian $\mathcal{T}$-group is rigid. More interestingly, if R is a finite $\mathbb{Z}$-algebra and S is another ring, then for any $\mathcal{T}$-group G we have

(5.1) $$(G^R)^S \cong G^{R \otimes_\mathbb{Z} S}.$$

Therefore G^R is not rigid if there are non-isomorphic rings S_1 and S_2 with $R \otimes S_1 \cong R \otimes S_2$. The simplest example is perhaps $R = \mathbb{Z}[\sqrt{-1}]$, $S_1 = \mathbb{Z}[\sqrt{p}]$ and $S_2 = \mathbb{Z}[\sqrt{-p}]$, for a prime p. Let us ask the following

Question. Which irreducible $\mathcal{T}$-groups are rigid?

We now concentrate on $\mathcal{T}$-groups of class 2 and shall call them $\mathcal{T}_2$-groups. It is easy to see that the polynomials defining multiplication and powers in a $\mathcal{T}_2$-group have integer coefficients, so G^R makes sense for any $\mathcal{T}_2$-group G and any ring R.

We fix a $\mathcal{T}_2$-group G with centre Z. We shall use $[-, -]$ to denote the bilinear map on $G^{ab} \times G^{ab}$ induced by commutation in G. Our aim is to construct a ring $\mathfrak{X}(G, S)$ using only the structure of G^S as an abstract group; if it so turns out that $\mathfrak{X}(G, S)$ determines S, then we shall be able to infer that G is rigid.

For an element x of G^{ab} we define

$$x^\perp = \{y \in G^{ab} \mid [x, y] = 1\}$$

and for a subset X of G^{ab} we set

$$X^\perp = \bigcap_{x \in X} x^\perp.$$

We shall call G *strongly isolated* if

(5.2) $$[-, g] : X^\perp \to Z$$

is bijective for some finite subset X and some element g of G^{ab}.

We now assume in addition that G is strongly isolated. It is clear that

$Z = G'$ and straightforward to verify that G^R is also strongly isolated for every finite $\mathbb{Z}$-algebra R.

For a ring S and subset X of $(G^S)^{ab}$, we write Φ_X for the set of all maps $[-, g]: X^\perp \to Z^S$ $(g \in (G^S)^{ab})$ and Ψ_X for the subset consisting of all such maps which are bijective. The ring $\mathfrak{X}(G, S)$ is defined to be the subring of $\text{End}_{\mathbb{Z}}(Z^S)$ generated by

$$\bigcup_X \{\psi^{-1}\varphi \mid \psi \in \Psi_X, \varphi \in \Phi_X\}$$

where X runs over all finite subsets of $(G^S)^{ab}$.

5.3 LEMMA $\mathfrak{X}(G, S)$ contains the image $S.\text{Id}$ of S in $\text{End}(Z^S)$ and $\mathfrak{X}(G, S)$ is contained in $\text{End}_S(Z^S)$. Thus $\mathfrak{X}(G, S)$ is an S-subalgebra of $\text{End}_S(Z^S)$.

Proof.   Let X and g be as in (5.2), and let $\psi: X^\perp \to Z^S$ be the associated map. For s in S put

$$\phi_s = [-, g^s]: X^\perp \to Z^S.$$

If $z \in Z^S$ then $z = h\psi = [h, g]$ for some h in $X^\perp$, and then $z\psi^{-1}\varphi_s = [h, g^s] = [h, g]^s = z^s$. Thus $\psi^{-1}\varphi_s \in \mathfrak{X}(G, S)$ is the endomorphism $s.\text{Id}$ of Z^S.

The second claim follows from the fact that Φ_X consists of S-linear mappings.

Since Z^S is the centre of G^S, it is obvious that the ring $\mathfrak{X}(G, S)$ depends only on the structure of G^S as an abstract group; the functorial nature of the construction is exhibited by

5.4 LEMMA   Let R and S be rings and $\alpha: G^R \to G^S$ an isomorphism. Then α restricts to an isomorphism $\alpha_1: Z^R \to Z^S$ and induces a ring isomorphism

$$\hat{\alpha}: \mathfrak{X}(G, R) \to \mathfrak{X}(G, S)$$

with $\hat{\alpha}(\theta) = \alpha_1^{-1}\theta\alpha_1$ for all $\theta \in \mathfrak{X}(G, R)$.

The verification is easy.

The relationship between $\mathfrak{X}(G, S)$ and S is unknown in general. We discuss some special cases.

5.5 THEOREM   If Z is cyclic then $\mathfrak{X}(G, S) \cong S$.

Proof. Since $Z^S \cong S$ as operator group,

$$S \, . \, \mathrm{Id} = \mathrm{End}_S(Z^S) \supseteq \mathfrak{X}(G, S) \supseteq S \, . \, \mathrm{Id}.$$

Observe that any $\mathcal{T}_2$-group whose derived subgroup coincides with its centre and is cyclic is necessarily strongly isolated: this comes at once from the normal form for alternating bilinear forms over $\mathbb{Z}$. Thus every such group is rigid.

Let $U_n(S)$ denote the group of upper uni-triangular $n \times n$ matrices over S.

5.6 THEOREM   *The group $U_n(\mathbb{Z})$ is rigid for every $n \geqslant 3$.*

As this is perhaps a rather surprising result, let us repeat what it means: if R, S are commutative rings with identity and $U_n(R) \cong U_n(S)$ for some $n \geqslant 3$ then $R \cong S$. To prove it, we consider the group

$$V_n = U_n(\mathbb{Z})/\gamma_3 U_n(\mathbb{Z}).$$

5.7 *If $n \geqslant 3$ and S is any ring then*

$$V_n^S \cong U_n(S)/\gamma_3 U_n(S).$$

This is easy to see, because the structure of V_n as an affine group is particularly transparent: in an obvious sense, V_n may be identified with the first two lines of entries above the diagonal in $U_n(\mathbb{Z})$. Thus 5.6 follows from

5.8 THEOREM   *For every $n \geqslant 3$ the centre of $\mathfrak{X}(V_n, S)$ is isomorphic with S. In particular, V_n is rigid if $n \geqslant 3$.*

Proof. First we check that V_n is strongly isolated by taking

$$X = \{(1, 0, \ldots, 0)\} \subseteq V_n^{ab} = \mathbb{Z}^{n-1}$$

and

$$g = (0, 1, \ldots, 1) \in V_n^{ab} = \mathbb{Z}^{n-1};$$

(5.2) is an isomorphism in this case.

The next step is to identify $\mathfrak{X}(V_n, S)$. Let Z be the centre of V_n. Identifying Z^S with S^{n-2} as usual, we have

$$S.1_{n-2} \subseteq \mathfrak{X}(V_n, S) \subseteq M_{n-2}(S).$$

It follows at once that $\mathfrak{X}(V_3, S) = S$. It is not hard to see that $\mathfrak{X}(V_4, S) = S.1_2$ and that for $n \geqslant 5$, $\mathfrak{X}(V_n, S)$ is the ring of all matrices of the form

$$\begin{pmatrix} * & 0 & \ldots & 0 & 0 \\ * & * & \ldots & * & * \\ \vdots & \vdots & & \vdots & \vdots \\ * & * & & * & * \\ 0 & 0 & \ldots & 0 & * \end{pmatrix}$$

in $M_{n-2}(S)$. In each case, then, one sees that the centre of $\mathfrak{X}(V_n, S)$ is isomorphic to S.

Rigid groups can be used to make counterexamples to the Hasse principle. In [17], K. Komatsu produces non-isomorphic algebraic number fields K_1 and K_2, with rings of intergers R_1, R_2 respectively, and a bijection $\varphi \colon W(R_1) \to W(R_2)$, where $W(R_i)$ is the set of prime ideals of R_i, such that $(R_1)_{\mathfrak{p}} \cong (R_2)_{\varphi(\mathfrak{p})}$ for each $\mathfrak{p}$ in $W(R_1)$ (here $R_{\mathfrak{p}}$ denotes the $\mathfrak{p}$-adic completion of a ring R). For each prime p,

$$R_i \otimes \mathbb{Z}_p \cong \bigoplus_{\mathfrak{p}|p} (R_i)_{\mathfrak{p}},$$

so that $R_1 \otimes \mathbb{Z}_p \cong R_2 \otimes \mathbb{Z}_p$. Hence if G is a $\mathscr{T}$-group such that G^{R_i} is defined $(i = 1, 2)$ then for each p we have

$$(G^{R_1})^{\mathbb{Z}_p} \cong G^{R_1 \otimes \mathbb{Z}_p} \cong G^{R_2 \otimes \mathbb{Z}_p} \cong (G^{R_2})^{\mathbb{Z}_p}.$$

But if G is rigid then

$$(G^{R_1})^{\mathbb{Q}} \cong G^{K_1} \not\cong G^{K_2} \cong (G^{R_2})^{\mathbb{Q}}.$$

Choosing the fields K_i carefully, one can obtain groups with particularly simple presentations. Let m be a square-free integer with $|m| \geqslant 3$ and $m \equiv 7$ or $15 \pmod{16}$. Put $\alpha_1 = m^{1/8}$ and $\alpha_2 = \sqrt{2\alpha_1}$; then $R_i = \mathbb{Z}[\alpha_i]$ is the ring intergers in $K_i = \mathbb{Q}(\alpha_i)$ for each i (this is a long but routine exercise), and these fields satisfy Komatsu's conditions. The discussion above shows that if $G_i = U_3(R_i)$ for each i then

5.9 $\mathscr{F}(G_1) = \mathscr{F}(G_2)$, but G_1 is not commensurable with G_2.

Thus G_1 and G_2 have weak class number $\geqslant 2$. They are both $\mathscr{T}_2$-groups of Hirsch length 24. Picking the most obvious set of generators for the groups one obtains the following presentation:

$$G_i = \langle x_0, x_1, \ldots, x_7, y_0, y_1, \ldots, y_7;$$
$$[x_k, x_l] = [y_k, y_l] \quad (\text{all } k, l),$$
$$[x_k, y_l] = [x_0, y_{k+l}] \quad (\text{all } k, l \text{ with } k+l < 8),$$
$$[x_k, y_l] = [x_0, y_{k+l-8}]^{m_i} \quad (\text{all } k, l \text{ with } k+l \geqslant 8),$$
$$[u, v, w] = 1 \quad (\text{all words } u, v, w)\rangle$$

where $m_1 = m$ and $m_2 = 16m$.

Groups of smaller Hirsch length which violate the Hasse principle are given in the next section.

Finally, we return briefly to the point of view of section 1. Let G be a $\mathscr{T}$-group. If R is an integral domain with field of fractions k and $(k: \mathbb{Q}) = d$ is finite, then

$$(G^R)^{\mathbb{Q}} \cong G^{R \otimes \mathbb{Q}} = G^{k \otimes \mathbb{Q}} \cong G^{\mathbb{Q}^d} \cong (G^{\mathbb{Q}})^d.$$

Thus the $\bar{\mathbb{Q}}$-isomorphism class of G^R depends only on $G^{\mathbb{Q}}$ and on d. To carry out the 'descent', we also need to know the algebraic group $\mathfrak{B}_{G^R}$. This does not seem to be easy to describe in general; for groups like those discussed above, however, there is a nice formula. Let R and k be as above. There is an algebraic group $\mathfrak{C}_k$ over $\mathbb{Q}$ defined by the stipulation that

$$\mathfrak{C}_k(K) = \text{Aut}(k \otimes_{\mathbb{Q}} K)$$

for every extension field K of $\mathbb{Q}$. (Thus $\mathfrak{C}_k(\mathbb{Q}) = \text{Gal}(k/\mathbb{Q})$, for example, if k is normal over $\mathbb{Q}$.) This is one ingredient in our formula; the second ingredient is the *restriction of scalars functor* $\mathscr{R}_{k/\mathbb{Q}}$, from algebraic k-groups to algebraic $\mathbb{Q}$-groups: for the definition see [7], §2.8.

5.10 THEOREM *Let G be a strongly isolated $\mathscr{T}_2$-group, and assume that $S \cdot \text{Id}_Z$ is a characteristic subring of $\mathfrak{X}(G, S)$ for every $\mathbb{Q}$-algebra S. Let R*

and k be as above. Then $\mathfrak{B}_{GR}$ *is isomorphic as algebraic group over* $\mathbb{Q}$ *to the group*

$$(\mathscr{R}_{k/\mathbb{Q}}\mathfrak{B}_G)\mathbin{]}\mathfrak{E}_k.$$

We omit the proof, which is based on 5.4. Examples of groups to which 5.10 applies are all $\mathscr{T}_2$-groups with cyclic centre, and the groups V_n defined above. In particular, 5.10 can be used to obtain a complete description of the automorphism group of $U_3(R)$, for every ring R of algebraic integers.

6. $\mathscr{T}_2$-Groups with Small Centre

To each nilpotent group G of class 2 with centre Z_G one can associate an alternating bilinear map

$$\varphi_G\colon (G/Z_G)\times(G/Z_G)\to Z_G$$

given by $(aZ_G, bZ_G)\mapsto [a, b]$. This map is non-degenerate, in an obvious sense, and provided G is a reasonable sort of group (for example if $\mathrm{Ext}^1_{\mathbb{Z}}(G/Z_G, Z_G) = 0$), it determines G up to isomorphism: if H is another such group then $H \cong G$ if and only if there exist isomorphisms $\alpha\colon G/Z_G \to H/Z_H$ and $\beta\colon Z_G \to Z_H$ such that

(6.1) $\varphi_G = (\alpha, \alpha)\circ\varphi_H\circ\beta^{-1}.$

(See R. B. Warfield Jr. [34], Ch. 5.) Thus the classification of such groups is equivalent to that of non-degenerate alternating bilinear maps under an equivalence relation like (6.1). It is easy to see that if G and H are $\mathscr{T}_2$-groups and k is a field of characteristic zero, then $G \cong_k H$ if and only if there exist k-linear isomorphisms $\alpha\colon (G/Z_G)^k \to (H/Z_H)^k$ and $\beta\colon (Z_G)^k \to (Z_H)^k$ satisfying (6.1).

The well known classification of alternating bilinear forms thus yields the classification of $\mathscr{T}_2$-groups with cyclic centre (cf. Grunewald and Scharlau [11]). In [14], $\mathscr{T}_2$-groups of Hirsch length 6 were classified by studying a certain invariant, the *Pfaffian form*, of an alternating bilinear map: in particular we showed there that the isomorphism classes of such groups G with $G' = Z_G \cong \mathbb{Z}^2$ correspond bijectively to projective equivalence classes of binary quadratic forms over $\mathbb{Z}$. Here we want to generalise this result to all $\mathscr{T}_2$-groups whose centres have rank 2; however, we only know how to do this up to $\mathbb{Q}$-isomorphism, so in effect we are classifying *torsion-free radicable nilpotent groups of class 2 and of finite rank with centres of rank 2*. For convenience we shall call such groups $\mathscr{D}^*$-groups. (For $\mathscr{D}^*$-groups G, we use the terms 'rank' and 'Hirsch length', interchangeably, to mean the sum of the ranks of G/Z_G and Z_G.)

To state the results, we need a

Definition. Let G be a $\mathscr{D}^*$-group. A central decomposition of G is a family $\{H_1, \ldots, H_m\}$ of subgroups of G such that (i) $Z_{H_i} = Z_G$ for each i; (ii) G/Z_G is the direct product of the subgroups H_i/Z_G; and (iii) $[H_i, H_j] = 1$ whenever $i \neq j$. The group G is (centrally) indecomposable if the only such decomposition is $\{G\}$.

6.2 THEOREM *Every $\mathscr{D}^*$-group G has a central decomposition into indecomposable constituents, and the decomposition is unique up to an automorphism of G. In particular, the constituents are unique up to isomorphism.*

6.3 THEOREM

(i) *Let G be an indecomposable $\mathscr{D}^*$-group of rank $n+2$. Then, w.r.t. a suitable basis of G/Z_G and a suitable basis $(\mathbf{e}, \mathbf{f})$ of Z_G, the map φ_G is represented by a matrix $\mathbf{A}$ as follows:*

(a) $n = 2m+1$.

$$
\mathbf{A} = \begin{pmatrix}
 & & & \mathbf{e} & \mathbf{f} & & 0 \\
 & 0 & & & \mathbf{e} & \mathbf{f} & \\
 & & & & 0 & \ddots & \ddots \\
 & & & & & & \mathbf{e} \quad \mathbf{f} \\
-\mathbf{e} & & 0 & & & & \\
-\mathbf{f} & \ddots & & & & & \\
 & \ddots & \ddots & -\mathbf{e} & & 0 & \\
0 & & -\mathbf{f} & & & &
\end{pmatrix}
$$

(b) $n = 2m$.

$$
\mathbf{A} = \begin{pmatrix} 0 & \mathbf{B} \\ -\mathbf{B}^t & 0 \end{pmatrix}
$$

where

$$\mathbf{B} = \begin{pmatrix} \mathbf{e} & \mathbf{f} & & & & \\ & \ddots & & \ddots & & 0 \\ & & \ddots & & \ddots & \\ & 0 & & \ddots & & \ddots \\ & & & & \mathbf{e} & \mathbf{f} \\ a_m\mathbf{f} & a_{m-1}\mathbf{f} & \cdots & a_2\mathbf{f} & \mathbf{e}+a_1\mathbf{f} \end{pmatrix}$$

and $X^m - a_1 X^{m-1} - \ldots - a_{m-1} X - a_m$ *is a primary polynomial over* $\mathbb{Q}$.

(ii) *If* G *is any* $\mathscr{D}^*$*-group, then w.r.t. suitable bases as above,* φ_G *is represented by the diagonal sum of matrices like* $\mathbf{A}$ *above.*

(As usual, we say that $\mathbf{A} = (\mathbf{a}_{ij})$ *represents* φ if

$$\varphi(r_1, \ldots, r_n; s_1, \ldots, s_n) = \sum_{i,j} r_i s_j \mathbf{a}_{ij}.$$

A polynomial is *primary* if it is a power of an irreducible polynomial.)
Note that in part (i) n has to be at least 2, so that $m \geqslant 1$.

6.4 THEOREM

(i) *Indecomposable* $\mathscr{D}^*$*-groups of equal odd rank are isomorphic.*

(ii) *The isomorphism classes of indecomposable* $\mathscr{D}^*$*-groups of rank* $2m+2$ *correspond bijectively with the projective equivalence classes of primary binary forms of degree* m *over* $\mathbb{Q}$.

The correspondence in (ii) associates to a group G with φ_G given by 6.3(b) the form $F(X, Y) = X^m - a_1 X^{m-1} Y - \ldots - a_m Y^m$; this is none other than the determinant of $\mathbf{B}$ if $\mathbf{e}, \mathbf{f}$ are replaced by X, Y respectively, hence F is (equivalent to) the Pfaffian form Pf_G of φ_G (cf. [14]).
    The results above do not quite settle the question of isomorphism between $\mathscr{D}^*$-groups, because a set of indecomposable groups can be formed into a central product in different ways (according to different identifications of their centres). The general classification is summed up in

6.5 *Suppose* $G = G_1 \ldots G_r$ *and* $H = H_1 \ldots H_s$ *are* $\mathscr{D}^*$*-groups with given central decompositions into indecomposable constituents* G_i, H_i. *Then* $G \cong H$ *if and only if the following hold:*

(i) $r = s$, *and after suitable re-ordering of the constituents one also has*

(ii) *for some* $t \leqslant r$, $G_1, \ldots, G_t$ *have even rank, and there exist* β *in* $GL_2(\mathbb{Q})$ *and* $\lambda_1, \ldots, \lambda_t \in \mathbb{Q}^\times$ *such that*

$$Pf_{H_i} = \lambda_i Pf^\beta_{G_i} \qquad for \; i = 1, \ldots, t,$$

and

(iii) G_i *and* H_i *have equal odd rank for* $i = t+1, \ldots, r$.

The results 6.2–6.5 all have analogues with 'k-isomorphism' replacing 'isomorphism', for any field k of characteristic zero; so in particular they can be used to study the classification over $\mathbb{C}$, or over p-adic fields. The proofs are based on the work of R. Scharlau [27].

Specific results along these lines depend on the classification of binary forms. This is an interesting subject, and we shall make some brief remarks about it in section 7. The applications we wish to mention depend on two new results (see 7.1 and 7.2).

(a) *There exist two irreducible binary octic forms over* $\mathbb{Q}$ *which are projectively equivalent over* $\mathbb{Q}_p$ *for all primes* p *(including* ∞*), but are not projectively equivalent over* $\mathbb{Q}$.

(b) *Binary cubic forms over* $\mathbb{Q}$ *which are projectively equivalent over every* p-adic field are projectively equivalent over $\mathbb{Q}$.

6.6 COROLLARY *There exist non-commensurable* $\mathcal{T}$-groups G and H of *class 2 and Hirsch length* 18 *such that* $G \cong_{\mathbb{Q}_p} H$ *for all primes* p *(including* ∞*). In particular,* $c'(G) \geqslant 2$.

6.7 COROLLARY $\mathcal{T}$-groups of class 2, with centre of rank 2, having *Hirsch length* 11 *or at most* 9 *all have weak class number one.*

This should be compared with the result of [11], that $\mathcal{T}$-groups of class 2 and Hirsch length at most 5 have class number one, while those of Hirsch length 6 need not. On the other hand, there is quite a gap between 6.6 and 6.7 which remains to be filled.

Corollary 6.6 follows from 6.4(ii) and from (a); given binary forms as in (a), one can use 6.3(i)(b) to write down explicit presentations for the groups G and H. To prove 6.7, suppose for example that G and H are $\mathcal{T}_2$-groups of Hirsch length 8, with centres of rank 2, and that $G^{\mathbb{Q}}$ and $H^{\mathbb{Q}}$ are indecomposable $\mathcal{D}^*$-groups. Then 6.4(ii) associates binary cubic forms Φ and Ψ, say, to $G^{\mathbb{Q}}$ and $H^{\mathbb{Q}}$ respectively; moreover, if $G \underset{\mathbb{Q}_p}{\cong} H$ then Φ and Ψ are projectively equivalent over $\mathbb{Q}_p$, while if Φ is projectively equivalent to Ψ over $\mathbb{Q}$ then $G \underset{\mathbb{Q}}{\cong} H$. Thus 6.7 in this case follows from (b). The other cases are dealt with similarly, though a little further argument is required to deal with decomposable groups.

As another application, let us consider the classification of $\mathscr{D}^*$-groups up to $\mathbb{C}$-isomorphism. With a linear form $aX - bY$ over $\mathbb{C}$ we associate the point $b/a \in \mathbb{C} \cup \{\infty\} = \mathbf{P}_1$, the complex projective line; the action of $GL_2(\mathbb{C})$ on linear forms then corresponds (contragrediently) to its action on $\mathbf{P}_1$ via Möbius transformations. In this way the projective equivalence classes of ordered r-tuples of linear forms are parametrised by points of $\mathbf{P}_1^r/PGL_2(\mathbb{C})$; except for a closed subset of smaller dimension, this object can be identified with $\mathbf{P}_1^{r-3}$ provided $r > 3$, because the Möbius transformations are triply transitive on $\mathbf{P}_1$, and a Möbius transformation is determined by its effect on three distinct points. Now what question 3.1 asks for, in the present context, is a description of the space W of isomorphism classes of $(n+2)$-dimensional nilpotent Lie algebras over $\mathbb{C}$, of class 2 and having 2-dimensional centres. It follows from 6.5 that W is the disjoint union of finitely many subspaces, one for each partition of n into parts $\geqslant 2$; the space W', say, corresponding to the partition

$$n = 2m_1 + \ldots + 2m_t + (2m_{t+1}+1) + \ldots + (2m_r+1)$$

may be identified with the set of all classes, under simultaneous projective equivalence, of sets $\{L_1^{m_1}, \ldots, L_t^{m_t}\}$ where $L_1, \ldots, L_t$ are binary linear forms over $\mathbb{C}$. The precise nature of this set depends on how many of the exponents m_i are distinct, and we shall not attempt to write down a general answer. The preceding paragraph shows that in general W' looks like a finite union of spaces $\mathbf{P}_1^f/\Gamma$, Γ being a finite group of transformations induced by the action of $\mathrm{Sym}(f+3)$ on $\mathbf{P}_1^{f+3}$ (this is to allow for the change from ordered to unordered r-tuples). When $t = 3$, W' is finite and consists of the equivalence classes of

$$\{X^{m_1}, (X-Y)^{m_2}, Y^{m_3}\}, \{X^{m_1}, X^{m_2}, Y^{m_3}\}, \{X^{m_1}, Y^{m_2}, Y^{m_3}\},$$

$$\{X^{m_1}, Y^{m_2}, X^{m_3}\}, \{X^{m_1}, X^{m_2}, X^{m_3}\};$$

some of these coalesce if any of the m_i are equal to each other. When $t = 2$, W' has one point or two, and when $t = 1$, W' has one point. If $t = 4$ and $m_1, \ldots, m_4$ are all distinct then $W' = \mathbf{P}_1/\Gamma \cup F$ where F is some finite set and Γ is the group generated by $z \mapsto 1/z$ and $z \mapsto 1-z$. Thus W' is infinite in this case.

To return to $\mathscr{D}^*$-groups, we also need to identify the points of our space W which correspond to Lie algebras defined over $\mathbb{Q}$. It is not clear to us how this should be done, but it is easy to see that the equivalence classes in question are exactly those containing a representative set of forms of the type

$$\{L_{11}^{e_1}, \ldots, L_{1r_1}^{e_1}, \ldots, L_{s1}^{e_s}, \ldots, L_{sr_s}^{e_s}\}$$

where $L_{ij} = X - \alpha_{ij}Y$ and for each i, $\{\alpha_{i1}, \ldots, \alpha_{ir_i}\}$ is a complete set of conjugate algebraic numbers over $\mathbb{Q}$.

For the sake of definiteness, let us at least state

6.8 COROLLARY   *Among $\mathscr{T}_2$-groups with centre of rank 2, those of Hirsch length at most 9, or equal to 11, lie in finitely many $\mathbb{C}$-isomorphism classes. There are infinitely many $\mathbb{C}$-isomorphism classes of such groups of Hirsch length h for every $h \geqslant 10$ except $h = 11$.*

The first claim follows from the preceding discussion. To verify the second statement, partition $h-2$ as $2+2+2+2+$ (some odd numbers $\geqslant 3$), and observe that the corresponding W' then contains the equivalence class of the set

$$\{X, X-Y, Y, X-\lambda Y\} = S_\lambda$$

for every $\lambda \in \mathbb{Q}$; these 4-tuples correspond to $\mathscr{D}^*$-groups, and they lie in infinitely many distinct equivalence classes. The group corresponding to S_λ is the central product of various indecomposable constituents of odd rank and the Mal'cev completion of the group G given by

$$\langle x_1, \ldots, x_4, y_1, \ldots, y_4, e, f; \quad \gamma_3(G) = 1,$$

$$[x_i, x_j] = [y_i, y_j] = 1 \quad (\text{all } i, j),$$

$$[x_i, y_j] = 1 \quad (\text{all } i, j \text{ with } i \neq j),$$

$$[x_1, y_1] = e, [x_2, y_2] = ef^{-1}, [x_3, y_3] = f, [x_4, y_4] = ef^{-\lambda}\rangle.$$

7.   Binary Forms

This is not the place for an essay on the classification of forms (nor are we qualified to write one). We shall confine ourselves to stating, baldly, two results, and saying a little about the proofs. Applications of these results have been given in sections 4 and 6. Recall that two forms are said to be *projectively equivalent* if each can be transformed into a constant multiple of the other by a linear substitution of variables.

Let $k = \mathbb{Q}(\omega)$ be a quadratic field, with Galois automorphism $x \mapsto x'$. For $v \in k^{\times}$ put

$$F_v(X, Y) = v'(\omega X + Y)^8 + v(\omega' X + Y)^8.$$

7.1 THEOREM   *Let $\omega = \sqrt{7}$, $v = 3 + \omega$ and $\mu = (3+\omega)^5$. Then F_v and F_μ are irreducible binary octic forms over $\mathbb{Q}$ which are projectively equivalent over $\mathbb{Q}_p$ for all p (including ∞) but not projectively equivalent over $\mathbb{Q}$.*

The proof depends in a rather delicate way on the arithmetic of the field extension $k(e^{2\pi i/8})/k$, where $k = \mathbb{Q}(\sqrt{7})$, and uses results from [20]. It also depends on the classification of finite subgroups of $PSL_2(\mathbb{C})$ (see [31]).

These and similar examples were found by a systematic calculation of Galois cohomology, similar in spirit to that involved in the following proof of

7.2 THEOREM   *Two binary cubic forms over $\mathbb{Q}$ which are projectively equivalent over $\mathbb{Q}_p$ for all (finite) primes p are projectively equivalent over $\mathbb{Q}$.*

The usual descent theory argument, of the kind described in section 1, shows that 7.2 is equivalent to

7.3 PROPOSITION   *Let $P \leqslant GL_2$ be the projective automorphism group of a cubic binary form F over $\mathbb{Q}$. Then the map*

$$\omega_P \colon H^1(\mathbb{Q}, P) \to \prod_{\text{finite } p} H^1(\mathbb{Q}_p, P)$$

has trivial kernel.

The remainder of this section is devoted to an outline of the proof; we shall use some basic properties of Galois cohomology (exact sequences, naturality) without special mention: they can all be found in any of the relevant texts mentioned in section 1 ([7], [29], [30], [35]).

7.4 *If F has 2 equal zeros then P is conjugate under $GL_2(\mathbb{Q})$ to the diagonal group. If F has 3 equal zeros then P is conjugate under $GL_2(\mathbb{Q})$ to the full lower triangular group. In either case, $H^1(\mathbb{Q}, P) = 0$.*

This is easy. Assume henceforth that F has no repeated zeros.

7.5 *There is an exact sequence*

$$1 \to Z \to P \to H \to 1$$

where Z is the group of scalar matrices in GL_2 and $PGL_2 \geqslant H = P/Z \cong S_3$ (the symmetric group on 3 letters).

Proof. Look at the action of PGL_2, by Möbius transformations, on the zeros of F.

7.6 *There is a commutative diagram with exact row*

$$
\begin{array}{ccccc}
H^1(\mathbb{Q}, Z) & \to & H^1(\mathbb{Q}, P) & \to & H^1(\mathbb{Q}, H) \\
\| & & \downarrow \omega_P & & \downarrow \omega_H \\
0 & & \prod_p H^1(\mathbb{Q}_p, P) & \to & \prod_p H^1(\mathbb{Q}_p, H).
\end{array}
$$

This follows from 7.5, and reduces the problem to showing that the kernel of ω_H is zero. There are two ingredients to the proof; the first is algebraic number theory:

7.7 *Let Γ be the group of automorphisms of H induced by $\mathrm{Gal}(\bar{\mathbb{Q}}/\mathbb{Q})$, and denote by $\mathscr{C}$ the set of all cyclic subgroups of Γ. Then $\ker \omega_H$ can be mapped injectively into the kernel of the restriction map*

$$H^1(\Gamma, H) \to \prod_{C \in \mathscr{C}} H^1(C, H).$$

Proof. Let Δ denote the centraliser of H in $\mathrm{Gal}(\bar{\mathbb{Q}}/\mathbb{Q})$, so that

$$1 \to \Delta \to \mathrm{Gal}(\bar{\mathbb{Q}}/\mathbb{Q}) \to \Gamma \to 1$$

is exact. For each prime p, we have a similar exact sequence

$$1 \to \Delta_p \to \mathrm{Gal}(\bar{\mathbb{Q}}_p/\mathbb{Q}_p) \to \Gamma_p \to 1,$$

where Γ_p is the image in Γ of the restriction map $\mathrm{Gal}(\bar{\mathbb{Q}}_p/\mathbb{Q}_p) \to \mathrm{Gal}(\bar{\mathbb{Q}}/\mathbb{Q})$. Since Δ and the Δ_p centralise H, there is a big commutative diagram with exact rows:

$$
\begin{array}{ccccccc}
 & & & & \ker \omega_H & & \\
 & & & & \downarrow & & \\
0 \to & H^1(\Gamma, H) & \to & H^1(\mathbb{Q}, H) & \to & H^1(\Delta, H) & \\
 & \downarrow & & \downarrow \omega_H & & \downarrow & \qquad (*)\\
0 \to & \prod_p H^1(\Gamma_p, H) & \to & \prod_p H^1(\mathbb{Q}_p, H) & \to & \prod_p H^1(\Delta_p, H). &
\end{array}
$$

Now we must apply the Frobenius density theorem (see [16], Ch. 5). Suitably interpreted, this tells us that if N is any normal subgroup of finite index in Δ then

$$
\Delta/N = \bigcup_p \Delta_p N/N. \qquad (\substack{* \\ *})
$$

Let $\delta : \Delta \to H$ be a cocycle which goes to zero in $H^1(\Delta_p, H)$ for every p. Then δ is actually a homomorphism, and taking $N = \ker \delta$ in $(\substack{* \\ *})$ we deduce that $\delta = 0$. Thus the right-most vertical map in $(*)$ is injective.

Since the map $H^1(\Gamma, H) \to H^1(\mathbb{Q}, H)$ is actually injective, it follows that $\ker \omega_H$ is bijective with the kernel of $H^1(\Gamma, H) \to \prod_p H^1(\Gamma_p, H)$. But a second application of the Frobenius theorem shows that every cyclic subgroup of Γ is a Γ_p for some prime p; the result follows.

The second ingredient in the proof that $\ker \omega_H = 0$ is finite group theory:

7.8 *Let $H = S_3$ and let $\mathscr{C}$ denote the set of all cyclic subgroups of H. If Γ is any subgroup of $\mathrm{Aut}\, H$ then the restriction map*

$$
H^1(\Gamma, H) \to \prod_{C \in \mathscr{C}} H^1(C, H)
$$

has trivial kernel.

Proof. Recall that $\mathrm{Aut}\, S_3 = \mathrm{Inn}\, S_3$, so that every proper subgroup of $\mathrm{Aut}\, H$ is cyclic. Thus if $\Gamma \neq \mathrm{Aut}\, H$ the result is trivial. Suppose that $\Gamma = \mathrm{Aut}\, H$, and let $* : H \to \Gamma$ be the isomorphism sending each element of H to the corresponding inner automorphism.

Let $\delta : \Gamma \to H$ be a cocycle which goes to zero in $H^1(C, H)$ for each

$C \in \mathscr{C}$. Define $\alpha: H \to H$ by

$$h\alpha = h \cdot h^*\delta.$$

One verifies trivially that α is an endomorphism of H. In fact, α is an automorphism: for if $h\alpha = 1$ then $1 = h\alpha = h \cdot b^{-h^*}b = b^{-1}hb$, where $\delta|_{\langle h^* \rangle}$ is the coboundary corresponding to $b \in H$, whence $h = 1$. Hence $\alpha = c^*$ for some $c \in H$, and we obtain

$$h^*\delta = h^{-1} \cdot h\alpha = h^{-1} \cdot h^{c^*} = (c^{-1})^h c = c^{-h^*}c$$

for all $h \in H$, i.e. for all $h^* \in \Gamma$. Thus δ is a coboundary as required.

The proof of 7.3 is now completed by taking 7.8, 7.7 and 7.6 together.

References

[1] Auslander, L., Baumslag, G., Automorphism groups of finitely generated nilpotent groups, *Bull. Amer. Math. Soc.* **73** (1967), 716–717.

[2] Ax, J., Solving Diophantine problems modulo every prime, *Ann. Math.* **85** (1967), 161–183.

[3] Ax, J., Kochen, S., Diophantine problems over local fields, II, *Amer. J. Math.* **87** (1965), 631–648.

[4] Baumslag, G., Lectures on nilpotent groups, *C.B.M.S. Regional Conf. Series* **2** (1971).

[5] Borel, A., Some finiteness properties of adele groups over number fields, *I.H.E.S. Pub. Math.* **16** (1963), 101–126.

[6] Borel, A., 'Linear Algebraic Groups', Benjamin, New York (1969).

[7] Borel, A., Serre, J-P., Théorèmes de finitude en cohomologie galoisienne, *Comment. Math. Helv.* **39** (1964), 111–164.

[8] Borevich, Z., Shafarevich, I. R., 'Number Theory', Academic Press, Orlando and New York (1966).

[9] Dyer, J. L., A nilpotent Lie algebra with nilpotent automorphism group, *Bull. Amer. Math. Soc.* **76** (1970), 52–56.

[10] Grunewald, F. J., Pickel, P. F., Segal, D., Polycyclic groups with isomorphic finite quotients, *Ann. Math.* **111** (1980), 155–195.

[11] Grunewald, F. J., Scharlau, R., A note on torsion-free finitely generated nilpotent groups of class 2, *J. Algebra* **58** (1979), 162–175.

[12] Grunewald, F. J., Segal, D., Some general algorithms I: arithmetic groups, *Ann. Math.* **112** (1980), 531–583.

[13] Grunewald, F. J., Segal, D., Some general algorithms II: nilpotent groups, *Ann. Math.* **112** (1980), 585–617.

[14] Grunewald, F. J., Segal, D., Sterling, L. S., Nilpotent groups of Hirsch length six, *Math. Z.* **179** (1982), 219–235.

[15] Hall, P., Nilpotent groups. (Lectures given at the Canadian Math. Congress, Summer Seminar, Univ. of Alberta, 1957.) *Queen Mary College Math. Notes* (1969).

158   F. GRUNEWALD AND D. SEGAL

[16] Janusz, G. J., 'Algebraic Number Fields', Academic Press, Orlando and New York (1973).
[17] Komatsu, K., On the adele rings and zeta functions of algebraic number fields, *Kodai Math. J.* **1** (1978), 394–400.
[18] Fogarty, J., 'Invariant Theory', Benjamin, New York (1969).
[19] Lang, S., 'Algebra', Addison-Wesley, Reading, Mass. (1965).
[20] Lorenz, F., Zur Theorie der Normenreste, *J. reine angew. Math.* **334** (1982), 157–170.
[21] Pickel, P. F., Finitely generated nilpotent groups with isomorphic finite quotients, *Bull. Amer. Math. Soc.* **77** (1971), 216–219.
[22] Pickel, P. F., Finitely generated nilpotent groups with isomorphic finite quotients, *Trans. Amer. Math. Soc.* **160** (1971), 327–341.
[23] Pickel, P. F., Nilpotent-by-finite groups with isomorphic finite quotients, *Trans. Amer. Math. Soc.* **183** (1973), 313–325.
[24] Remeslennikov, V. N., Conjugacy of subgroups in nilpotent groups, *Algebra i Logika* **6** (1967), no. 2, 61–76 (Russian).
[25] Sarkisian, R. A., Algorithmic questions for linear algebraic groups, II, *Mat. Sbornik* **113** (155) (1980), 400–436 (Russian); *Math. USSR Sbornik* **41** (1982), 329–359.
[26] Sarkisian, R. A., On a problem of equality in Galois cohomology, *Algebra i Logika* **19** (1980), 707–725 (Russian); *Algebra and Logic* **19** (1980), 459–472.
[27] Scharlau, R., Paare alternierender Formen, *Math. Z.* **147** (1976), 13–19.
[28] Segal, D., 'Polycyclic groups', Cambridge Univ. Press (1983).
[29] Serre, J-P., 'Local Fields', Springer, New York (1979).
[30] Springer, T. A., Galois cohomology of linear algebraic groups, *Proc. Sympos. Pure Math.* **9** (1966), 149–158.
[31] Springer, T. A., Invariant theory, *Springer Lecture Notes* **585** (1977).
[32] Tarski, A., 'A Decision Method for Elementary Algebra and Geometry', 2nd ed., Univ. of California, Berkeley and Los Angeles (1951).
[33] Vergne, M., Variétés d'algèbres de Lie nilpotentes, Thèse à la Faculté des Sciences de l'Université de Paris (1966).
[34] Warfield, R. B., Jr., Nilpotent groups, *Springer Lecture Notes in Math.* **513** (1976).
[35] Waterhouse, W. C., 'Introduction to Affine Group Schemes', Springer, New York, Heidelberg and Berlin (1979).

Additional reference

[36] Bryant, R. M., Groves, J. R. J., Algebraic groups of automorphisms of nilpotent groups and Lie algebras, *in preparation*.

Added in proof: in fact there exist irreducible binary quartic forms over $\mathbb{Q}$ which violate the Hasse principle. Thus groups G and H as in Corollary 6.6 may be found having Hirsch length 10; this closes the gap between 6.6 and Cor. 6.7.

Finiteness, Solubility and Nilpotence

DEREK J. S. ROBINSON

1. Introduction 159
2. Generalized nilpotent groups 165
3. Generalized soluble groups 175
4. Finiteness conditions on commutators and conjugates ... 180
5. Locally finite groups 188
6. Embedding theorems 194
References 201

1. Introduction

Infinite groups first arose at the end of the last century as groups specified by a set of generators and defining relations. The motivation came from geometry and topology. In particular the work of Max Dehn on manifolds led directly to the decision problems of modern combinatorial group theory.

The later structural school of infinite group theory began with the publication in 1916 of O. J. Schmidt's book 'Abstract Theory of Groups' [80]. Schmidt and such other early workers as R. Baer, S. N. Černikov and A. G. Kuroš were strongly influenced by Emmy Noether's view of mathematics.

Much effort was invested in attempts to create theories of infinite groups analogous to the successful theory of finite groups. This led on the one hand to the study of finiteness conditions, properties weaker than finiteness, and on the other to the introduction of group-theoretical properties which were known to be equivalent for finite groups but which turned out to be distinct for infinite groups. In this way the numerous classes of generalized soluble and nilpotent groups emerged.

By the 1940's a considerable body of knowledge had accumulated. In

GROUP THEORY: essays for Philip Hall
ISBN 0-12-304880-X

order to organize this material, Kuroš and Černikov wrote their well known survey article [52]. This recognized the existence of

> ...a new branch of group theory...whose program it is: to study groups which are in one sense or another close to Abelian groups, under restrictions which are in one sense or another close to the finiteness of the groups.

It is our purpose here to examine the state of the theory some thirty-five years after the Kuroš–Černikov survey and to record some of the principal accomplishments, especially in areas to which Philip Hall made important contributions. During the remainder of this section we shall review some of the basic concepts. (For further details and all unexplained notation and terminology we refer the reader to Robinson [73].)

Classes of groups and closure operations

By the 1950's the number of classes of infinite groups that had been recognized as significant had become very large. Many of these classes were given alphabetic labels such as *FC, SI, SN*; the resulting notation was unwieldy and occasionally confusing. In the late 1950's Philip Hall proposed a scheme for expressing classes of groups in terms of certain basic classes by means of closure operations (see especially Hall [34]). This systemization proved to be suggestive as well as succinct, and it has been widely adopted over the last two decades.

By a *class of groups* we shall understand a class in the usual sense whose members are groups, with the additional stipulation that the class contain a unit group and be closed under isomorphism. Basic examples of classes are

$$\mathfrak{F}, \quad \mathfrak{A}, \quad \mathfrak{N},$$

the classes of all finite groups, abelian groups and nilpotent groups respectively.

If $\mathfrak{X}$ and $\mathfrak{Y}$ are classes of groups, the *product* class $\mathfrak{X}\mathfrak{Y}$ consists of all groups G with a normal subgroup N such that $N \in \mathfrak{X}$ and $G/N \in \mathfrak{Y}$. In words, $\mathfrak{X}\mathfrak{Y}$ is the class of all ($\mathfrak{X}$-by-$\mathfrak{Y}$)-*groups*. Thus we have a non-associative, non-commutative binary operation on group classes.

A *closure operation* A is a function associating to each class of groups $\mathfrak{X}$ a class $A\mathfrak{X}$ such that A fixes the class of unit groups (1) and

$$\mathfrak{X} \subseteq A\mathfrak{X} = A^2\mathfrak{X} \subseteq A\mathfrak{Y}$$

whenever $\mathfrak{X} \subseteq \mathfrak{Y}$. A class $\mathfrak{X}$ is A-*closed* if $\mathfrak{X} = \text{A}\mathfrak{X}$. Examples of closure operations abound, some of the more common ones being

$$\text{S}, \text{S}_n, \text{H (or Q)}, \text{P}, \text{L}, \text{R}.$$

Here $\text{S}\mathfrak{X}$ and $\text{S}_n\mathfrak{X}$ are the classes of groups isomorphic with subgroups and subnormal subgroups of $\mathfrak{X}$-groups respectively; $\text{H}\mathfrak{X}$ is the class of homomorphic images of $\mathfrak{X}$-groups, $\text{P}\mathfrak{X}$ consists of groups having a series of finite length with factors in $\mathfrak{X}$, while $\text{R}\mathfrak{X}$ is the class of residually $\mathfrak{X}$-groups and $\text{L}\mathfrak{X}$ is the class of locally $\mathfrak{X}$-groups (each finite subset is contained in an $\mathfrak{X}$-subgroup).

Closure operations may be partially ordered by means of the rule $\text{A} \leqslant \text{B}$ if $\text{A}\mathfrak{X} \subseteq \text{B}\mathfrak{X}$ for all classes $\mathfrak{X}$. The composite of two closure operations A and B need not be a closure operation. Therefore we define $\langle \text{A}, \text{B} \rangle$ to be the smallest closure operation containing A and B (in the above ordering). Note that $\langle \text{A}, \text{B} \rangle = \text{AB}$ if $\text{BA} \leqslant \text{AB}$. There is a natural definition of intersection: $(\text{A} \cap \text{B})(\mathfrak{X}) = \text{A}\mathfrak{X} \cap \text{B}\mathfrak{X}$. The analogy between the algebra of closure operations and the lattice of subgroups of a group is apparent.

A closure operation A is called *unary* if

$$\text{A}\mathfrak{X} = \bigcup_{G \in \mathfrak{X}} \text{A}(G)$$

where (G) is the class of unit groups and groups isomorphic with G. For a unary operation A we can define a class $\mathfrak{X}^{\text{A}}$ as the union of all A-closed subclasses of $\mathfrak{X}$; this is the largest A-closed subclass of $\mathfrak{X}$. To take a well known example, if $\mathfrak{G}$ is the class of finitely generated groups, then $\mathfrak{G}^{\text{S}}$ is the class of groups which satisfy the maximal condition.

If $\mathfrak{X}^{-}$ denotes the class of groups not in $\mathfrak{X}$ together with (1), we define

$$\mathfrak{X}^{-\text{A}} = (\mathfrak{X}^{-})^{\text{A}},$$

the largest A-closed class which intersects $\mathfrak{X}$ in (1). For example $\mathfrak{A}^{-\text{H}}$ is the class of perfect groups.

We shall see later that many complex classes of groups can be expressed concisely by means of closure operations.

Series

The concept of a series of finite length is a very old one in group theory and goes back to Galois. Series of general order type, on the other hand, first arose during the 1930's when G. Birkhoff and Kuroš were searching for versions of the Jordan–Hölder Theorem valid for infinite groups. We shall recall the definition in the form proposed by Hall [34].

A *series* of general order type in a group G is a set of subgroups **S** which is linearly ordered (i.e. totally ordered) by inclusion and which satisfies the following conditions:

(i) If $1 \neq x \in G$, then the union of all members of **S** which do not contain x is a subgroup V_x in **S**;

(ii) if $1 \neq x \in G$, then the intersection of all members of **S** which contain x is a subgroup Λ_x in **S**;

(iii) $V_x \lhd \Lambda_x$;

(iv) every member of **S** is of the form V_x or Λ_x for some $x \neq 1$ in G.

The V_x and Λ_x are the *terms* of **S** and the Λ_x/V_x are the *factors* of **S**.

The factors of **S** may be linearly ordered by the rule $\Lambda_x/V_x \ll \Lambda_y/V_y$ if and only if $\Lambda_x \leqslant V_y$. Note that given $x, y \neq 1$ in G, then $\Lambda_x \leqslant V_y$ or $\Lambda_y \leqslant V_x$. The order-type Σ of **S** is defined to be that of the factors with this ordering. Thus we can write $\mathbf{S} = \{\Lambda_\sigma, V_\sigma \mid \sigma \in \Sigma\}$. If **S** has order-type an ordinal or the reverse order-type of an ordinal, then **S** is an *ascending* or a *descending series* respectively.

If G is an Ω-*operator* group, we obtain the concept of an Ω-*series* in G by requiring all terms to be Ω-subgroups. If $\Omega = \mathrm{Inn}\, G$, then an Ω-series is called a *normal series*, all terms being normal in G. If all terms of a series are merely subnormal, we speak of a *subnormal series*.

Two new closure operations are useful: if $\mathfrak{X}$ is a class of groups, define $\hat{P}\mathfrak{X}$ and $\hat{P}_{\mathrm{sn}}\mathfrak{X}$ to be the classes of groups having a series and a subnormal series respectively with factors in $\mathfrak{X}$. We can also define $\hat{P}_n\mathfrak{X}$ to be the class of groups having a normal series with $\mathfrak{X}$-factors; but $\hat{P}_n$ is not a closure operation since $\hat{P}_n \neq \hat{P}_n^2$. Further operations are obtained by using ascending or descending series, in which case we write $\acute{P}$ or $\grave{P}$, respectively, instead of $\hat{P}$.

A subgroup which occurs in a series, an ascending series or a descending series of a group G is said to be *serial*, *ascendant* or *descendant* in G: these are, of course, generalizations of subnormal subgroups. We may also define $\hat{s}\mathfrak{X}$, $\acute{s}\mathfrak{X}$, $\grave{s}\mathfrak{X}$ to be the classes of groups isomorphic with serial subgroups, ascendant subgroups and descendant subgroups of $\mathfrak{X}$-groups.

These closure operations are particularly useful in sorting out the various classes of generalized soluble groups.

Composition series. If **S** and **S*** are Ω-series in an Ω-group G, then **S*** is said to be a *refinement* of **S** if every term of **S** is also a term of **S***, and it is clear then that $V_x \leqslant V_x^* \leqslant \Lambda_x^* \leqslant \Lambda_x$ for all non-trivial x in G, in the obvious notation. An Ω-*composition series* in G is an Ω-series in G which has no proper refinements. Taking Ω to be the empty set it is evident that the composition series are precisely the series whose factors are *absolutely simple*, i.e., have no proper non-trivial serial subgroups. An Inn G-composition series of G is called a *principal* or *chief series*.

It is an easy exercise in the use of Zorn's Lemma to show that every Ω-series can be refined to an Ω-composition series; thus Ω-composition series always exist in an Ω-group. However, two composition series need not have isomorphic factors, as the examples $\ldots 4\mathbb{Z} < 2\mathbb{Z} < \mathbb{Z}$ and $\ldots 9\mathbb{Z} < 3\mathbb{Z} < \mathbb{Z}$ show.

Finiteness conditions

A *finiteness condition* is usually held to be any group-theoretical property which is possessed by all-finite groups. We shall now review some of the more useful of these.

(A) *Finitely generated groups and finitely presented groups.* Perhaps the most familiar finiteness condition, apart from finiteness itself, is the property of being finitely generated. It is well known that there are $2^{\aleph_0}$ non-isomorphic finitely generated groups, and even $2^{\aleph_0}$ finitely generated soluble groups (B. H. Neumann [63], P. Hall [26]). On the other hand, there can be only countably many *finite presented groups*, i.e. groups which have a presentation with finitely many generators and relations.

The fact that 2-generator groups could have a very complicated structure was established by G. Higman, B. H. Neumann and H. Neumann in 1949 in [44]. They showed that *every countable group can be embedded in a 2-generator group*. It is a much harder problem to decide which countable groups occur as subgroups of finitely presented groups. We shall return to this question in section 6.

(B) *Torsion groups and locally finite groups.* A *torsion* or *periodic group* is a group in which every element has finite order. Almost nothing is known about the structure of torsion groups which are not locally finite. That locally finite groups form a proper subclass of the class of torsion groups was first shown by E. S. Golod [17] and P. S. Novikov and S. I. Adjan [68]. Recently simpler examples of finitely generated infinite p-groups have been given by R. I. Grigorčuk [22] and N. D. Gupta and S. Sidki [25]: the last two authors have constructed a 2-generator infinite p-group (with p odd) which is realized as a group of automorphisms of the regular tree of degree p.

(C) *Chain conditions.* If $\mathscr{S}$ is a collection of subgroups of a group G, then G is said to satisfy the *ascending chain condition* on $\mathscr{S}$-subgroups if there is no strictly ascending infinite chain of subgroups in $\mathscr{S}$. This is equivalent to the *maximal condition* on $\mathscr{S}$, i.e. every non-empty set of subgroups in $\mathscr{S}$ has at least one maximal element. The descending chain condition and the minimal condition on $\mathscr{S}$ are defined analogously and are also equivalent.

Chain conditions are among the most commonly encountered finiteness conditions in group theory. The basic examples are max and min, the maximal and minimal conditions on subgroups, $\mathscr{S}$ being taken to be the set of all subgroups. It had long been an open question whether max and min together imply finiteness. However, this possibility has been disproved by recent examples of A. Ju. Ol'šankiĭ [69] and E. Rips (unpublished); indeed, Rips has found an infinite group in which every proper non-trivial subgroup has order p (here p is some large prime); these are the so-called *Tarski p-groups.*

Further familiar properties are max-s_n, min-s_n and max-n, min-n, the maximal and minimal conditions on subnormal and normal subgroups respectively. Since non-abelian simple groups can have extremely complex structures, these properties are usually studied in connection with some form of solubility or nilpotence.

Finally we mention max-*ab* and min-*ab*, the maximal and minimal conditions on abelian subgroups. These are quite weak conditions in general—for example Rips' infinite p-group satisfies both of them—so they too have been studied mainly for classes of generalized soluble groups in which one can reasonably expect to find many abelian subgroups. The basic results here are due to A. I. Mal'cev [56] and Schmidt [81] who proved respectively that for soluble groups max is equivalent with max-*ab* and min with min-*ab*. There have been a whole series of results in the same vein. For a detailed account with simplified proofs see Robinson [74], where crucial use is made of near-splitting techniques that split a group over a normal subgroup to within finite index and finite intersection.

(D) *Groups with finite rank.* There are many definitions of 'finite rank'. We shall consider a few of the most important. A group has *finite Prüfer rank* if there is an upper bound for the minimum number of generators of its finitely generated subgroups. This notion has been used mainly for soluble groups, since the *abelian* groups with finite Prüfer rank are well known: they are exactly the groups in which the p-ranks are boundedly finite for all p, where p is 0 or a prime. However, nothing is known in general about groups with finite Prüfer rank: notice that Rips' group is of this type.

A group has *finite abelian subgroup rank* if each abelian subgroup has finite p-rank for all p, where p is 0 or a prime. Such groups form a larger class than groups with finite Prüfer rank. An intermediate class is formed by groups with *finite abelian section rank*: these, of course, are groups in which each abelian section has finite p-rank for all p, where again p is 0 or a prime.

These finiteness conditions have been widely applied to generalized soluble groups. For example, R. Baer and H. Heineken [9] have shown that, for radical groups, finite abelian subgroup rank and finite abelian section rank are the same property (see Theorem 3.4 below). This is not true in general since free groups obviously have finite abelian subgroup rank.

(E) *Finiteness conditions on commutators and conjugates.* Many finiteness conditions have been studied which restrict the set of commutators or conjugates of a group in some way. Abelian groups will usually satisfy these conditions and so will finite groups. We shall discuss such properties in detail in section 4.

2. Generalized Nilpotent Groups

By a generalization of nilpotence is meant any group theoretical property which for finite groups is equivalent to nilpotence. Thus the several well known characterizations of finite nilpotent groups give rise to classes of generalized nilpotent groups. It is convenient to divide such classes into two families, those which are contained in the class of locally nilpotent groups and the remainder, most of which contain all the locally nilpotent groups.

Locally nilpotent groups

Local nilpotence is perhaps the best known generalization of nilpotence, yet relatively little is known about the structure, or even the normal structure, of locally nilpotent groups. Of course the elements of finite order in a locally nilpotent group G form a characteristic subgroup T which is a direct product of locally finite p-groups; also G/T is torsion-free. Thus attention is primarily directed at locally finite p-groups and torsion-free locally nilpotent groups.

The fundamental result for the entire theory is the Hirsch–Plotkin Theorem: the product of two normal locally nilpotent subgroups is locally nilpotent. This has as a consequence

THEOREM 2.1 (cf. [73], I, p. 57) *In any group G there is a unique*

maximal normal locally nilpotent subgroup (the Hirsch–Plotkin radical).
It contains all ascendant locally nilpotent subgroups of G.

Almost our only information about the general structure of locally nilpotent groups comes from

THEOREM 2.2 (cf. [73], I, pp. 154, 166)

(i) (Baer [4], D. H. McLain [59]) *A maximal subgroup of a locally nilpotent group is normal.*

(ii) (Mal'cev [55]) *A principal factor of a locally nilpotent group is central.*

Under most finiteness restrictions the structure of a locally nilpotent group is much simpler, as the following results indicate.

THEOREM 2.3 (McLain [59]; cf. [73], I, pp. 154, 166)

(i) *If a locally nilpotent group satisfies* max-n, *it is a finitely generated nilpotent group.*

(ii) *If a locally nilpotent group satisfies* min-n, *it is a Černikov group.*

(A *Černikov group* is a finite extension of an abelian group with min.)

THEOREM 2.4 (Mal'cev [56]; cf. [73], II, p. 38) *If a locally nilpotent group has finite abelian subgroup rank, its primary components are Černikov groups and the quotient group by the torsion-subgroup is a torsion-free nilpotent group of finite rank.*

Apart from the information provided by Theorems 2.2 and 2.3, we have little knowledge of the subnormal structure of locally nilpotent groups. For example, nothing is known about locally nilpotent groups whose subnormal subgroups have bounded indices of subnormality. In particular it is not known if a locally nilpotent group in which normality is a transitive relation need be soluble.[‡]

McLain's characteristically simple groups. Among the most famous examples of locally nilpotent groups are the characteristically simple groups constructed by McLain in 1954. These are infinite analogues of unitriangular matrix groups, and they have been extensively employed in the construction of counterexamples. We shall briefly recall the definition.

‡ It need not be, as has recently been shown by S. Thomas.

Let F be a field and Λ a linearly ordered set with more than one element. We consider an F-vector space V with a basis $\{v_\lambda \mid \lambda \in \Lambda\}$. Denote by $e_{\lambda\mu}$ the usual elementary linear transformation of V where

$$v_\lambda e_{\lambda\mu} = v_\mu \quad \text{and} \quad v_\nu e_{\lambda\mu} = 0 \quad \text{if } \nu \neq \lambda.$$

Then McLain's group is

$$M \equiv M(\Lambda, F) = \langle 1 + a e_{\lambda\mu} \mid a \in F, \lambda < \mu \in \Lambda \rangle.$$

The most important properties of these groups can be summarized as follows.

THEOREM 2.5 (McLain [58]; cf. [73], II, p. 15)

(i) *M is the product of its normal abelian subgroups and so is locally nilpotent.*

(ii) *M is either torsion-free or a p-group according as the characteristic of F is 0 or a prime p.*

(iii) *Assume that $|\Lambda| > 2$ and that given $\lambda_i < \mu_i$, $i = 1, 2$, in Λ, there is an order-automorphism σ of Λ such that $\lambda_1 \sigma = \lambda_2$ and $\mu_1 \sigma = \mu_2$. Then M is a characteristically simple group. Thus M is a perfect group with trivial centre.*

Notice that the hypothesis (iii) forces Λ to be infinite. For example, Λ could be the set of all rational numbers in natural order.

We consider next some special classes of locally nilpotent groups.

Fitting groups, Baer groups and Gruenberg groups. A group which coincides with its Fitting subgroup is called a *Fitting group*; these are exactly the groups in which every finite subset is contained in a normal nilpotent subgroup. For example, every section of a direct product of nilpotent groups is a Fitting group; so are the McLain groups.

Groups in which each finite subset is contained in a subnormal nilpotent subgroup form a wider class; these are called *Baer groups*. It is a fundamental theorem of Baer [7] that two finitely generated subnormal nilpotent subgroups always generate a subnormal nilpotent subgroup. It follows from this that Baer groups are the groups which are generated by their subnormal abelian subgroups.

A still wider class is formed by those groups in which each finite

subset lies in an ascendant nilpotent subgroup, the *Gruenberg groups.*
There is an alternative description of these as the groups which are
generated by their ascendant abelian subgroups. This is based on the
important result of K. W. Gruenberg [24] that two finitely generated
ascendant nilpotent subgroups generate an ascendant nilpotent subgroup.

We define closure operations N and Ń by making NX and ŃX consist
of all groups generated by their subnormal X-subgroups and their
ascendant X-subgroups respectively. Thus N𝔄 is the class of Baer
groups and Ń𝔄 that of Gruenberg groups.

It follows from Theorem 2.1 that every Gruenberg group is locally
nilpotent. Moreover, it is easy to see that a *countable* locally nilpotent
group is a Gruenberg group. However, these classes are distinct, as was
shown by M. I. Kargapolov [46] and L. Kovács and B. H. Neumann
(unpublished). Let $G_0 = 1$, $G_{\alpha+1} = C \smallsetminus G_\alpha$ and $G_\lambda = \bigcup_{\beta < \lambda} G_\beta$, where C
is a cyclic group of order p, α is an ordinal and λ is a limit ordinal. If ρ
is the first uncountable ordinal, then $G = G_\rho$ is a locally finite p-group
with no non-trivial ascendant abelian subgroups.

Hypercentral groups and the normalizer condition.   A property that is
well known to be equivalent to nilpotence for finite groups is the
normalizer condition: every proper subgroup is properly contained in
its normalizer. However for infinite groups this remains a relatively
mysterious property. By forming successive normalizers it is easily
seen that the normalizer condition is equivalent to every subgroup
being ascendant. Hence every group with the normalizer condition is a
Gruenberg group.

It is easy to see that every nilpotent group satisfies the normalizer
condition and, more generally, so does every *hypercentral group:* this is
a group with an ascending central series, that is, one in which every
factor is central.

Obviously, if every subgroup of a group is subnormal, the group
satisfies the normalizer condition. Groups with the former property
form another obscure class about which almost nothing is known. Its
members can be non-nilpotent, as was discovered by H. Heineken and
I. J. Mohamed [41]; they can even be non-nilpotent and hypercentral
(H. Smith [110]). But a hypercentral member has to be soluble (C. J. B.
Brookes [98]). An open problem is whether there are any insoluble
members at all.

J. E. Roseblade [79] has shown that a group in which every subgroup
is subnormal of defect at most d is nilpotent of class at most $f(d)$ for
some function f. A classical theorem of Dedekind shows that $f(1)$ may

be taken to be 2 and Heineken [101] proved that $f(2)$ can be taken to be 4 (see also S. K. Mahdavianary [107]), but realistic bounds for $f(d)$ when $d > 2$ are not known.

The relative positions of the seven classes discussed above are indicated in Fig. 1. It is known that all classes are distinct.

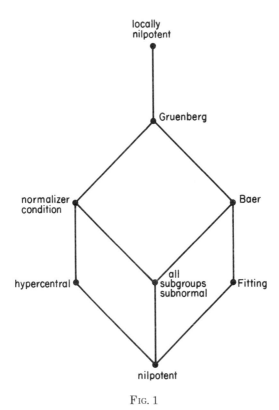

F IG. 1

Non-locally nilpotent classes

We shall now discuss some types of generalized nilpotent groups which need not be locally nilpotent.

Z-groups, Z̄-groups, hypocentral groups. A group G is called a *Z-group* if it has a *central series* $\{\Lambda_\sigma, V_\sigma \mid \sigma \in \Sigma\}$: this means that $[\Lambda_\sigma, G] \leqslant V_\sigma$ for all σ in Σ. The class of Z-groups is an immensely wide one. Somewhat narrower is the class of *Z̄-groups*, i.e. groups in which every principal factor is central. It is easy to see that

$$\bar{Z} = Z^{\mathrm{H}},$$

the largest image closed subclass of Z. By Theorem 2.2 every locally nilpotent group is a $\bar{Z}$-group.

Groups with a descending central series are called *hypocentral* (or *ZD*-groups): very little is known about them. More familiar is the subclass of *residually nilpotent groups*: these have a descending central series of type ω. By a famous theorem of Magnus, *every free group is residually nilpotent*. There are numerous non-nilpotent polycyclic groups which are residually nilpotent. However, no example seems to be known of a residually nilpotent group with max which is insoluble.

Since every group is an image of a free group, it is clear that even residually nilpotent groups can be very far from being nilpotent. Another indication of this is the following remarkable result of P. Hall and B. Hartley [37].

THEOREM 2.6 (cf. [73], II, p. 92)   *Every group can be embedded in the normal product of two free groups.*

Thus there is no analogue of the Hirsch–Plotkin theorem for residually nilpotent groups.

Hall's $\bar{Z}$-groups.   In 1964 Hall [35] made an important contribution to the theory of generalized nilpotent groups by exhibiting a $\bar{Z}$-group with non-cyclic free subgroups. This shows that $\bar{Z}$ is not a subgroup closed class, so that

$$\bar{\bar{Z}} = \bar{Z}^{\mathrm{S}} = Z^{\mathrm{HS}}$$

is a strictly smaller class than $\bar{Z}$.

Hall's example is a group of 2×2 matrices over the ring of p-adic integers. Define

$$M = M(h, k, l)$$

to be the set of all matrices

$$\begin{pmatrix} a & b \\ c & d \end{pmatrix}$$

where the p-adic integers a, b, c, d satisfy $ad - bc = 1$, $b \equiv 0 \bmod p^h$, $c \equiv 0 \bmod p^k$ and $d \equiv 1 \bmod p^l$; here $h \geqslant 0$, $k > 0$, $l > 0$ and $h + k \geqslant l$. It is easy to see that M is a group. Hall showed by direct matrix calculations that if $1 \neq N \triangleleft M$, then M/N is a finite p-group. In addition he proved that M has no minimal normal subgroups. From these facts it follows that M is a $\bar{Z}$-group. However, if r is a sufficiently large positive integer, M will contain the matrices

$$\begin{pmatrix} 1 & p^r \\ 0 & 1 \end{pmatrix} \quad \text{and} \quad \begin{pmatrix} 1 & 0 \\ p^r & 1 \end{pmatrix}$$

and these generate a free group of rank 2 (cf., for example, [108], p. 48).

A very similar group was produced at about the same time by Ju. I. Merzljakov [60]; his example is linear of degree 2 over the ring of rational numbers with odd denominators.

Residually central groups.   A group G is called *residually central* if for each non-trivial element x there is a normal subgroup N such that $x \notin N$ and xN belongs to the centre of G/N: equivalently $x \notin [x, G]$ for every $x \neq 1$ in G. It is easy to see that every Z-group is residually central. It is also not hard to establish the following fact.

THEOREM 2.7 (C. Ayoub [2], J. R. Durbin [16]; cf. [73], II, p. 7)   *A residually central group with* min-n *is a hypercentral Černikov group.*

Thus for finite groups residual centrality is equivalent to nilpotence. It is surprising that this natural characterization of finite nilpotent groups was not discovered until 1968.

The first example of a residually central group which is not a Z-group was given by R. E. Phillips and J. E. Roseblade [71] on the basis of work of P. A. Linnell [105] on the zero divisor problem for group algebras.

THEOREM 2.8 [71]   *Let G be an infinite torsion-free polycyclic group which is residually nilpotent and abelian by finite, and has G/G' finite. Let p be a prime not dividing $|G : G'|$. Then the standard wreath product $W = C \wr G$, where C is a cyclic group of order p, is a residually central group but it is not a Z-group.*

Proof.   The base group B of W is essentially the group algebra FG where F is the field of p elements. Suppose that $1 \neq x \in [x, W]$; then x must belong to B since G is residually nilpotent. Hence $x \in [x, G]$ and this means that in additive notation $0 \neq x = x\eta$ where η belongs to I, the augmentation ideal of FG. But Linnell has shown that FG has no divisors of zero; thus $\eta = 1$, which is impossible. Therefore W is residually central.

However, W cannot be a Z-group; for if it were, it would be hypocentral since by a well known theorem of P. Hall [26] W satisfies max-n. On the other hand, it follows from the choice of p that $I = I^2$, so that $[B, G] = [B, {}_2G]$, which contradicts hypocentrality.

It is not hard to find such polycyclic groups G, for example

$$\langle x, y \mid (x^2)^y = x^{-2}, (y^2)^x = y^{-2} \rangle,$$

which is even supersoluble.

Baer nilpotent groups, Engel groups. A group in which every finite section is nilpotent is said to be *Baer-nilpotent*. This is a very complex class. It contains all locally nilpotent groups, but also the Tarski p-groups of Rips, which are simple.

A much better known class is the class of *Engel groups*: a group G is Engel if to each pair of elements x, y there corresponds a positive integer $n = n(x, y)$ such that $[x, {}_ny] = 1$. It is well known that finite Engel groups are nilpotent, so Engel groups are Baer-nilpotent; it is obvious that locally nilpotent groups are Engel groups. A celebrated example of E. S. Golod [17] is an Engel p-group which is not locally nilpotent.

A great deal has been written on the subject of Engel groups, particularly in conjunction with a finiteness condition or some generalization of solubility. Thus T. A. Peng [70] has shown that an Engel group with max-ab is finitely generated and nilpotent. However, the structure of Engel groups with min-ab has still not been completely clarified; partial results have been given by D. Held [42] and J. E. Martin and J. A. Pamphilon [57]. Another unsolved problem is to describe the structure of an Engel group with finite abelian subgroup rank. For information on this see [73].

Groups in which every subgroup is serial. It has been shown by Kuroš (see [73], II, p. 10) that every subgroup of a group is serial if and only if $H' \leqslant \text{Frat } H$ for every subgroup H. Hence by Theorem 2.2 every subgroup of a locally nilpotent group is serial. It is easy to see that a group in which all subgroups are serial is Baer-nilpotent; but it need not be locally nilpotent. J. S. Wilson [95] showed this by means of an adaptation of the Golod construction mentioned above.

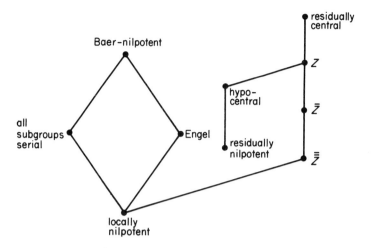

F$_{\text{IG}}$. 2 *Diagram of group classes*. All ten of the classes are distinct.

Stability groups

Let $\mathbf{S} = \{\Lambda_\sigma, V_\sigma \mid \sigma \in \Sigma\}$ be a series in a group G. The set of all auto-morphisms γ of G such that $[\Lambda_\sigma, \gamma] \leqslant V_\sigma$ for all σ is a subgroup of Aut G called the *stability group of* $\mathbf{S}$. If Γ is a group acting faithfully on G and if $[\Lambda_\sigma, \Gamma] \leqslant V_\sigma$ for all σ, then Γ is said to *stabilize* $\mathbf{S}$ *faithfully*; of course Γ is isomorphic with a subgroup of the stability group of $\mathbf{S}$. The main problem of stability theory is to identify the groups which can faith-fully stabilize a series of given type.

There is a strong connection between stability groups and central series. Indeed suppose that H is a group with a central series of order type Σ and let G be the standard wreath product

$$H \wr T$$

where $T = \langle t \rangle$ has order 2. If $\{\Lambda_\sigma, V_\sigma \mid \sigma \in \Sigma\}$ is the given central series in H, define $\bar{\Lambda}_\sigma = \Lambda_\sigma \Lambda_\sigma^t$ and $\bar{V}_\sigma = V_\sigma V_\sigma^t$. Let $\Sigma + 1$ be the ordered set obtained by adjoining a symbol λ to Σ with the rule $\sigma < \lambda$ for all σ in Σ. Define $\bar{\Lambda}_\lambda$ to be G and $\bar{V}_\lambda$ to be HH^t. Then $\{\bar{\Lambda}_\sigma, \bar{V}_\sigma \mid \sigma \in \Sigma + 1\}$ is a series in G with order type $\Sigma + 1$. Moreover, $[\bar{\Lambda}_\sigma, H] \leqslant \bar{V}_\sigma$ since $[H^t, H] = 1$. Also $H \cap Z(G) = 1$. Hence H faithfully stabilizes a series of order type $\Sigma + 1$. This furnishes examples of groups that faithfully stabilize series of any ordinal type which is not a limit ordinal.

Stability groups of series of finite length were first studied by L. A. Kalužnin [45]. Hall [29] proved the following basic result.

THEOREM 2.9 *The stability group of a series of finite length $n > 0$ is nilpotent of class at most $\binom{n}{2}$. For normal series the bound may be reduced to $n - 1$.*

Combining this with the conclusion of the previous paragraph, we obtain the

COROLLARY *The groups which faithfully stabilize a series (or even a normal series) of finite length are exactly the nilpotent groups.*

The study of stability groups of series of infinite order type was initiated in an important paper of Hall and Hartley [37]. Let us consider ascending series first: two main theorems are proved about the stability groups of such series.

THEOREM 2.10 [37] *The stability group of an ascending series is hypo-central.*

THEOREM 2.11 [37] *A finitely generated group can faithfully stabilize an ascending series if and only if it is residually nilpotent.*

From the second of these theorems we see that *a finite group faithfully stabilizes an ascending series (or even an ascending normal series) if and only if it is nilpotent.* Thus the property of faithfully stabilizing an ascending series is a generalization of nilpotence. Groups with this property are evidently locally residually nilpotent and hypocentral; however, no internal characterization is known. We mention a partial result which is given in [37]: *the groups that faithfully stabilize an ascending (normal) series of type $\leqslant \omega$ are exactly the residually nilpotent groups.*

However, when we turn to descending series, we find a totally different situation.

THEOREM 2.12 [37] *Every group faithfully stabilizes some descending series.*

Hall and Hartley established this theorem by means of a collecting process for free products similar to the well known collecting process for free groups. The key result that emerges from this theory is

THEOREM 2.13 [37] *If $G = \operatorname*{Fr}_{\lambda \in \Lambda} G_\lambda$ is a free product of groups, then G has a descending series whose factors are isomorphic with the free factors G_λ.*

This is a result which is likely to have other applications.

3. Generalized Soluble Groups

By a generalization of solubility is meant any group-theoretical property which is equivalent to solubility for finite groups.

Locally soluble groups

Although locally soluble groups are probably the most natural type of generalized soluble group, very little is known about their structure. This is largely due to our ignorance of finitely generated soluble groups, except in the abelian-by-polycyclic case.

One obstacle to progress is the lack of a theorem of the Hirsch–Plotkin type. Indeed *the product of two normal locally soluble subgroups need not be locally soluble.* An example was given by Philip Hall in lectures at Cambridge in 1963, and a very similar one by Baumslag, Kovács and Neumann [13] (cf. [73], II, p. 90).

The first step in the construction is to observe that if G/N is an infinite cyclic group, then G may be embedded in a normal product of two copies of the standard wreath product $N \wr (G/N)$. This is applied with N equal to the McLain group $M(\mathbb{Z}, GF(p))$ and G the semi-direct product $\langle \tau \rangle \ltimes N$ where τ is the automorphism of N given by $1 + e_{ij} \mapsto 1 + e_{i+1, j+1}$. Since G is finitely generated (by $1 + e_{01}$ and τ), it cannot be locally soluble. On the other hand, it is easy to see that $N \wr (G/N)$ is locally soluble.

Our meagre knowledge of the structure of locally soluble groups is based on

THEOREM 3.1 (Mal'cev [55], cf. [73], I, p. 154) *A principal factor of a locally soluble group is abelian.*

This at least indicates that there are no non-cyclic simple, locally soluble groups.

Series with abelian factors

One of the widest classes of generalized soluble groups to have received attention is the class $\grave{P}\mathfrak{A}$ of groups having a series with abelian factors; these groups are called *SN-groups*. Successively smaller sub-classes are $\hat{P}_{s_n} \mathfrak{A}$ and $SI = \hat{P}_n \mathfrak{A}$, the classes of groups with a subnormal and normal series with abelian factors respectively.

Groups having a descending series with abelian factors are called *hypoabelian*; these are exactly the groups whose transfinite derived series reach the identity, and they include, of course, *residually soluble groups*. Little is known of groups of these types.

More tractable are the classes $\acute{P}\mathfrak{A}$, $\acute{P}_{s_n} \mathfrak{A}$, $\acute{P}_n \mathfrak{A}$ of groups with an ascending series, an ascending subnormal series, an ascending normal series with abelian factors. The second and third of these types are also called *subsoluble groups* and *hyperabelian groups*.

A group all of whose composition factors are abelian is called an $\overline{SN}$-*group*. It is easy to see that $\overline{SN}$ is the largest subclass of SN which is closed under the formation of homomorphic images and serial sub-groups; thus

$$\overline{SN} = (SN)^{H\hat{S}} = (\acute{P}\mathfrak{A})^{H\hat{S}}.$$

We can define an $\overline{SI}$-*group* to be a group all of whose principal factors are abelian; it is easily seen that

$$\overline{SI} = (SI)^{H} = (\hat{P}_n \mathfrak{A})^{H}.$$

The example of a $\bar{Z}$-group with non-cyclic free subgroups given by Philip Hall (see section 2) shows that $\overline{SI}$ is not subgroup closed. Hence there is a smaller class

$$\overline{\overline{SI}} = (\overline{SI})^{S} = (\hat{P}_n \mathfrak{A})^{HS}.$$

(In fact $\overline{\overline{SI}}$ is not even s_n-closed, as J. S. Wilson [93] has shown.) Notice that every locally soluble group is an $\overline{\overline{SI}}$-group by Theorem 3.1.

The class $\overline{SN}$ is not subgroup closed. This was shown by Wilson using an example similar to those of Hall and Merzljakov mentioned in section 2.

Let G be the group of 2×2 matrices X over the ring $\mathbb{Z}_{(p)}$ of rational numbers with denominators prime to p (the local ring at p over $\mathbb{Z}$), which satisfy $X \equiv 1 \pmod{p}$ and $\det X = 1$; thus G is a subgroup of $SL(2, \mathbb{Z}_{(p)})$. Define G_r to be the kernel of the homomorphism $G \to SL(2, \mathbb{Z}/p^r\mathbb{Z})$ arising from the canonical homomorphisms

$$\mathbb{Z}_{(p)} \to \mathbb{Z}_{(p)}/p^r\mathbb{Z}_{(p)} \xrightarrow{\sim} \mathbb{Z}/p^r\mathbb{Z}.$$

Wilson established the following facts about G.

THEOREM 3.2 (Wilson [94])   *The non-central subnormal subgroups of G are precisely the subgroups that contain some G_r. The non-subnormal serial subgroups of G are metabelian and each is contained in infinitely many subnormal subgroups.*

From this and the fact that $G_r \cap Z(G) = 1$ one can deduce that G is an $\overline{SN}$-group. But G has non-cyclic free subgroups, so $\overline{SN}$ is not subgroup-closed and hence

$$\overline{\overline{SN}} = (\overline{SN})^{\mathrm{S}} = (\hat{\mathfrak{p}}\mathfrak{A})^{\mathrm{HS}}$$

is a proper subclass of $\overline{SN}$.

In 1961 Philip Hall showed that the class $\overline{\overline{SN}}$ contains non-cyclic simple groups—we shall discuss the construction later in this section. This shows that $\overline{\overline{SN}}$, and hence $\overline{SN}$, is not a subclass of $\overline{SI}$. It was later shown by Wilson that $\overline{SI}$ is not a subclass of $\overline{SN}$. To achieve this Wilson again used linear groups but this time of infinite degree.

Let F be the rational function field over $\mathbb{Q}$ in indeterminates $\{x_i \mid i \in \mathbb{Z}\}$ and let σ be the field automorphism of F which maps each x_i to x_{i+1}. Define R to be the skew polynomial ring $F[t:\sigma]$; thus $ft = tf^\sigma$ for f in F. Write $GL(R)$ for the group of all invertible countably infinite matrices over R which are row and column finite. If I is the ideal Rt, the canonical homomorphism $R \to R/I$ induces a homomorphism $GL(R) \to GL(R/I)$ with kernel G say.

THEOREM 3.3 (Wilson [92])

(i) *Every proper quotient of G is nilpotent.*

(ii) *The intersection of all the non-trivial normal subgroups of G is the identity.*

(iii) *There is a descendant subgroup of G which has a free quotient of countably infinite rank.*

It is clear from (i) and (ii) that G is an $\overline{SI}$-group, while (iii) shows that G is not an $\overline{SN}$-group.

Radical groups

A group which has an ascending series with locally nilpotent factors is called a *radical group*. It is easy to see that these are precisely the groups which coincide with a term of their iterated ascending series of Hirsch–Plotkin radicals. The class of radical groups is useful since it contains all soluble groups and all locally nilpotent groups, while it is better behaved than the class of locally soluble groups. For example, a product of normal radical subgroups is always radical. It is clear from Theorem 2.1 that every $\acute{P}\mathfrak{A}$-group is radical. Also radical groups are *SN*-groups since every locally nilpotent group is a *Z*-group; but the class of radical groups is subgroup and image closed, so in fact *every radical group is in* $(\acute{P}\mathfrak{A})^{\text{HS}} = \overline{\overline{SN}}$.

Residually commutable groups. Finally we mention a type of group akin to residually central groups. Call a group *G residually commutable* if given $a \neq 1$ and $b \neq 1$ in *G*, at least one of *a* and *b* fails to belong to $[a, b]^G$. This property was introduced by Ayoub [2] in 1969.

It is easy to see that every *SI*-group is residually commutable. However, no example seems to be known of a residually commutable group which is not an *SI*-group. Ayoub has shown that a finite residually commutable group is soluble, so residual commutability is a generalization of solubility.

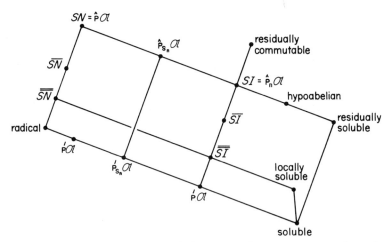

FIG. 3   *Diagram of group classes.* All classes are distinct except perhaps for *SI* and the class of residually commutable groups.

The effect of finiteness conditions

Since it has proved difficult to obtain non-trivial results about the structure of generalized soluble groups, most effort has been expended in studying these groups when finiteness conditions are imposed. For example, it is a theorem of Černikov (see [73]) that an *SN*-group with min is a soluble Černikov group. In addition Ayoub [2] showed that a residually commutable group with min is also of this type. Conse-

quently *all sixteen classes above coincide under* min. The effect of max
is a good deal less certain. For example, no residually soluble groups
with max are known which are not soluble. Nor apparently is it known
if $\overline{SI}$-groups with max are soluble. In addition the existence of infinite
simple $\overline{SN}$-groups shows that little can be expected of the properties
max-n and min-n.

The deepest results have been obtained in the case of finiteness
conditions on abelian subgroups. We shall quote a theorem of Baer and
Heineken to illustrate this.

THEOREM 3.4 (Baer and Heineken [9])   *Let G be a radical group and let
T be its maximal normal torsion subgroup. Assume that G has finite
abelian subgroup rank. Then*

(i) *G/T is a soluble group with an abelian series of finite length whose
factors are finite or torsion-free;*
(ii) *Sylow subgroups of G are Černikov groups;*
(iii) *G is countable and hyperabelian;*
(iv) *G has finite abelian section rank.*

For a comprehensive account of this and related results see [74].

Simple generalized soluble groups

In 1961 Philip Hall published the startling result that the class SN
contains non-cyclic simple groups. This demonstrated how far removed
SN and similar classes are from soluble groups. It also provided
examples of non-strictly simple groups: recall that a group is *strictly
simple* if it is non-trivial and contains no proper non-trivial ascendant
subgroups. However, it is still not known if there are strictly simple
groups which are not absolutely simple.

Hall's construction is based on his previous work on wreath products
of linearly ordered sets of permutation groups (see [33]).

Let H be any non-trivial group and let W be the restricted standard
wreath power

$$W = \operatorname{Wr} H^{\mathbb{Z}};$$

this is generated by isomorphic copies H_n of H. The order-automorphism
$n \mapsto n+1$ of $\mathbb{Z}$ gives rise to an automorphism τ of W such that
$H_n^\tau = H_{n+1}$. Define

$$f(H) = \langle \tau \rangle \ltimes W,$$

the semi-direct product. We identify H with the subgroup H_0 of $f(H)$.

Hall showed in [33] that W' is a minimal normal subgroup of $f(H)$. This result is already of interest: if we take $|H| = p$, then W' is a characteristically simple locally finite p-group with trivial Baer radical.

The next step is to iterate the construction by defining

$$G_0 = H, \ G_{\alpha+1} = f(G_\alpha) \text{ and } G_\lambda = \bigcup_{\beta < \lambda} G_\beta$$

where α is an ordinal and λ a limit ordinal. Write

$$g_\alpha(H) = G_\alpha'.$$

Then the following holds.

THEOREM 3.5 (Hall [34])   *If H is a non-trivial group and $\lambda > 0$ is a limit ordinal, then $g_\lambda(H)$ is a simple group with an ascending series of ordinal type λ.*

It can be verified that $g_\lambda(H)$ is an $\overline{\overline{SN}}$-group provided H is. Hence we derive the

COROLLARY   *There exist non-cyclic simple $\overline{\overline{SN}}$-groups.*

4. Finiteness Conditions on Commutators and Conjugates

We shall examine finiteness conditions which restrict the set of commutators or conjugates of elements in a group. The class of groups satisfying such a condition will usually contain all abelian groups and all finite groups.

Centre-by-finite groups

Almost all results in this area depend ultimately on the following classical theorem of I. Schur [83].

THEOREM 4.1   *If G is a group with $|G\colon Z(G)|$ finite of order m, where $Z(G)$ denotes the centre of G, then G' is finite and $(G')^m = 1$.*

This is equivalent to the fact that the Schur multiplicator of a finite group is finite and is annihilated by the group order. The theorem can easily be proved by using the transfer into $Z(G)$.

In 1951 Baer gave a generalization of Schur's theorem, replacing the centre by a term $\zeta_i(G)$, $i = 1, 2, 3, \ldots$, of the upper central series.

THEOREM 4.2 (Baer [6]; cf. [73], I, p. 113) *If G is a group such that $G/\zeta_i G$ is finite, then $\gamma_{i+1}G$ is finite.*

Here $\gamma_{i+1}G$ is the $(i+1)$th term of the lower central series of G. Philip Hall considered the converse problem: what conclusion can be drawn from the finiteness of $\gamma_{i+1}G$? Using a simple lemma on commutator subgroups, he proved

THEOREM 4.3 (Hall [27]; cf. [73], I, p. 117) *If G is a group such that $\gamma_{i+1}G$ is finite, then $|G:\zeta_{2i}G|$ is finite.*

Combining the theorems of Baer and Hall, we see that *a group G is finite-by-nilpotent if and only if some $\zeta_i(G)$ has finite index.* If G is a finite-by-nilpotent group, define $k(G)$ and $l(G)$ to be the smallest nonnegative integers i and j such that $\gamma_{i+1}G$ and $|G:\zeta_j G|$ are finite. Then by Theorems 4.2 and 4.3

$$0 \leqslant k(G) \leqslant l(G) \leqslant 2k(G).$$

Hall showed that these are the only restrictions on $k(G)$ and $l(G)$: if k and l are integers satisfying

$$0 \leqslant k \leqslant l \leqslant 2k, \qquad\qquad (*)$$

then there exists a finite-by-nilpotent group G such that $k(G) = k$ and $l(G) = l$.

Hall's construction is as follows. Let P be the central product of a countably infinite number of non-abelian groups of order p^3: these are to have exponent p if $p > 2$ and to be quaternion groups if $p = 2$. Now let G be the standard wreath product of P and a group $C = \langle c \rangle$ of order p,

$$G = P \wr C.$$

This is a nilpotent p-group of class $2p$. It is not difficult to see that $\zeta_p G = B' = \gamma_{p+1}G$ where B is the base group of the wreath product. Hence $k(G) = p$ and $l(G) = 2p$; this realizes the pair $(p, 2p)$. To realize a pair (k, l) satisfying $(*)$ we choose $p > 2k$ and consider the group

$$\bar{G} = \langle c, \zeta_{p+k}G \rangle / \zeta_{2k-l}G.$$

For this group $k(\bar{G}) = k$ and $l(\bar{G}) = l$.

Philip Hall's problems on words

In the course of attempts to generalize the Baer–Hall theorems to verbal subgroups the following problems were formulated: they are usually attributed to Hall (see [28]).

If w is a word and G is any group, let $w(G)$ and $w^*(G)$ be the corresponding verbal and marginal subgroups of G.

   I If $G/w^*(G)$ is finite, is $w(G)$ finite?
  II If $w(G)$ is finite and G satisfies max, does it follow that $G/w^*(G)$ is finite?
 III If w is finite-valued on G, does it follow that $w(G)$ is finite?

(The extra hypothesis 'G has max' is inserted in II since otherwise very simple examples would defeat the conjecture.) For example, if $w = [x_1, x_2, \ldots, x_{k+1}]$, the first problem has a positive answer by Baer's theorem.

All three problems have positive answers if w is an *outer commutator word*. Recall that the outer commutator words of weight 1 are $x_1, x_2, \ldots$, while the outer commutator words of weight $n > 1$ are defined inductively by $w = [w_1, w_2]$ where w_i is an outer commutator word of weight $n_i < n$ and $n = n_1 + n_2$. The solutions to the problems in the case of outer commutator words were obtained by R. F. Turner-Smith [90] (I and II) and J. Wilson [91] (III).

However, all three problems remain open in the general case. We refer the reader to [73], I, p. 111 ff. for a detailed account.

Groups with finite automorphism group

Groups with finite automorphism group form a special class of centre-by-finite groups and much has been written about them. Such groups were first studied by Baer [8] who proved that *a torsion group with finite automorphism group is finite*. However, it is known that there are many torsion-free abelian groups with automorphism group of order 2: a useful classification of such groups is not to be expected. On the other hand, J. Alperin [1] showed that *a finitely generated group has finite automorphism group if and only if it is a finite central extension of a cyclic group*.

The first general result about the structure of groups with finite automorphism group was

THEOREM 4.4 (V. I. Nagrebeckiĭ [62]) *The elements of finite order in a group with finite automorphism group form a finite fully invariant subgroup containing the derived subgroup.*

This of course implies Baer's theorem. More recently homological techniques have been used effectively in studying groups with finite automorphism group, a point of view adopted by Robinson [75]. We shall explain the basis for this.

Let G be a group with centre C and central quotient group Q. Write Δ for the cohomology class of the central extension $1 \to C \to G \to Q \to 1$. There is a natural left action of Aut Q and a natural right action of Aut C on the second cohomology group $H^2(Q, C)$. These actions yield necessary and sufficient conditions for an element (γ, κ) of Aut $C \times$ Aut Q to be induced by an automorphism of G, namely $\kappa\Delta = \Delta\gamma$. (A detailed account of automorphisms of group extensions can be found in Robinson [77].) If $C(\Delta)$ is the set of pairs (γ, κ) satisfying this equation, we have a natural exact sequence

$$0 \to \mathrm{Hom}(Q/Q', C) \to \mathrm{Aut}\, G \to C(\Delta) \to 1.$$

This sequence may be used to give a criterion for a group to have finite automorphism group. Using the above notation we may state

THEOREM 4.5 (Robinson [75])   *The group G has finite automorphism group if and only if Q, $\mathrm{Hom}(Q/Q', C)$ and $C_{\mathrm{Aut}\, C}(\Delta)$ are all finite.*

The sequence can also be used to characterize those abelian groups which are isomorphic with the centre of a group with finite automorphism group. In general the centre can have infinite automorphism group; in fact *there is a group G such that* Aut $G \simeq S_4$ *but* Aut$(\zeta(G))$ *is uncountable.* On the other hand, there is

THEOREM 4.6 (Robinson [75])   *A nilpotent group has finite automorphism group if and only if its centre has finite index and finite automorphism group.*

There is evidence to suggest that the finite groups which are automorphism groups of infinite groups are of rather special type. The most striking results are due to J. T. Hallett and K. A. Hirsch [39], who show that finite groups which are the automorphism groups of a torsion-free group have exponent dividing 12. Moreover, they give a detailed classification of such groups. It seems possible that groups having locally finite automorphism groups should be of restricted type. Certainly there are several wide classes of locally finite groups which cannot be the automorphism groups of any group, as the following result indicates.

THEOREM 4.7 (Robinson [76])   *No group of the following types can be the automorphism group of any group:*

  (i) *infinite Černikov groups;*
 (ii) *nilpotent torsion groups of infinite exponent;*
(iii) *infinite nilpotent torsion groups without elements of order 2 or 3.*

To this list we may add countable torsion *FC*-groups of infinite exponent which involve elements of only a finite number of prime orders (J. Zimmerman [114]).

Groups with finite conjugacy classes[‡]

A group which has finite conjugacy classes is called an *FC-group*. These groups were introduced in 1948 by Baer [5]; a later paper of B. H. Neumann [64] completed the elementary theory. The basic results (cf. [73], I, p. 121 ff.) are summed up as follows.

THEOREM 4.8 (Baer, B. H. Neumann)

  (i) *If G is an FC-group, then $G/Z(G)$ is a residually finite, torsion FC-group.*

 (ii) *The elements of finite order in an FC-group form a fully invariant subgroup containing the derived subgroup.*

(iii) *A torsion group is FC if and only if every finite subset is contained in a finite normal subgroup.*

(iv) *Every FC-group can be embedded in an FC-group which is the direct product of a torsion FC-group and a torsion-free abelian group.*

Some special types of *FC*-groups were studied in later papers by B. H. Neumann. Two of his most important results are

THEOREM 4.9 (B. H. Neumann [65])   *A group G has boundedly finite conjugacy classes if and only if G' is finite.*

THEOREM 4.10 (B. H. Neumann [66])   *A group has finite conjugacy classes of subgroups if and only if it is centre-by-finite.*

Torsion FC-groups.   It is clear from Theorem 4.8 that it is the torsion *FC*-groups which are of greatest interest. These include all sections of direct products of finite groups. In a paper published in 1959 Hall

---

‡ *FC*-groups are also important in the study of group rings; see Passman's essay.

raised the natural question of which torsion FC-groups can be sub-groups, quotient groups or sections of direct products of finite groups? Of course in the case of subgroups it is clear that the torsion FC-group must at least be residually finite. However, it is far from obvious what further conditions must be met. By a simple and elegant argument Hall was able to settle the countable case. He proved

THEOREM 4.11 (P. Hall [30])   *Every countable, residually finite, torsion FC-group is isomorphic with a subgroup of a direct (i.e., restricted) product of finite groups.*

This has the amusing consequence that the countable, residually finite, torsion FC-groups are exactly the subgroups of the group

$$S_2 \times S_3 \times S_4 \times \ldots. \qquad (*)$$

Hall also considered the more difficult question of groups that are not residually finite and obtained a definitive result.

THEOREM 4.12 (P. Hall [30])   *Every countable torsion FC-group is isomorphic with a section of a direct product of finite groups.*

This implies that all countable FC-groups are to be found among the sections of the direct product of the group $(*)$ and a rational vector space of countably infinite dimension.

Covering groups

Hall established Theorem 4.12 by introducing the *covering group* of a group generated by normal subgroups. This construction is of independent interest.

Let G be a group which is generated by a family of normal subgroups $\{G_\lambda \mid \lambda \in \Lambda\}$ where Λ is a linearly ordered set. To each λ in Λ and x in G_λ we associate a symbol $u_\lambda(x)$ and we write U for the set of all such symbols. If $y \in G_\mu$, we define a permutation $\pi_\mu(y)$ of U; this is to send $u_\lambda(x)$ to

$$u_\lambda(y^{-1}xy), \quad u_\lambda(xy) \quad \text{or } u_\lambda(x)$$

according as $\lambda < \mu$, $\lambda = \mu$ or $\lambda > \mu$. The permutation group

$$\bar{G} = \langle \pi_\mu(y) \mid \mu \in \Lambda, y \in G_\mu \rangle$$

is the covering group. Hall showed that the assignment $\pi_\mu(y) \mapsto y$, $(y \in G_\mu)$, determines an epimorphism $\bar{G} \to G$.

Let $\bar{K}_\lambda$ be the kernel of the induced transitive representation of $\bar{G}$ on the orbit $\{u_\lambda(x) \,|\, x \in G_\lambda\}$. Then $\bar{G}$ is a subcartesian product of the groups $\bar{G}/\bar{K}_\lambda$.

If G is a countable torsion FC-group, we may choose the G_λ to be suitable finite normal subgroups of G. Then $\bar{G}/\bar{K}_\lambda$ is finite and $\bar{G}$ is residually finite. It is not difficult to see that $\bar{G}$ is a countable torsion FC-group; since it is also residually finite, $\bar{G}$ embeds in a direct product of finite groups by Theorem 4.11.

Hall constructs a number of interesting groups in his paper. For example, he exhibits a countable torsion FC-group that is not an image of a direct product of finite groups; this contrasts with the easily established result that every abelian torsion group is an image of a direct product of finite cyclic groups.

In addition Hall gives an example which shows how much more complex the situation is for uncountable groups.

THEOREM 4.13 (P. Hall [30]) *There is an uncountable torsion FC-group which is not isomorphic with any section of a direct product of finite groups.*

This group is less well known than it ought to be, so we shall describe its construction. Let S be any infinite set. We associate to each subset X and each finite subset F of S symbols a_X and b_F. Consider a group G whose elements are the triples (ε, a_X, b_F) with $\varepsilon = \pm 1$; the group operation is given by

$$(\varepsilon, a_X, b_F)(\varepsilon', a_{X'}, b_{F'}) = (\varepsilon'', a_{X+X'}, b_{F+F'})$$

where $X + Y$ denotes the symmetric difference $(X \cup Y) \backslash (X \cap Y)$ and where

$$\varepsilon'' = \varepsilon\varepsilon'(-1)^{|F \cap X'|}.$$

It is easy to check that G is a group and that $C \equiv G' = Z(G)$ has order 2, while G/C is an elementary abelian 2-group with cardinality $2^{|S|}$. Thus G is an extra-special 2-group and is clearly an FC-group.

Hall observed that if M is an infinite subgroup of a section L of some direct product of finite groups, then

$$|L : C_L(M)| \leqslant |M|.$$

However, this condition is violated in the group G. Indeed, if A and B are the subgroups generated by the a_X's and b_F's respectively, then

$C_G(B) = BC$ and
$$|G: C_G(B)| = |ABC: BC| = |A| > |B|.$$

Further developments

Philip Hall's work on torsion FC-groups has been continued in a sequence of papers by Gorčakov. We record some of his results.

THEOREM 4.14 (Yu. M. Gorčakov [20]) *If G is any FC-group, then $G/\zeta_2 G$ can be embedded in a direct product of finite groups.*

In general $G/Z(G)$ cannot be embedded in a direct product of finite groups, although this is possible if G is residually finite (Gorčakov [20]).

In particular, *an FC-group with trivial centre can be embedded in a direct product of finite groups.* Gorčakov [18] has also established the dual result: *if G is a perfect, residually finite, torsion FC-group, it has such an embedding.*

THEOREM 4.15 (Gorčakov [19]) *A residually finite, torsion FC-group is isomorphic with a section of a direct product of finite groups.*

Compare this with Hall's example, which is not residually finite. In a recent paper M. J. Tomkinson has given the following characterization of residually finite torsion FC-groups in terms of centrally restricted products.

THEOREM 4.16 (Tomkinson [89]) *A group is a residually finite, torsion FC-group if and only if it is isomorphic with a subgroup of a centrally restricted product of finite groups.*

Here the *centrally restricted product* of a set of torsion groups $\{G_i \mid i \in I\}$ is the subgroup of $\underset{i \in I}{\mathrm{Cr}}\, G_i$ (the Cartesian product) consisting of all elements of finite order such that for almost all i the i-component belongs to $Z(G_i)$.

However, the general problem remains open: which torsion FC-groups are realizable as subgroups, quotients or sections of direct products of finite groups?

Groups with finite automorphism classes

The *automorphism classes* of elements of a group G are the Aut G-orbits under the natural action of Aut G on G. Groups with finite automorphism classes form a natural subclass of the class of FC-groups, containing all groups with finite automorphism group. However, no satisfactory characterization of groups with finite automorphism classes is known.

Groups whose automorphism classes are boundedly finite are more tractable. In a forthcoming paper the following result is established.

THEOREM 4.17 (D. J. S. Robinson and J. Wiegold [78]) *Let T be the torsion subgroup of the centre of a group G. Then the automorphism classes of G are boundedly finite if and only if T is finite and $\mathrm{Aut}\,G$ induces a finite group of automorphisms in G/T.*

Thus if the centre of G is torsion-free and G has boundedly finite automorphism classes, then $\mathrm{Aut}\,G$ is finite. However, there are groups with boundedly finite automorphism classes which have infinite automorphism group.

Finally we note a corresponding result for automorphism classes of subgroups, meaning $\mathrm{Aut}\,G$-orbits on the set of subgroups of G.

THEOREM 4.18 (Robinson and Wiegold [78]) *A group G has boundedly finite automorphism classes of subgroups if and only if either $\mathrm{Aut}\,G$ is finite or $G = G_1 \times G_2$, where G_1 is a locally cyclic torsion group, G_2 is a finite central extension of a direct product of finitely many groups of type p^∞ for different primes p, and G_1 and G_2 do not have elements of the same prime order.*

5. Locally Finite Groups

We shall not attempt to describe the whole range of work in the very extensive theory of locally finite groups. Instead we consider two topics to which Philip Hall made notable contributions.

Universal locally finite groups

A locally finite group U is said to be *universal* if

(i) each finite group can be embedded in U, and
(ii) isomorphic finite subgroups of U are conjugate.

In [31] Hall proves the existence and uniqueness of countable universal locally finite groups. His method of establishing existence is remarkably simple. One observes first that for any finite group H the right regular representation $H \to \mathrm{Sym}\,H$ maps isomorphic subgroups of

H to conjugate subgroups of $\operatorname{Sym} H$. Now choose G_1 to be any finite group of order at least 3 and define inductively $G_{i+1} = \operatorname{Sym} G_i$, with G_i embedded in G_{i+1} by means of the right regular representation. Let $U = \varinjlim G_i$ be the direct limit of the sequence of groups $G_1, G_2, \ldots$. It follows from the embedding of G_i in G_{i+1} that U is a countable universal locally finite group.

Hall also established the following property of U:

(iii) if $m > 1$, the elements of U with order m form a conjugacy class $U[m]$ and $U = U[m]U[m]$.

This is based on an elementary result: given integers $m > 1$ and $n > 1$, there is a finite group $\langle x, y \rangle$ such that x and y have order m and xy has order n. We deduce at once from (ii) and (iii) that

(iv) U is a simple group.

We record two further properties of U:

(v) U contains as subgroups $2^{\aleph_0}$ copies of every infinite countable locally finite group.

(vi) (K. K. Hickin, unpublished) If π is a proper set of primes and P is a countable infinite locally finite π-group, then P can be embedded as a maximal π-subgroup of U. In particular, if p is a prime, then every countable infinite locally finite p-group can be embedded in U as a Sylow p-subgroup of U.

We state Hall's uniqueness result as

THEOREM 5.1   *Up to isomorphism there is only one countable universal locally finite group.*

Hall's work was extended to uncountable groups by O. H. Kegel and B. A. F. Wehrfritz in [49]. They establish the existence of universal locally finite groups of arbitrary infinite cardinality and they also prove that such groups possess the properties (iii), (iv) and (v). We mention also the stronger result (see [49], p. 182): *every infinite locally finite group can be embedded in a universal locally finite group of the same cardinality.*

Kegel and Wehrfritz left open the question: are there non-isomorphic universal locally finite groups of the same cardinality $\mathfrak{c} > \aleph_0$? This was answered by A. Macintyre and S. Shelah in a remarkable paper [54] published in 1976. Using the method of indiscernibles in infinitary logic they proved the following

THEOREM 5.2   *For each cardinal $\mathfrak{c} \geqslant \aleph_1$ the set of isomorphism classes of universal locally finite groups of cardinality $\mathfrak{c}$ has cardinality $2^{\mathfrak{c}}$.*

In [102] Hickin shows that, in the case $\mathfrak{c} = \aleph_1$ of Theorem 5.2, there is a collection of $2^{\aleph_1}$ locally finite groups such that given two non-isomorphic universal groups in it, neither embeds in the other.

It is all too apparent from these results that countable groups are the exception, not the rule, among universal locally finite groups. The uncountable universal locally finite groups are far from being satisfactorily understood and remain a topic where further research can be expected.

We conclude by mentioning an open problem: which locally finite groups H embed in every universal locally finite group of cardinality $\geqslant|H|$? Macintyre [106] calls such groups 'inevitable'. He shows that there are no inevitable abelian groups of cardinality $\aleph_1$; and Hickin proves in [102] that no locally finite group of cardinality $\aleph_1$ is inevitable. S. Thomas has recently proved that no uncountable soluble group is inevitable [113].

Generalizations of Hall's construction.   Hall's countable universal locally finite group has suggested similar, more general, constructions. These give considerable insight into the possible finite subgroup structures of countable locally finite simple groups. The possibility of such constructions was first noted by Hickin; further additions and refinements are due to R. E. Phillips. I am indebted to him for the remarks which follow.

The notion of an existentially closed group plays an essential role. The key ingredients of the theory may be found in J. Hirschfeld and W. H. Wheeler [104]; here we explain only what we need. Let $\mathfrak{L}$ be a given class of groups and G a group in $\mathfrak{L}$. The symbol $W(x_j, g_k)$ denotes a word in a finite number of variables x_j and a finite number of elements g_k in G. A finite set of equations and inequations

$$W_r(x_j, g_k) = 1$$
$$W_s(x_j, g_k) \neq 1 \tag{$*$}$$

is said to be $\mathfrak{L}$-*consistent with* G if there is a group H in $\mathfrak{L}$ with $G \leqslant H$ such that $(*)$ has a solution in H. The $\mathfrak{L}$-group G is *existentially closed in* $\mathfrak{L}$ if every system $(*)$ which is $\mathfrak{L}$-consistent with G has a solution in G itself.

Suppose G is in $\mathfrak{L}$ and $\mathfrak{c} = \max\{|G|, \aleph_0\}$. If $\mathfrak{L}$ is closed under taking ascending unions of cardinality $\leqslant\mathfrak{c}$, there is an existentially closed group K in $\mathfrak{L}$ with $G \leqslant K$ and $|K| = \mathfrak{c}$ [104].

Macintyre and Shelah [54] have pointed out that the existentially closed groups in the class of locally finite groups are precisely the universal groups. Thus *Hall's group U may be viewed as the unique existentially closed countable locally finite group.*

Let A be a countable periodic abelian group and $\mathfrak{L}_A$ consist of all countable locally finite groups G whose centre contains A. Since $\mathfrak{L}_A$ is closed under ascending countable unions, there exist existentially closed groups in $\mathfrak{L}_A$.

THEOREM 5.3   *Up to isomorphism there is only one existentially closed group in $\mathfrak{L}_A$.*

Let E_A denote an existentially closed member of $\mathfrak{L}_A$. The group E_A has properties similar to those of Hall's group U. We list some of these.

(i)$_A$ If $H \in \mathfrak{L}_A$ and H/A is finite, then there is an embedding of H in E_A which is the identity on A.

(ii)$_A$ If H and K are subgroups of E_A with H/A and K/A finite, then H and K are conjugate in E_A if and only if there is an isomorphism of H onto K restricting to the identity on A.

There are also exact analogues of the properties (v) and (vi) but none is known of (iii). Property (iv) takes the form

(iv)$_A$ E_A is perfect and E_A/A is simple.

Theorem 5.3, (i)$_A$ and (ii)$_A$ show that $E_1 \cong U$; also if A is a finite abelian subgroup of U, then $E_A \cong C_U(A)$, the centralizer of A in U.

There are a great number of simple groups of the form E_A/A. This follows from the result that E_A/A *has Schur multiplicator A.* Thus A is *not isomorphic to B implies E_A/A is not isomorphic to E_B/B.* Also, the case $A = 1$ gives the fact that U has trivial Schur multiplicator.

Although similar to U, the groups E_A/A can have a more complex structure. For example, if p is a prime, there are $|A/p^nA|$ conjugacy classes of cyclic subgroups of order p^n in E_A/A. The principal result about the finite subgroups of E_A/A is

THEOREM 5.4   *If the Prüfer rank of A is at least 3, then E_A/A is not locally-(finite simple).*

Like the other statements about E_A/A this is not at all easy to prove. If A is finite however, it follows without difficulty from the fact that the Schur multiplicator of any finite simple non-abelian group has Prüfer rank $\leqslant 2$ (R. L. Griess [99]). The first example of a countable locally finite simple group that is not locally-(finite simple) was given by V. N. Serežkin and A. E. Zalesskiĭ [109].

The existence of infinite abelian subgroups

A celebrated question of uncertain antiquity in the theory of groups is: does every infinite group possess an infinite abelian subgroup? Obviously it must do if elements of infinite order are present, so the question is only of interest for infinite periodic groups.

There is a similar well known problem which is generally associated with the name of O. J. Schmidt: if an infinite group has all of its proper subgroups finite, does it follow that the group is of Prüfer type?

We know now that both questions have negative answers, thanks to work of P. S. Novikov and S. I. Adjan [68], A. Ju. Ol'šanskiĭ [69] and E. Rips (unpublished). Indeed Rips' example is an infinite group all of whose proper non-trivial subgroups have order p where p is some large prime. This group is a finitely generated infinite simple p-group.

That the situation is quite different for locally finite groups was shown independently by P. Hall and C. R. Kulatilaka [38] and M. I. Kargapolov [47].

THEOREM 5.5 *Every infinite locally finite group has an infinite abelian subgroup.*

The proof given by Hall and Kulatilaka begins by supposing the result to be false: a short but clever argument reduces to the study of an infinite locally finite group G such that $C_G(g)$ is finite for every $g \neq 1$ in G. A careful analysis of the structure of G leads to a contradiction. The most troublesome case, when G is residually finite, is disposed of by an appeal to Burnside's theorem on the structure of Frobenius groups.

A significant feature of the proof (and also of Kargapolov's) is the use of the Feit–Thompson Theorem.

The work of Hall–Kulatilaka and Kargapolov marked the beginning of a period of heightened interest in locally finite groups during which deep results and powerful techniques from finite group theory have been used increasingly. Particularly noteworthy is the work of V. P. Šunkov, in a long series of papers, and of Kegel and Wehrfritz, whose book gives an excellent account of the subject up to 1973. We mention two high points of the theory.

THEOREM 5.6 (Kegel and Wehrfritz [48], Šunkov [85]) *A locally finite group satisfying the minimal condition on abelian subgroups is a Černikov group.*

The proof of this depends on deep results of Bender on finite groups with strongly embedded subgroups.

THEOREM 5.7 (Šunkov [86]) *A locally finite group with finite Prüfer rank has a locally soluble subgroup of finite index.*

Here it should be said that the structure of locally soluble groups with finite Prüfer rank is reasonably well understood (see [73], §10.3).

Theorem 5.7 has been strengthened by V. V. Belyaev [96]: *If G is a locally finite group with* min-p *for all primes p, then G has a locally soluble subgroup of finite index.* Note the corollary: if G is infinite, it cannot be simple.

It is reasonable to expect that simple locally finite groups will be studied more intensively using the machinery of the classification of finite simple groups. For example, a very recent result in this direction asserts that a simple locally finite *linear* group is a Chevalley group or a group of twisted type over a locally finite field [100, 111, 112, 97, 96].

Centralizers of involutions

The role of centralizers in the proof of Theorem 5.5 has been noted. In a finite simple group the centralizers of involutions exert a strong influence over the structure of the group. The best known illustration of this is the theorem of Brauer and Fowler, according to which there are only finitely many finite simple groups in which the centralizer of an involution has bounded order.

It has been found that also in infinite torsion groups the centralizers of involutions play an important role. Consider for example, the following result.

THEOREM 5.8 (Šunkov [87]) *Let G be a torsion group containing an involution whose centralizer is finite. Then G is soluble by finite, and hence locally finite.*

This has recently been refined by B. Hartley and T. Meixner: Using Šunkov's theorem and a theorem of Fong on finite groups, they prove

THEOREM 5.9 (Hartley and Meixner [40]) *Let G be a torsion group with an involution i such that $C_G(i)$ has finite order m. Then G contains a nilpotent subgroup of class at most 2 and index not exceeding some number f(m).*

By Šunkov's theorem the group G is locally finite and it is routine to reduce to the case where G is finite. The problem then becomes one for finite groups: let H be a finite soluble group and let α be an automorphism of H with order 2 such that $|C_H(\alpha)| = m$; it has to be shown that H has a nilpotent subgroup of class at most 2 with index bounded by a function of m.

6.   Embedding Theorems

The classical example of an embedding theorem is the result of Higman, Neumann and Neumann that every countable group can be embedded in a 2-generator group. However, the first embedding theorem to appear in infinite group theory is much earlier: it is

THEOREM 6.1   *An arbitrary group G can be embedded in a simple group G^* such that G^* is finite if G is finite and $|G^*| = |G|$ if G is infinite.*

If G is infinite, this may be deduced from the determination by Baer [3] in 1934 of the composition series of the infinite symmetric groups: for another proof (with the assertion about cardinality) see [73], I, p. 144.

There are many embedding theorems to be found in the papers of Philip Hall. Indeed his last published work [36] contains many far-reaching generalizations of the theorem of Higman, Neumann and Neumann. He also makes considerable progress in the problem of embedding in finitely generated simple groups. We shall examine some of these results and discuss recent developments.

Embeddings in finitely generated groups

The first embedding theorem to be discovered by Philip Hall is the fact that *every non-trivial countable abelian group can be realized as the centre of a finitely generated centre-by-metabelian group.* This was a seminal result in the theory of finitely generated soluble groups and has been invoked many times.

One application occurs in the work of G. Baumslag, F. B. Cannonito and C. F. Miller [12], who use Hall's construction to show that the class of finitely generated soluble groups with soluble word problem is not extension closed.

In [32] Hall took up the problem of embeddings in finitely generated groups in a very general setting. Shortly before, B. H. Neumann and H. Neumann had extended the original Higman–Neumann–Neumann embedding theorem by establishing

THEOREM 6.2 (Neumann and Neumann [67]) *A countable group in a variety $\mathfrak{V}$ can be embedded in a 2-generator group in the variety $\mathfrak{V}\mathfrak{A}^2$.*

This shows, for example, that every countable soluble group of derived length l embeds in a 2-generator soluble group of derived length at most $l+2$, a result that does much to explain why finitely generated soluble groups of derived length 3 are so much harder than those of length 2. In [32] Hall adapted the Neumanns' construction, making it possible to embed certain groups—actually those that are derived groups—as normal subgroups of finitely generated groups. He then put the techniques to extraordinary use.

Let $a_0, a_1, a_2, \ldots$ be a countable sequence of generators for a group H. Let C and D denote the cartesian and direct powers of $\aleph_0$ copies of H; thus

$$C = \operatorname{Cr} H^{\mathbb{Z}} \quad \text{and} \quad D = \operatorname{Dr} H^{\mathbb{Z}}.$$

Let t be the shift operator on C, so that $(c^t)_i = c_{i-1}$ where $c \in C$ and $i \in \mathbb{Z}$. Also define an element u of C by $u_{2^r} = a_r$ and $u_n = 1$ if $n \neq 2^r$ for any $r \geqslant 0$. We construct the semi-direct product $\langle t \rangle \ltimes C$, and define the subgroup $G = \langle t, u \rangle$. An easy argument reveals that if M is the normal closure of u in G, then $M' = D' = \operatorname{Dr}(H')^{\mathbb{Z}}$. This fact has many applications, of which the following are a sample.

(a) The theorem of Neumann and Neumann quoted above may be deduced as follows. Let K be a countable group in a variety $\mathfrak{V}$. Then it suffices to embed K in the derived subgroup of a countable group H in the variety $\mathfrak{V}\mathfrak{A}$. For, upon applying the construction, we shall have $K \leqslant H' \leqslant G$ and $G \in \mathfrak{V}\mathfrak{A}^2$ since $C \in \mathfrak{V}\mathfrak{A}$.

But such an embedding is easily achieved. For each x in K we define elements $x^{(0)}$ and $x^{(1)}$ in $\operatorname{Cr} K^{\mathbb{Z}}$ by the rules

$$(x^{(0)})_i = x \quad \text{and} \quad (x^{(1)})_i = x^i.$$

Then, with t the shift operator as before, we have $(x^{(0)})^t = x^{(0)}$ and $[x^{(1)}, t] = (x^{(0)})^{-1}$, which shows that t normalizes the subgroup $\langle x^{(0)}, x^{(1)} \rangle$, so we may form the semi-direct product

$$H = \langle t \rangle \ltimes \langle x^{(0)}, x^{(1)} \rangle.$$

Finally the mapping $x \mapsto x^{(0)}$ is an embedding of K into H'.

(b) Hall's original purpose in introducing the construction was to give examples of finitely generated soluble groups of derived length 3 with non-nilpotent Frattini subgroups (see [32]).

To make this application one chooses primes p and q such that $q \equiv 1 \pmod{p^2}$, so there is a primitive q-adic p^2th root of unity ω. Let N be a group of type q^∞; then the mapping $x \mapsto x^\omega$ is an automorphism α of N. Consider the semi-direct product

$$H = \langle \alpha \rangle \ltimes N.$$

We apply the construction to the group H, obtaining a finitely generated group $G = \langle t, u \rangle$ which is easily seen to be soluble with derived length 3. It is also easy to prove that the Frattini subgroup of G contains the subgroup $\langle \alpha^p, N \rangle$, which is not even locally nilpotent.

(c) Somewhat later R. S. Dark [15] used the same technique to prove that *every countable group is isomorphic with a subnormal subgroup of a 2-generator group.*

To prove this one first takes a countable group $K = \langle a_1, a_2, \ldots \rangle$ and embeds K as the direct factor K_0 of $C = \operatorname{Cr} K^{\mathbb{Z}}$. Let t be the shift operator on C and define elements $b(i)$ in C by the rule

$$(b(i))_n = \begin{cases} a_i & \text{if } n > 0 \\ 1 & \text{if } n \leqslant 0. \end{cases}$$

Let H be the subgroup $\langle t, b(i) \mid i = 1, 2, \ldots \rangle$ of $\langle t \rangle \ltimes C$. Then $[b(i), t^{-1}]$ has its 0-component equal to a_i and all other components trivial. Hence $K = K_0 \leqslant H'$. Of course K is subnormal in $\langle t \rangle \ltimes C$ and so, *a fortiori*, in H'. Dark's theorem now follows on applying the Hall construction to the group H.

Yet another generalization of the Higman–Neumann–Neumann Theorem was discovered by F. Levin [53]. He showed that *every countable group embeds in a 2-generator group in which the generators have arbitrary orders m and n, where $m \leqslant n$, provided only that $m > 1$ and $n > 2$,* conditions which are clearly indispensible. Later C. F. Miller and P. E. Schupp [61] proved that the 2-generator group can be made complete and hopfian.

Results such as these suggest the general problem of embedding in a join of given groups. This is the subject of Hall's paper [36]. In it the following far-reaching generalization of the theorems of Dark and Levin is established.

THEOREM 6.3 *Let* H, K, L *be groups such that* $|H| > 1$, $|K| > 2$ *and* $|L| \leqslant |H * K|$. *Then* L *is isomorphic with a subnormal subgroup of defect at most* 2 *in a group* $J = \langle H, K \rangle$, *generated by copies of* H *and* K.

Here again it is clear that the conditions on H, K, L are indispensable if $|H| \leqslant |K|$.

The proof of the theorem yields extra information about the case where H, K, L are soluble groups. If the derived lengths of H, K, L are h, k, l respectively, then in fact J can be chosen to have derived length at most $h + k + l + 1$. By taking L to be countable and H and K infinite cyclic we obtain the

COROLLARY *A countable soluble group of derived length* l *can be subnormally embedded with defect at most* 2 *in a* 2-*generator soluble group of derived length at most* $l + 3$.

This would lead one to suspect that the subnormal structure of finitely generated soluble groups of derived length 4 is much more complicated than that of groups of derived length 3. The suspicion is partly confirmed by the fact that finitely generated soluble groups of derived length 3 have the subnormal join property, whereas those of derived length 4 do not in general (see Robinson [72]).

The proof of Theorem 6.3 is easier if H or K is large. Suppose that in fact $|H| \geqslant |K * L|$ with $|K| > 2$. Define

$$Q = K_1 * K_2 / [K_1, K_2]'$$

where $K_i \cong K$. Now let

$$W = (L \wr Q) \wr H,$$

where $\wr$ denotes the unrestricted wreath product. Hall shows that there is a subgroup K^* of the base group of the outer wreath product in W such that $K \cong K^*$ and $L \leqslant [H, K^*]$. Now L is 2-step subnormal in W and hence also in $\langle H, K^* \rangle$.

Embeddings in simple groups

In the same paper [36] Hall also studied embeddings of groups in simple joins of given groups. For example, he proved that if K_1, K_2, K_3 are non-trivial groups and L is a group satisfying $|L| \leqslant |K_1 * K_2 * K_3|$, then L may be embedded in a simple group of the form $\langle K_1, K_2, K_3 \rangle$ unless $|K_1| = |K_2| = |K_3| = 2$ or two of the K_i have order 2 and the third is uncountable. This implies (on taking $|K_1| = |K_2| = 2$ and $|K_3| = 3$) that *every countable group can be embedded in a 3-generator simple group.*

This last result was improved by A. P. Goryuškin [21] who proved that *every countable group can be embedded in a 2-generator simple group.* Goryuškin's argument is amazingly short and involves an ingenious sequence of free products with amalgamation. We give an outline of the steps of his proof.

In the first place it is sufficient to perform the embedding for a countable *simple* group, say G_1; this is because every countable group embeds in a countable simple group. Goryuškin embeds G_1 in a group $G_2 = \langle a, b \rangle$ where a and b have infinite orders, $\langle a \rangle \cap \langle b \rangle = 1$ and G_1 is generated by certain conjugates of a. Then he embeds G_2 in a 2-generator group G_3 such that the normal closure of a in G_3 coincides with G_3. Now consider a normal subgroup M of G_3 which is maximal subject to $M \cap G_1 = 1$. If $M < N \triangleleft G_3$, then $N \cap G_1 \neq 1$ and so $G_1 \leqslant N$ by the simplicity of G_1. Hence $a \in N$ and therefore $N = G_3$. Thus M is actually a maximal normal subgroup of G_3. Therefore $G = G_3/M$ is a 2-generator simple group with a subgroup MG_1/M which is isomorphic with G_1.

Shortly after the appearance of Goryuškin's paper, Schupp [82], in an impressive demonstration of the power of small cancellation theory, established the following theorem, which seems close to the best possible.

THEOREM 6.4   *Let H and K be groups such that $|H| > 1$ and $|K| > 2$. Let L be a group satisfying $|L| \leqslant |H * K|$. Then L can be embedded in a simple group of the form $\langle H, K \rangle$.*

Hall's techniques have been used by Hickin and Phillips [103] for embeddings of periodic groups in 2-generator periodic groups, a situation where small cancellation theory is not appropriate. However, Hall's paper [36] is not as widely known as it deserves to be. It contains an immense variety of material of independent interest.

Embeddings in finitely presented groups
In conclusion, we consider the problem of embedding countable groups in finitely presented groups. This is a subject of much interest at present,[‡] partly because of the connection with the word problem. The basic result is the following celebrated theorem of Higman.

‡ cf. Strebel's essay.

THEOREM 6.5 (Higman [43]) *A countable group with a recursive presentation can be embedded in a finitely presented group.*

We recall that a presentation of a group is said to be *recursive* if it has a recursively enumerable set of generators and a recursively enumerable set of relators in these generators.

It is very easy to show that a finitely generated subgroup of a finitely presented group must have a recursive presentation. We obtain therefore the following characterization of finitely generated subgroups of finitely presented groups.

THEOREM 6.6 (Higman [43]) *A finitely generated group can be embedded in a finitely presented group if and only if it has a recursive presentation.*

No such characterization exists for the infinitely generated subgroups of a finitely presented group. However, numerous classes of groups which can be embedded in a finitely presented group are known; we mention some examples.

THEOREM 6.7 (Baumslag, Cannonito and Miller [12])

(i) *Every countable locally free group can be embedded in a finitely presented group.*

(ii) *Every countable locally polycyclic-by-finite group can be embedded in a finitely presented group.*

Special cases of (ii) are that every countable abelian group and every countable locally finite group may be embedded in a finitely presented group; these results had already been noticed by Higman. Since there are countable abelian groups which have no recursive presentations, it follows that to have a recursive presentation is not a prerequisite for embeddability in a finitely presented group.

Higman's theorem gives another proof of the insolubility of the word problem for finitely presented groups. For there are finitely generated recursively presented groups with insoluble word problem: a finitely presented group containing such a group as a subgroup cannot have soluble word problem.

A further connection with the word problem is indicated by the following remarkable theorem.

THEOREM 6.8 (W. W. Boone and G. Higman [14]) *A finitely generated group G has soluble word problem if and only if there are embeddings of G in a simple group G_1 and of G_1 in a finitely presented group G_2.*

In particular, every finitely presented simple group has soluble word problem. From this it is clear that there are finitely presented groups which cannot be embedded in any finitely presented simple group. This is in contrast to the results of Hall, Goryuškin and Schupp above.

A question of current interest is a relative version of the embedding problem solved by Higman's Theorem (6.6): Given a class of groups $\mathfrak{X}$, which finitely generated recursively presented $\mathfrak{X}$-groups can be embedded in a finitely presented $\mathfrak{X}$-group? For example, Baumslag established

THEOREM 6.9 (G. Baumslag [10]) *Every finitely generated metabelian group can be embedded in a finitely presented metabelian group.*

Here it should be noted that finitely generated metabelian groups have soluble word problem and so possess recursive presentations.
However, the situation is less straightforward for the variety of centre-by-metabelian groups.

THEOREM 6.10 (J. R. G. Groves [23], R. Strebel [84]) *A finitely generated centre-by-metabelian group can be embedded in a finitely presented centre-by-metabelian group if and only if it is abelian-by-polycyclic.*

The necessity of the condition is due to Groves, while the sufficiency is a recent result of Strebel. Another very interesting result in this direction is

THEOREM 6.11 (M. W. Thomson [88]) *A finitely generated soluble linear group over a field can be embedded in a finitely presented soluble linear group over the same field.*

This has the corollary that every countable free nilpotent group can be embedded in a finitely presented soluble linear group.
A major unsolved problem is to determine which finitely generated recursively presented soluble groups can be embedded in a finitely presented soluble group. The same question for finitely generated soluble groups of finite (Prüfer) rank is also open; however, there is a partial result in this case.

THEOREM 6.12 (G. Baumslag and R. Bieri [11]) *Every finitely generated residually finite soluble group of finite rank can be embedded in a constructible soluble group.*

Here a soluble group is said to be *constructible* if it can be built up from the trivial group by a finite number of *HNN*-extensions and finite extensions. Constructible soluble groups are finitely presented, residually finite and of finite rank. D. Kilsch has generalized Theorem 6.12 by proving

THEOREM 6.13 (Kilsch [50])   *The groups which can be embedded in a constructible soluble group are precisely the residually finite, soluble minimax groups.*

It should be observed that a finitely generated soluble group of finite rank is a minimax group ([73], II, p. 176).

We conclude with a result of G. P. Kukin:

THEOREM 6.14 (Kukin [51])   *A finitely generated soluble group of derived length l, with a recursive presentation, can be embedded in a group which is finitely presented in the variety* $\mathfrak{U}^{l+2}$.

Here a group G is said to be *finitely presented in a variety* $\mathfrak{V}$ if it can be generated by a finite set of elements subject to a finite set of relations in addition to the identical relations of $\mathfrak{V}$.

References

[1] Alperin, J., Groups with finitely many automorphisms, *Pacific J. Math.* **12** (1962), 1–5.

[2] Ayoub, C., On properties possessed by solvable and nilpotent groups, *J. Austral. Math. Soc.* **9** (1969), 218–227.

[3] Baer, R., Die Kompositionsreihe der Gruppe aller eineindeutigen Abbildungen einer unendliche Menge, *Studia Math.* **5** (1934), 15–17.

[4] Baer, R., Nilpotent groups and their generalizations, *Trans. Amer. Math. Soc.* **47** (1940), 393–434.

[5] Baer, R., Finiteness properties of groups, *Duke Math. J.* **15** (1948), 1021–1032.

[6] Baer, R., Endlichkeitskriterien für Kommutatorgruppen, *Math. Ann.* **124** (1952), 161–177.

[7] Baer, R., Nilgruppen, *Math. Z.* **62** (1955), 402–437.

[8] Baer, R., Finite extensions of abelian groups with the minimum condition, *Trans. Amer. Math. Soc.* **79** (1955), 521–540.

[9] Baer, R., Heineken, H., Radical groups with finite abelian subgroup rank, *Illinois J. Math.* **16** (1972), 533–580.

[10] Baumslag, G., Subgroups of finitely presently metabelian groups, *J. Austral. Math. Soc.* **14** (1973), 98–110.

[11] Baumslag, G., Bieri, R., Constructable soluble groups, *Math. Z.* **151** (1976), 249–257.

[12] Baumslag, G., Cannonito, F. B., Miller, C. F. III., Infinitely generated subgroups of finitely presented groups I, *Math. Z.* **153** (1977), 117–134.

[13] Baumslag, G., Kovács, L. G., Neumann, B. H., On products of normal subgroups, *Acta Sci. Math. (Szeged.)* **26** (1965), 145–147.

[14] Boone, W. W., Higman, G., An algebraic characterization of groups with soluble word problem, *J. Austral. Math. Soc.* **18** (1974), 41–53.

[15] Dark, R. S., On subnormal embedding theorems for groups, *J. London Math. Soc.* **43** (1968), 387–390.

[16] Durbin, J. R., Residually central elements in groups, *J. Algebra* **9** (1968), 408–413.

[17] Golod, E. S., On nil-algebras and residually finite p-groups, *Izv. Akad. Nauk SSSR Ser. Mat.* **28** (1964), 273–276.

[18] Gorčakov, Yu. M., On locally normal groups, *Mat. Sb.* **67** (1965), 244–254.

[19] Gorčakov, Yu. M., Locally normal groups, *Sibirsk. Mat. Ž.* **6** (1971), 1259–1272.

[20] Gorčakov, Yu. M., Subgroups of direct products, *Algebra i Logika* **15** (1976), 622–627.

[21] Goryuškin, A. P., Embedding countable groups in two generator simple groups, *Mat. Zametki* **16** (1974), 231–235.

[22] Grigorčuk, R. I., On Burnside's problem for periodic groups, *Funkcional. Anal. i Prilozen* **14** (1980), 53–54.

[23] Groves, J. R. G., Finitely presented centre-by-metabelian groups, *J. London Math. Soc.* (2) **18** (1978), 65–69.

[24] Gruenberg, K. W., The Engel elements of a soluble group, *Illinois J. Math.* **3** (1959), 151–168.

[25] Gupta, N. D., Sidki, S., Some infinite p-groups, *Algebra i Logika* **22** (1983), 584–589.

[26] Hall, P., Finiteness conditions for soluble groups, *Proc. London Math. Soc.* (3) **4** (1954), 419–436.

[27] Hall, P., Finite-by-nilpotent groups, *Proc. Cambridge Phil. Soc.* **52** (1956), 611–616.

[28] Hall, P., Nilpotent groups. (Lectures given at the Canadian Math. Congress, Summer Seminar, Univ. of Alberta, 1957.) *Queen Mary College Math. Notes* (1969).

[29] Hall, P., Some sufficient conditions for a group to be nilpotent, *Illinois J. Math.* **2** (1958), 787–801.

[30] Hall, P., Periodic *FC*-groups, *J. London Math. Soc.* **34** (1959), 289–304.

[31] Hall, P., Some constructions for locally finite groups, *J. London Math. Soc.* **34** (1959), 305–319.

[32] Hall, P., The Frattini subgroup of finitely generated groups, *Proc. London Math. Soc.* (3) **11** (1961), 327–352.

[33] Hall, P., Wreath powers and characteristically simple groups, *Proc. Cambridge Phil. Soc.* **58** (1962), 170–184.

[34] Hall, P., On non-strictly simple groups, *Proc. Cambridge Phil. Soc.* **59** (1963), 531–553.

[35] Hall, P., A note on $\overline{SI}$-groups, *J. London Math. Soc.* **39** (1964), 338–344.

[36] Hall, P., On the embedding of a group in a join of given groups, *J. Austral. Math. Soc.* **17** (1974), 434–495.

[37] Hall, P., Hartley, B., The stability group of a series of subgroups, *Proc. London Math. Soc.* (3) **16** (1966), 1–39.

[38] Hall, P., Kulatilaka, C. R., A property of locally finite groups, *J. London Math. Soc.* **39** (1964), 235–239.

[39] Hallett, J. T., Hirsch, K. A., Torsion-free groups having finite automorphism groups I, *J. Algebra* **2** (1965), 287–298.

[40] Hartley, B., Meixner, T., Periodic groups in which the centralizer of an involution has bounded order, *J. Algebra* **64** (1980), 285–291.

[41] Heineken, H., Mohamed, I. J., A group with trivial centre satisfying the normalizer condition, *J. Algebra* **10** (1968), 368–376.

[42] Held, D., On bounded Engel elements in groups, *J. Algebra* **3** (1966), 360–365.

[43] Higman, G., Subgroups of finitely presented groups, *Proc. Roy. Soc. Ser. A* **262** (1961), 455–475.

[44] Higman, G., Neumann, B. H., Neumann, H., Embedding theorems for groups, *J. London Math. Soc.* **24** (1949), 247–254.

[45] Kalužnin, L. A., Über gewisse Beziehungen zwischen einer Gruppe und ihren Automorphismen, *Berliner Math. Tagung* (1953), 164–182.

[46] Kargapolov, M. I., Some problems in the theory of nilpotent and soluble groups, *Dokl. Akad. Nauk SSSR* **127** (1959), 1164–1166.

[47] Kargapolov, M. I., On a problem of O. J. Schmidt, *Sibirsk. Math. Ž.* **4** (1963), 232–235.

[48] Kegel, O. H., Wehrfritz, B. A. F., Strong finiteness conditions in locally finite groups, *Math. Z.* **117** (1970), 309–324.

[49] Kegel, O. H., Wehrfritz, B. A. F., 'Locally Finite Groups', North-Holland, Amsterdam (1973).

[50] Kilsch, D., On minimax groups which are embedded in constructible groups, *J. London Math. Soc.* (2) **18** (1978), 472–474.

[51] Kukin, G. P., On the embedding of recursively presented algebras and groups, *Dokl. Akad. Nauk SSSR* **251** (1980), 37–39.

[52] Kuroš, A. G., Černikov, S. N., Soluble and nilpotent groups, *Uspehi Mat. Nauk* **2** (1947), 18–59; *Amer. Math. Soc. Translations* (1) **80** (1953).

[53] Levin, F., Factor groups of the modular groups, *J. London Math. Soc.* **43** (1968), 195–203.

[54] Macintyre, A., Shelah, S., Uncountable universal locally finite groups, *J. Algebra* **43** (1976), 168–175.

[55] Mal'cev, A. I., On a general method for obtaining local theorems in group theory, *Ivanov. Gos. Ped. Inst. Učen. Zap.* **1** (1941), 3–9.

[56] Mal'cev, A. I., On certain classes of infinite soluble groups, *Mat. Sb.* **28** (1951), 567–588.

[57] Martin, J. E., Pamphilon, J. A., Engel elements in groups with the minimal condition, *J. London Math. Soc.* (2) **6** (1973), 281–285.

[58] McLain, D. H., A characteristically-simple group, *Proc. Cambridge Phil. Soc.* **50** (1954), 641–642.

[59] McLain, D. H., On locally nilpotent groups, *Proc. Cambridge Phil. Soc.* **52** (1956), 5–11.

[60] Merzljakov, Ju. I., On the theory of generalized soluble and nilpotent groups, *Algebra i Logika* **2** (1963), 29–36.

[61] Miller, C. F. III, Schupp, P. E., Embeddings into Hopfian groups, *J. Algebra* **17** (1971), 171–176.

[62] Nagrebeckiĭ, V. I., On the periodic part of a group with a finite number of automorphisms, *Dokl. Akad. Nauk SSSR* **205** (1972), 519–521.

[63] Neumann, B. H., Some remarks on infinite groups, *J. London Math. Soc.* **12** (1937), 120–127.

[64] Neumann, B. H., Groups with finite classes of conjugate elements, *Proc. London Math. Soc.* (3) **1** (1951), 178–187.

[65] Neumann, B. H., Groups covered by permutable subsets, *J. London Math. Soc.* **29** (1954), 236–248.

[66] Neumann, B. H., Groups with finite classes of conjugate subgroups, *Math. Z.* **63** (1955), 76–96.

[67] Neumann, B. H., Neumann, H., Embedding theorems for groups, *J. London Math. Soc.* **34** (1959), 465–479.

[68] Novikov, P. S., Adjan, S. I., Commutative subgroups and the conjugacy problem in free periodic groups of odd order, *Izv. Akad. Nauk. SSSR Ser. Mat.* **32** (1968), 1176–1190.

[69] Ol'šankiĭ, A. Ju., An infinite simple torsion-free noetherian group, *Izv. Akad. Nauk. SSSR Ser. Mat.* **43** (1979), 1328–1393.

[70] Peng, T. A., Engel elements of groups with maximal condition on abelian subgroups, *Nanta Math.* **1** (1966), 23–28.

[71] Phillips, R. E., Roseblade, J. E., A residually central group that is not a Z-group, *Michigan Math. J.* **25** (1978), 233–234.

[72] Robinson, D. J. S., Joins of subnormal subgroups, *Illinois J. Math.* **9** (1965), 144–168.

[73] Robinson, D. J. S., 'Finiteness Conditions and Generalized Soluble Groups', Springer, Berlin (1972).

[74] Robinson, D. J. S., A new treatment of soluble groups with finiteness conditions on their abelian subgroups, *Bull. London Math. Soc.* **8** (1976), 113–129.

[75] Robinson, D. J. S., A contribution to the theory of groups with finitely many automorphisms, *Proc. London Math. Soc.* (3) **35** (1977), 34–54.

[76] Robinson, D. J. S., Infinite torsion groups as automorphism groups, *Quart. J. Math.* (2) **30** (1979), 351–364.

[77] Robinson, D. J. S., Applications of cohomology to the theory of groups, in 'Groups—St. Andrews 1981', *London Math. Soc. Lecture Notes* **71** (1982), 46–80.

[78] Robinson, D. J. S., Wiegold, J., Groups with boundedly finite automorphism classes, *Rend. Sem. Mat. Univ. Padova* **71** (1984) (to appear).

[79] Roseblade, J. E., On groups in which every subgroup is subnormal, *J. Algebra* **2** (1965), 402–412.

[80] Schmidt, O. J., 'Abstract Theory of Groups', Kiev (1916); 2nd Ed. Moscow (1933).

[81] Schmidt, O. J., Infinite soluble groups, *Mat. Sb.* **17** (1945), 145–162.

[82] Schupp, P. E., Embeddings into simple groups, *J. London Math. Soc.* (2) **13** (1976), 90–94.

[83] Schur, I., Neuer Beweis eines Satzes über endliche Gruppen, *Sitzber. Akad. Wiss. Berlin* (1902), 1013–1019.

[84] Strebel, R., Subgroups of finitely presented centre-by-metabelian groups, *J. London Math. Soc.* (2) **28** (1983), 481–491.

[85] Šunkov, V. P., On locally finite groups with the minimal condition for abelian subgroups, *Algebra i Logika* **9** (1970), 579–615.

[86] Šunkov, V. P., Locally finite groups of finite rank, *Algebra i Logika* **10** (1971), 199–225.

[87] Šunkov, V. P., On periodic groups with an almost regular involution, *Algebra i Logika* **11** (1972), 470–493.

[88] Thomson, M. W., Subgroups of finitely presented solvable linear groups, *Trans. Amer. Math. Soc.* **231** (1977), 133–142.

[89] Tomkinson, M. J., A characterization of residually finite periodic FC-groups, *Bull. London Math. Soc.* **13** (1981), 133–137.

[90] Turner-Smith, R. F., Marginal subgroup properties for outer commutator words, *Proc. London Math. Soc.* (3) **14** (1964), 321–341.

[91] Wilson, J., On outer-commutator words, *Canad. J. Math.* **26** (1974), 608–620.

[92] Wilson, J. S., An $\overline{SI}$-group which is not an $\overline{SN}$-group, *Bull. London Math. Soc.* **5** (1973), 192–196.

[93] Wilson, J. S., On normal subgroups of $\overline{SI}$-groups, *Arch. Math.* **25** (1974), 574–577.

[94] Wilson, J. S., $\overline{SN}$-groups with non-abelian free subgroups, *J. London Math. Soc.* (2) **7** (1974), 699–708.

[95] Wilson, J. S., On periodic generalized nilpotent groups, *Bull. London Math. Soc.* **9** (1977), 81–85.

Additional references

[96] Belyaev, V. V., Locally finite groups with Černikov Sylow *p*-subgroups, *Algebra i Logika* **20** (1981), 605–619.

[97] Borovik, A. V., Periodic linear groups of odd characteristic, *Dokl. Akad. Nauk SSSR* **266** (1982), 1289–1291.

[98] Brookes, C. J. B., Groups with every subgroup subnormal, *Bull. London Math. Soc.* **15** (1983), 235–238.

[99] Griess, Jr., R. L., Schur multipliers of the known finite simple groups II, *Proc. Sympos. Pure Math.* **37** (1980), 279–282.

[100] Hartley, B., Shute, G., Monomorphisms and direct limits of finite groups of Lie type, *Quart. J. Math.* (to appear).

[101] Heineken, H., A class of three-Engel groups, *J. Algebra* **17** (1971), 341–345.

[102] Hickin, K. K., Complete universal locally finite groups, *Trans. Amer. Math. Soc.* **239** (1978), 213–227.

[103] Hickin, K. K., Phillips, R. E., Joins of periodic groups, *Proc. London Math. Soc.* **24** (1979), 176–192.

[104] Hirschfeld, J., Wheeler, W. H., 'Forcing, Arithmetic and Division rings', *Springer Lecture Notes in Math.* **454** (1975).

[105] Linnell, P. A., Zero divisors and idempotents in group rings, *Math. Proc. Cambridge Phil. Soc.* **81** (1977), 365–368.

[106] Macintyre, A., Existentially closed structures and Jensen's principle, *Israel J. Math.* **25** (1976), 202–210.

[107] Mahdavianary, S. K., A special class of three-Engel groups, *Arch. Math.* **40** (1983), 193–199.

[108] Robinson, D. J. S., 'A Course in the Theory of Groups', Springer-Verlag, New York (1982).

[109] Serežkin, V. N., Zaleskii, A. E., Linear groups generated by pseudo-reflections, *Vesci Akad. Nauk BSSR Ser. Fiz.-Mat. Navuk* (1977), no. 5, 9–16, **137**.

[110] Smith, H., Hypercentral groups with all subgroups subnormal, *Bull. London Math. Soc.* **15** (1983), 229–234.

[111] Thomas, S., An identification theorem for the locally finite non-twisted Chevalley groups, *Arch. Math.* **40** (1983), 21–31.

[112] Thomas, S., The classification of the simple periodic linear groups, *Arch. Math.* **41** (1983), 103–116.

[113] Thomas, S., Complete universal locally finite groups of large cardinality, preprint (1983).

[114] Zimmerman, J., The occurrence of certain types of groups as automorphism groups, Ph.D. Dissertation, Univ. of Illinois (1983).

Group Rings of Polycyclic Groups

DONALD S. PASSMAN

1. Introduction 207
2. Groups with max-n 208
3. Primitive group rings 209
4. Residually finite groups 215
5. The Nullstellensatz 218
6. The Artin–Rees property 223
7. Nilpotence of the Frattini subgroup 226
8. Zalesskii-type subgroups 228
9. Orbitally sound groups 234
10. Orbital primes in abelian group rings 239
11. Prime ideals in polycyclic group rings 243
12. Irreducible modules 245
13. Concluding remarks 248
References 252

1. Introduction

Between 1954 and 1961, Philip Hall wrote a series of three papers on finitely generated solvable groups. The first paper [35] discussed groups with max-n, the maximal condition on normal subgroups, the second [36] studied residual finiteness and the third paper [37] considered the nilpotency of the Frattini subgroup. As it turned out, the proofs of these diverse results hinged upon certain properties of the integral group ring $\mathbb{Z}G$ of a polycyclic group G. Indeed these three group theory papers initiated the ring-theoretic study of group rings of polycyclic groups.

In this essay we discuss group rings of polycyclic-by-finite and certain related groups. Since this subject has been growing steadily for over 25 years, we can no longer hope to be all-inclusive. The topics considered in sections 2–7 stem most directly from the work of Philip Hall, while those of sections 8–12 depend on the work of J. E. Roseblade

GROUP THEORY: essays for Philip Hall
ISBN 0-12-304880-X

in [74]. In particular, these include irreducible modules, primitive and prime ideals and the Artin–Rees property. Furthermore, we emphasize material not contained in the standard references on group rings, namely the books [64] by I. B. S. Passi, [66, 68] by D. S. Passman and [80] by S. K. Sehgal in English and the books [5] by A. A. Bovdi and [59] by A. V. Mihalev and A. E. Zalesskii in Russian. We recommend to the reader the alternate survey by D. R. Farkas [100].

The role of polycyclic group rings in the subject of general group rings is analogous to the role played by polynomial rings within general ring theory. These rings are ever-present, they are concrete and understandable and therefore the results obtained for them are far stronger and more detailed than can ever be achieved in the general case. In addition, there are several very good reasons for their study. First, there are the group-theoretic applications. Second, they afford interesting examples for the Goldie theory [34] of noncommutative Noetherian rings. Finally, there seems to be a close analogy between these group rings and the universal enveloping algebras of finite-dimensional Lie algebras (see J. Dixmier [14]).

2. Groups with max-n

Let RG denote the group ring of G over the ring R. Usually R is either a field or the ring $\mathbb{Z}$ of integers and G is a polycyclic-by-finite group. We use $\mathfrak{P}$ to denote the class of all polycyclic groups and $\mathfrak{F}$ the class of all finite groups. Recall that R is a (right) Noetherian ring if it satisfies the ascending chain condition or maximal condition on its right ideals. The following result is crucial since it allows the entry of the rich theory of Noetherian rings into this subject.

THEOREM 2.1 (Hall [35]) *If R is Noetherian and G is in $\mathfrak{P}\mathfrak{F}$, then the group ring RG is Noetherian.*

Proof. By definition, G has a chain $1 = G_0 \lhd G_1 \lhd \ldots \lhd G_n = G$ with each quotient G_i/G_{i-1} either infinite cyclic or a finite group. By induction, we may assume $S = RG_{n-1}$ is Noetherian. If G_n/G_{n-1} is infinite cyclic, then RG is close to being a skew polynomial ring over S. More precisely, if $G = \langle G_{n-1}, x \rangle$, then RG is a free left and right S-module with basis $\{x^{\pm k} \mid k = 0, 1, 2, \ldots\}$ and with $x^{-1}Sx = S$. The argument of the Hilbert basis theorem now implies that RG is Noetherian. On the other hand, if G/G_{n-1} is finite, then RG is a finitely generated S-module and we obtain the same conclusion.

Even this beginning result has numerous group-theoretic consequences. A group G has max-n if its normal subgroups satisfy the
maximal condition. Certainly $\mathfrak{PF}$-groups satisfy max-n, since they
satisfy the stronger property max, namely the maximal condition on all
subgroups. Other groups which will be of interest to us are the
abelian-by-(polycyclic-by-finite) groups. We shall write $\mathfrak{APF}$ for this
class of groups.

THEOREM 2.2 (Hall [35])   *A finitely generated $\mathfrak{APF}$-group satisfies*
max-n.

Proof.   Let H be a finitely generated group with A a normal abelian
subgroup and with $G = H/A$ in $\mathfrak{PF}$. Then H acts on A by conjugation
with A acting trivially, so A is in fact a G-module. More properly, A
viewed additively is a module for the Noetherian ring $\mathbb{Z}G$. Since H is
finitely generated and G is finitely presented, we conclude that A is a
finitely generated $\mathbb{Z}G$-module and hence A satisfies the maximal condition on $\mathbb{Z}G$-submodules. Since these submodules are precisely the
normal subgroups of H contained in A, it follows that H satisfies
max-n.

Another example of interest comes from wreath products. Suppose H
and G are groups and form $W = \Pi_{g \in G} H_g$, the weak (restricted) direct
product of copies of H indexed by the elements of G. Then G acts on W
by permuting its factors and the wreath product $H \wr G$ is the semidirect
product of W by G. The following is a consequence of Theorem 2.1; the
result after it has a similar, but simpler, proof.

THEOREM 2.3 (Hall [35])   *Let H satisfy* max-n *and let G be in* $\mathfrak{PF}$. *Then*
the wreath product $H \wr G$ satisfies max-n.

THEOREM 2.4 (Hall [35])   *Let H and G satisfy* max-n *and assume that H*
has no nontrivial central factor. Then the wreath product $H \wr G$ satisfies
max-n.

3.   Primitive Group Rings

In the previous section we obtained sufficient conditions for the
group ring RG to be Noetherian. It is natural to ask whether these are
also necessary. More generally, we wish to determine when the group
ring RG satisfies any of a number of ring-theoretic conditions. Problems

of this sort have been extensively and perhaps excessively studied. Nevertheless, certain properties are clearly both interesting and important. We start by considering prime and semiprime group algebras.

A ring R is *prime* if for any two nonzero ideals I, J of R we have $IJ \neq 0$. A ring R is *semiprime* if it is a subdirect product of prime rings (i.e., if it is residually prime). It can be shown that R is semiprime if and only if for all nonzero ideals I of R we have $I^2 \neq 0$.

For G an arbitrary group, the (finite conjugate) *FC-center* of G, written $\Delta(G)$, is the set of those elements having only finitely many conjugates. Equivalently

$$\Delta(G) = \{x \in G \mid |G : C_G(x)| < \infty\}.$$

Then $\Delta(G)$ is a characteristic subgroup of G, the set $\Delta^+(G)$ of elements of $\Delta(G)$ of finite order is generated by all finite normal subgroups of G and $\Delta(G)/\Delta^+(G)$ is torsion-free abelian.

For H a subgroup of G, there is a natural projection map $\pi_H : RG \to RH$ given by

$$\pi_H(\Sigma_{x \in G} \, r_x x) = \Sigma_{x \in H} \, r_x x.$$

If I is a nonzero ideal of RG, then $\pi_H(I)$ is a nonzero ideal of RH. We reserve the symbol θ for the projection to RH when $H = \Delta(G)$.

While many of the results of this section hold for more general coefficient rings, we will assume for the most part that $R = K$ is a field. The following is proved using a coset counting argument known as *the Δ-method*.

LEMMA 3.1   *Let I, J be ideals of KG with $IJ = 0$. Then $\theta(I) \cdot \theta(J) = 0$.*

This allows questions of primeness and semiprimeness to be reduced to the subgroup $\Delta(G)$ where the problem is easily solved. Thus we have

THEOREM 3.2   *If char $K = 0$, then KG is semiprime.*

THEOREM 3.3 (Passman [65])   *If char $K = p > 0$, then KG is semiprime if and only if G has no finite normal subgroup of order divisible by p.*

THEOREM 3.4 (I. G. Connell [13])   *KG is prime if and only if G has no nontrivial finite normal subgroup.*

As a consequence we have

COROLLARY 3.5 [13]   *KG is Artinian if and only if G is finite.*

Proof. If G is finite, then KG is Artinian. Conversely, suppose KG is Artinian. We proceed by induction on the composition length n of KG. If KG is not prime, then G has a nontrivial finite normal subgroup N and the natural epimorphism $KG \to K[G/N]$ has a non-zero kernel. Thus $K[G/N]$ is Artinian of smaller composition length so G/N is finite and hence so is G. On the other hand, if KG is prime (this includes the possibility $n = 1$), then the Wedderburn theorems imply that KG is simple. Since there is a natural epimorphism $KG \to K$, we deduce that $G = 1$.

Even though the Artinian problem is solved, the characterization of Noetherian group rings has so far eluded solution. We do know that KG Noetherian implies that G satisfies max. In particular, if G is solvable then it is polycyclic. In general we may assume that $\Delta(G) = 1$. Then KG is a prime Noetherian ring and hence has a classical ring of quotients which is a full matrix ring over a division ring. But many non-Noetherian group rings also have this property. Furthermore, if K is algebraically closed of characteristic 0, then the periodic elements of G have bounded period. However, the existence of Tarski monsters and of other counterexamples to the Burnside problem seem to preclude a group-theoretic approach here.

Another problem of interest is to determine when KG is primitive, that is has a faithful irreducible module. For this to occur it is necessary that KG is semiprimitive, that is its Jacobson radical $J(KG)$ is zero, and that KG is prime so that $\Delta^+(G) = 1$. The first nontrivial examples of primitive group rings were given by E. Formanek and R. L. Snider [32]. In the following result, part (i) is a sharpened version of the above found in Fisher and Snider [28], part (ii) is contained in Formanek [30] and part (iii) occurs in Passman [67].

THEOREM 3.6   *Let KG be a group algebra.*

(i) *If G is a countable locally finite group, then KG is primitive if and only if it is prime and semiprimitive.*

(ii) *If $G = H_1 * H_2$ with $|H_1| > 1$ and $|H_2| > 2$, then KG is primitive.*

(iii) *If G is a torsion-free solvable group with $\Delta(G) = 1$, then there exists a field F containing K with FG primitive.*

At the end of this section we will see that (i) above does not hold for nondenumerable locally finite groups. A consequence of (ii) is that for any group H there exists $G \supseteq H$ with KG primitive for all K. Finally the hypothesis that $\Delta(G) = 1$ in part (iii) is necessary since it was also shown in [67] that if KG is primitive, for any group G, and if $|K| > |G|$, an equation of cardinal numbers, then $\Delta(G) = 1$. The first example of a primitive group ring with $\Delta(G) \neq 1$ was constructed by Formanek [30]. We offer another example based on work of R. S. Irving [43].

LEMMA 3.7 *Let $G = C \wr C$ with C infinite cyclic. If R is any countable commutative integral domain, then RG is primitive.*

Proof. We will construct a concrete faithful irreducible right RG-module. Since $G = C \wr C$ we have $G = A\langle x \rangle$ where A is free abelian with free generators $\ldots, y_{-2}, y_{-1}, y_0, y_1, y_2, \ldots$ and $x^{-1}y_i x = y_{i-1}$, for all i.

Embed R in a countably infinite field K and let V be the vector space over K with basis $\ldots, v_{-2}, v_{-1}, v_0, v_1, v_2, \ldots$. We define an action of G on V by setting

$$v_i x = v_{i+1} \quad \text{and} \quad v_i y_0 = \lambda_i v_i,$$

with each λ_i some fixed element of $K^\times = K \setminus \{0\}$. Since

$$v_i y_j = v_i x^j y_0 x^{-j} = v_{i+j} y_0 x^{-j} = \lambda_{i+j} v_{i+j} x^{-j} = \lambda_{i+j} v_i, \qquad (*)$$

the action of G on V is well defined. Because $R \subseteq K$, we conclude that V is an RG-module.

We choose the λ_i as follows. Since K is countable, the set of all finite sequences $k_0, k_1, \ldots, k_n$ (for all $n \geqslant 0$ and all $k_i \in K^\times$) is also countable. Laying them end to end, we can find a sequence $(\ldots, \lambda_{-2}, \lambda_{-1}, \lambda_0, \lambda_1, \lambda_2, \ldots)$ so that every finite sequence occurs as a segment somewhere in (λ_i). We claim that V is faithful and irreducible.

First we prove irreducibility. Let W be any nonzero RG-submodule of V. By applying a suitable power of x to a nonzero element of W, we can assume that W contains $w = t_0 v_0 + t_1 v_1 + \ldots + t_n v_n$ with t_i in K and $t_0 \neq 0$. Fix any k in $K^\times$ and consider the $n+1$ long sequence $k, 1, 1, \ldots, 1$. This occurs somewhere in (λ_i) with say $k = \lambda_j$. By $(*)$ we obtain $wy_j = kt_0 v_0 + t_1 v_1 + \ldots + t_n v_n$ and hence $wy_j - w = (k-1)t_0 v_0 \in W$. It follows by varying k that $W \supseteq Kv_0$. Since $v_i x = v_{i+1}$ we deduce that $W = V$.

To show that V is faithful, suppose $\alpha = \Sigma_i \alpha_i x^i \in RG$ acts trivially on V where each $\alpha_i \in RA$. Then $0 = v_j \alpha = \Sigma_i (v_j \alpha_i) x^i$ for each j. But $(v_j \alpha_i) x^i \in Kv_{j+i}$ so that each $v_j \alpha_i = 0$ and hence each α_i also acts trivially on V. Thus it suffices to show that RA acts faithfully.

Let α in RA act trivially on V. Then $x^m \alpha x^{-m}$ also acts trivially, and so, by choosing m suitably, we can assume that $\alpha \in R\langle y_0, y_1, \ldots, y_n \rangle$. Furthermore, by multiplying α by a suitable power of $(y_0 y_1 \cdots y_n)$ we can assume that all exponents in α are nonnegative. Hence $\alpha = f(y_0, y_1, \ldots, y_n)$ is a polynomial over R in $y_0, y_1, \ldots, y_n$. Again using $(*)$, we see that, for any j,

$$0 = v_j \alpha = v_j f(y_0, y_1, \ldots, y_n) = f(\lambda_j, \lambda_{j+1}, \ldots, \lambda_{j+n}) v_j.$$

Since every sequence of $n+1$ nonzero elements of K occurs as $\lambda_j, \lambda_{j+1}, \ldots, \lambda_{j+n}$ for some j, it follows that f vanishes on $K^{\times(n)}$. Because K is infinite we conclude that f is identically zero.

As a consequence, following [30], one constructs primitive group rings with $\Delta(G) \neq 1$. Indeed we have

COROLLARY 3.8   *Let* $G = C \times (C \setminus C)$ *with* C *infinite cyclic. Then* KG *is primitive if and only if* K *is countable.*

Proof.   If K is uncountable, then $|K| > |G|$ so KG is not primitive since $\Delta(G) \neq 1$ (cf. remarks after Theorem 3.6). Conversely, if K is countable, then $R = KC$ is a commutative countable integral domain. Since $K[C \times (C \setminus C)] \simeq R[C \setminus C]$, Lemma 3.7 yields the result.

At a certain point in this study it was conjectured that if KG is semi-primitive, $\Delta(G) = 1$ and K is big enough, then KG is primitive. However, the following theorem and example put an end to this.

THEOREM 3.9 (O. I. Domanov [15])   *Let* A *be a normal abelian subgroup of* G. *If* KA *contains more than* $2^{|G/A|}$ *idempotents, then* KG *is not primitive.*

Proof.   Suppose that KG is primitive and let the faithful irreducible module be KG/M with M a maximal right ideal. Then $M \cap KA \neq KA$ so we can choose L, a maximal right ideal of KA with $L \supseteq M \cap KA$. Since KA is commutative, KA/L is a field and we let $\lambda: KA \to KA/L$ be the natural map. Note that λ takes idempotents to 0 or 1.

Let e be a non-zero idempotent of KA and set $I = \Sigma_{x \in G} (KG) e^x$. Since I is a non-zero two-sided ideal of KG and KG acts faithfully on KG/M, we have $I \nsubseteq M$ and hence $M + I = KG$. Thus we can write

$$1 = \mu + \Sigma_{x \in S} \alpha_x e^x$$

with $\mu \in M$, $\alpha_x \in KG$ and S some finite subset of G. We multiply this equation on the right by $\Pi_{x \in S}(1 - e^x)$ and since KA is commutative we obtain

$$\Pi_{x \in S}(1 - e^x) = \mu . \Pi_{x \in S}(1 - e^x) \in KA \cap M \subseteq L.$$

Since L is a maximal ideal of KA, this yields $1 - e^x \in L$ for some x and hence $e^x \notin L$.

Let $\bar{G} = G/A$. To each idempotent e we associate the sequence $\Pi_{\bar{x} \in \bar{G}} \lambda(e^{\bar{x}})$ of 0's and 1's. There are $2^{|\bar{G}|}$ such sequences and KA has more than this many idempotents, by assumption. Thus there exist e, f distinct nonzero idempotents of KA with $\Pi_{\bar{x} \in \bar{G}} \lambda(e^{\bar{x}}) = \Pi_{\bar{x} \in \bar{G}} \lambda(f^{\bar{x}})$. The idempotent $e - ef$ then satisfies $\lambda((e - ef)^x) = 0$ for all x in G, and by the preceding paragraph it must be zero. Similarly $f = ef$, a contradiction since $e \neq f$.

Now for the example:

COROLLARY 3.10 [15]   *Let p be a prime and let N and H be elementary abelian p-groups with $|H| = \aleph_0$ and $|N| > 2^{\aleph_0}$. If $G = N \setminus H$, then G is a locally finite, metabelian p-group with $\Delta(G) = 1$. Furthermore, for all fields K, KG is not primitive even though $J(KG) = 0$ for char $K \neq p$.*

Proof.   The group-theoretic properties of G are clear. If char $K = p$, then $J(KG)$ is the augmentation ideal of KG, namely the kernel of the natural homomorphism $KG \to K$. Thus KG is not semiprimitive and hence not primitive. On the other hand, if char $K \neq p$, then since G is a locally finite p-group we have $J(KG) = 0$. Furthermore, G is the semi-direct product of $A = \Pi_{h \in H} N_h$ by H and KA contains more than $2^{\aleph_0}$ idempotents. Theorem 3.9 now implies that KG is not primitive.

Corollary 3.10 should be viewed in conjunction with the examples in Theorem 3.6 (i), (iii). There is at present no viable conjecture as to when KG is primitive for arbitrary groups G. The case of G in $\mathfrak{PF}$ has been settled and will be discussed later. In particular, if G is a non-identity finitely generated nilpotent group, then KG is never primitive. However, group rings of nonfinitely generated nilpotent groups can be primitive.

THEOREM 3.11 (K. A. Brown [99])   *Let G be a torsion-free nilpotent group with center Z. If G has an abelian subgroup B of infinite rank r with $B \cap Z = 1$ and $|KZ| \leqslant r$, then KG is primitive.*

A close look at the examples makes it seem possible that KG is primitive if KG is semiprimitive, K is large and if $|G: C_G(x)| = |G|$ for all $x \neq 1$ in G.

4. Residually Finite Groups

A group G is *residually finite* if it is a subdirect product of finite groups. Alternately, this happens if and only if for each $x \in G$, $x \neq 1$ there exists a normal subgroup N_x of finite index in G with $x \notin N_x$.

Let G be a $\mathfrak{PF}$-group and let $1 = G_0 \triangleleft G_1 \triangleleft \ldots \triangleleft G_n = G$ be a subnormal series with factors G_i/G_{i-1} which are either infinite cyclic or finite. Then the *Hirsch number* $h(G)$, the number of infinite cyclic factors here, is independent of the particular series and hence an invariant of the group.

It follows by induction on $h(G)$ that any $\mathfrak{PF}$-group is residually finite. It is also true that every finitely generated $\mathfrak{APF}$-group is residually finite. The proof of this was begun by Hall [36] and finished some fifteen years later by Jategaonkar [45] and Roseblade [71]. We consider it in the next three sections, finally proving it as Theorem 6.6. We begin with a few examples to see what is involved.

Let H be in $\mathfrak{PF}$ and let A be an irreducible module for the integral group ring $\mathbb{Z}H$, with A written multiplicatively. Then we can form $G = AH$, the semidirect product of A by H, and we observe that G is a finitely generated $\mathfrak{APF}$-group. Suppose G is residually finite. Then because $A \neq 1$, there exists a normal subgroup N of G of finite index with $N \not\supseteq A$. But then $N \cap A \triangleleft G$ is clearly a proper $\mathbb{Z}H$-submodule of A, so we must have $N \cap A = 1$ and hence A is embedded isomorphically in G/N. Thus we see that a necessary aspect of the proof must involve showing that all irreducible $\mathbb{Z}H$-modules are finite. This clearly also applies to irreducible $\mathbb{F}_pH$-modules.

There is more to be done. An R-module V is said to be *monolithic* if V has a nonzero submodule V_0, called the *lith* of V, that is contained in all nonzero submodules of V. Thus a monolithic module is merely an essential extension of an irreducible module. Now let us make the same construction as in the preceding paragraph, but this time with A a finitely generated monolithic $\mathbb{Z}H$-module with lith A_0. If $G = AH$ is residually finite, then we can choose $N \triangleleft G$ with G/N finite and $A_0 \not\subseteq N$. Thus $A \cap N$ is a $\mathbb{Z}H$-submodule of A not containing A_0, so we must have $A \cap N = 1$ and again A is finite. Therefore, we must be able to show that any monolithic $\mathbb{Z}H$-module is finite. In fact this is the key step in the proof and the final result follows easily because of

LEMMA 4.1 *Let V be an R-module. Then V is residually monolithic.*

Proof. Let $0 \neq v \in V$ and let W be a submodule of V maximal with the property that $v \notin W$. Then V/W is clearly monolithic with $v + W$ contained in its lith.

We begin by considering irreducible modules. A field K is *absolute* if it is algebraic over a finite field. Thus K is absolute if and only if the multiplicative group $K^{\times}$ is periodic. For K nonabsolute we can state the following

THEOREM 4.2 (Hall [36]) *Let G be a $\mathfrak{P}\mathfrak{F}$-group and let K be a nonabsolute field. Then all irreducible KG-modules are finite-dimensional over K if and only if G is abelian-by-finite.*

There are numerous extensions of this result but they require a more appropriate concept of finite-dimensional. An irreducible R-module V has *finite endomorphism dimension* if it is finite-dimensional as a vector space over the commuting ring $\mathrm{End}_R(V)$. It follows ([68], Chapter 5) that if G is abelian-by-finite, then all irreducible KG-modules have finite endomorphism dimension. Furthermore, the converse holds for suitable locally finite groups (Farkas and Snider [24], B. Hartley [40, 41]) and for suitable solvable groups (Snider [88, 103], B. A. F. Wehrfritz [106, 107, 108]). Indeed suppose K is a nonabsolute field of characteristic $p \geqslant 0$ and G is solvable-by-finite. If all irreducible KG-modules have finite endomorphism dimension, then $G/O_p(G)$ is abelian-by-finite.

It is an intriguing question whether the complex group algebra $\mathbb{C}G$ can have all primitive images finite-dimensional over their centres without there being a bound on these dimensions. By the above, this cannot happen if G is solvable or locally finite and it presumably cannot happen in general. Since this finite-dimensional property is inherited by subgroups and quotient groups, the problem splits into two halves. First, study finitely generated groups G with this property. Since finitely generated linear groups are residually finite, the same is true for G. Second, study arbitrary G knowing that each finitely generated subgroup is abelian-by-finite.

The goal now is to show that if $R = \mathbb{Z}$ or $\mathbb{F}_p$ and if G is in $\mathfrak{P}\mathfrak{F}$, then the irreducible RG-modules are finite. To understand what this amounts to, suppose that G is a finitely generated abelian group. Then KG is a finitely generated commutative K-algebra and every irreducible KG-module V is of the form $V = KG/M$ for some maximal ideal M. Thus V is the additive subgroup of the field $F = KG/M$ and it follows from the Hilbert Nullstellensatz that $|F:K|$ is finite and hence that V is finite-dimensional over K.

Thus what is required is a generalization of the Nullstellensatz. To see how this might be achieved, we consider another special situation. Let G be a finitely generated group and let V be an irreducible $\mathbb{Z}G$-module. Then V must be countable and, as an abelian group, characteristically simple. Thus V must be one of the following:

(1) a finite elementary abelian p-group for some prime p;
(2) a countably infinite elementary abelian p-group;
(3) a finite direct sum of copies of $\mathbb{Q}$, the additive group of rationals;
(4) a countably infinite direct sum of copies of $\mathbb{Q}$.

While (1), (2) and (4) are possible, it turns out that (3) is not.

To see this, suppose V is an n-dimensional vector space over $\mathbb{Q}$ with basis $\{v_1, v_2, \ldots, v_n\}$ and let G be generated by $x_1, x_2, \ldots, x_m$. Then the action of each $x_i^{\pm 1}$ on V can be described by an $n \times n$ matrix over $\mathbb{Q}$, and if d is a common denominator for all these finitely many entries, then $\Sigma_1^n v_i \mathbb{Z}[1/d]$ is a proper nonzero $\mathbb{Z}G$-submodule of V, a contradiction. Observe that for this argument we need to know that $\mathbb{Z}[1/d] \neq \mathbb{Q}$, a consequence of the fact that $\mathbb{Z}$ is a principal ideal domain with infinitely many primes. This latter property for certain rings R will prove to be of crucial importance.

Let R be commutative and let A be a normal abelian subgroup of G. Then G acts by conjugation on the group ring RA. If $\alpha \in RA$, $\alpha \neq 0$, then an RA-module M is said to be α^G-free if there exists a free RA-submodule M_0 of M such that for all $v \in M$ we have $v\alpha^{g_1}\alpha^{g_2} \ldots \alpha^{g_r} \in M_0$ for suitable $g_1, g_2, \ldots, g_r \in G$.

THEOREM 4.3 (Hall [36], Roseblade [71]) *Let R be a commutative Noetherian domain and let A be a torsion-free normal abelian subgroup of the $\mathfrak{PF}$-group G. Then any cyclic RG-module is an α^G-free RA-module for some nonzero $\alpha \in RA$.*

For the remainder of this section, we assume A is central in G. Then G acts trivially on RA and α^G-free RA-modules are merely said to be α-free. Suppose M is such a module. If RA_α denotes the ring RA localized at α, then

$$M_\alpha = M \otimes RA_\alpha = M_0 \otimes RA_\alpha$$

is a free RA_α-module. This concept is now known as *generic flatness*. It was shown to hold for enveloping algebras of finite-dimensional Lie algebras by D. Quillen [70] (see also M. Duflo [18] and R. S. Irving [44]). Its relationship to the Nullstellensatz is apparent from

LEMMA 4.4   *Let A be a torsion-free central subgroup of the $\mathfrak{PF}$-group G and assume that RA is a principal ideal domain with infinitely many primes. If M is an irreducible RG-module, then there exists a prime $\pi \in RA$ with $M\pi = 0$.*

Proof. If no such π exists, then $S = RA$ embeds isomorphically in $D = \mathrm{End}_{RG}(M)$. Hence D also contains F, the field of fractions of S. Since M is a vector space over F, it follows that M has an S-submodule isomorphic to F. By Theorem 4.3, M is an α-free S-module for a suitable α and hence, since S is a principal ideal domain, the submodule F is also α-free. This forces all primes of S to divide α and contradicts the fact that S has infinitely many primes.

As consequences we have

COROLLARY 4.5 [36]   *Let M be an irreducible $\mathbb{Z}G$-module with G a $\mathfrak{PF}$-group. Then there exists a prime $p \in \mathbb{Z}$ with $Mp = 0$. Hence M is an irreducible $\mathbb{F}_pG$-module.*

COROLLARY 4.6 [36]   *Let G be a finitely generated nilpotent group and let K be an absolute field. If M is an irreducible KG-module, then M is finite-dimensional over K.*

Proof.   For the first corollary take $A = 1$ in Lemma 4.4. For the second, we assume that G acts faithfully on M. If $A = \langle x \rangle$ is an infinite cyclic central subgroup of G, then Lemma 4.4 with $RA = KA$ implies that the image of x in $\mathrm{End}\,M$ is algebraic over K. Since K is absolute, we conclude that the image of x has finite order, a contradiction. Thus the center of G is periodic and G is finite.

5.   The Nullstellensatz

In this section we indicate how to prove that the irreducible $\mathbb{Z}G$-modules, for a $\mathfrak{PF}$-group G, are all finite. By Corollary 4.5, we already know that such modules are in fact irreducible $\mathbb{F}_pG$-modules for some prime p. Thus we now study irreducible KG-modules for K an absolute field. It is interesting to observe that in order to prove this generalized Nullstellensatz, we require a preliminary result which is an operator group analog of the ordinary commutative situation.

If G acts on a group H, then G permutes the elements of H, the subgroups of H and the ideals of KH. Any such object is said to be *orbital*, or more precisely G-orbital, if it has only finitely many G-conjugates. In particular, the set of orbital elements of H is

$$D_H(G) = \{x \in H \mid |G: C_G(x)| < \infty\}.$$

Note that when $G = H$ acts by conjugation, then $D_G(G) = \Delta(G)$, the *FC*-center.

If G acts on the finitely generated torsion-free abelian group A, then it also acts on the rational vector space $V = A \otimes \mathbb{Q}$. We say that G acts *rationally irreducibly* on A if V is an irreducible $\mathbb{Q}G$-module, or, equivalently, if the nonidentity G-invariant subgroups of A all have finite index in A. We say that A is a *plinth* for G if G and all its subgroups of finite index act rationally irreducibly on A. The first goal is to study the G-invariant prime ideals of KA.

Suppose now that A is a finitely generated abelian group and that G is any group of automorphisms of A. Then G acts on the group algebra KA. Let I be an ideal of KA, let U be a subgroup of A, and assume that $I > (I \cap KU) . KA$. A subgroup $W > U$ of A is said to be a $\log_U$ *subgroup* for I if and only if for all $\alpha \in I \setminus (I \cap KU) . KA$ there exist x, y in the support of α with $xy^{-1} \in W \setminus U$. These subgroups, for $U = 1$, were introduced in Bergman [3] as an analog over arbitrary fields of a certain logarithmic limit over the real or complex numbers.

LEMMA 5.1   *Let* $I > (I \cap KU) . KA$. *Then*

(i) *any* $\log_U$ *subgroup for* I *contains a minimal* $\log_U$ *subgroup for* I;
(ii) *there are only finitely many minimal* $\log_U$ *subgroups;*
(iii) *if* I *and* U *are* G-*orbital, then so are the minimal* $\log_U$ *subgroups for* I.

Proof.   For each $\alpha \in I \setminus (I \cap KU) . KA$ set

$$T_\alpha = \{xy^{-1}U \in A/U \mid x, y \in \operatorname{supp}\alpha, \ xy^{-1} \notin U\}.$$

Since $\alpha \notin (I \cap KU) . KA$, it is clear that each T_α is a finite nonempty subset of A/U. Furthermore, $W \supseteq U$ is a $\log_U$ subgroup for I if and only if $(W/U) \cap T_\alpha \neq \emptyset$ for all such α. Parts (i) and (ii) now follow by an easy induction argument. One shows that if T_λ $(\lambda \in \Lambda)$ are finite nonempty subsets of a finitely generated abelian group, then the collection of all subgroups which intersect each T_λ nontrivially has finitely many minimal members. Finally, if I and U are orbital, they are normalized by a subgroup of G of finite index which then permutes the finitely many minimal $\log_U$ subgroups of I.

Let P be a prime ideal of KA and let $\bar{\ } : KA \to KA/P$ be the natural map to the domain KA/P.

LEMMA 5.2   *Suppose that $P > (P \cap KU) . KA$ and let v be a valuation on the domain $\overline{KA}$. Let O_v be the associated valuation ring and M_v its unique maximal ideal. Assume $\overline{KU} \subseteq O_v$ but $\overline{KU} \cap M_v = 0$. If $W_v = \{a \in A \mid v(\bar{a}) = 1\}$, then A/W_v is torsion-free and W_v is a $\log_U$ subgroup for P.*

Proof.   Since A/W_v is a subgroup of the value group, it is torsion-free. Also since $\overline{KU} \subseteq O_v$, we have $W_v \supseteq U$. Let $\alpha \in P \backslash (P \cap KU) . KA$ and write $\alpha = \Sigma_1^n \alpha_i x_i$ with $\alpha_i \in KU$, $1 \in \text{supp } \alpha_i$ and $x_1, x_2, \ldots, x_n$ in distinct cosets of U in A. Assume the numbering so chosen that $\alpha_1, \alpha_2, \ldots, \alpha_r \notin P \cap KU$ while $\alpha_{r+1}, \ldots, \alpha_n \in P \cap KU$. Clearly $r \geqslant 1$ and we assume that $\bar{x}_1$ has minimum valuation among $\bar{x}_1, \bar{x}_2, \ldots, \bar{x}_r$. Then $\alpha x_1^{-1} \in P$ so

$$0 = \overline{\alpha x_1^{-1}} = \Sigma_1^n \bar{\alpha}_i \bar{x}_i \bar{x}_1^{-1} = \Sigma_1^r \bar{\alpha}_i \bar{x}_i \bar{x}_1^{-1}.$$

Note that $\bar{\alpha}_i$ and $\bar{x}_i \bar{x}_1^{-1}$ are contained in O_v and, since $\overline{KU} \cap M_v = 0$, that each $\bar{\alpha}_i$ is a unit of O_v for $i = 1, 2, \ldots, r$. Thus the term $\bar{\alpha}_1 \bar{x}_1 \bar{x}_1^{-1}$ is not contained in M_v so there must be a second such term, say $\bar{\alpha}_j \bar{x}_j \bar{x}_1^{-1}$, not in M_v. We conclude that $v(\bar{x}_j \bar{x}_1^{-1}) = 1$. Since x_j, $x_1 \in \text{supp } \alpha$ with $x_j x_1^{-1} \notin U$ and $x_j x_1^{-1} \in W_v$, the result follows.

Let B be a proper subgroup of A. We say that B is a *maximal isolated orbital* subgroup if B is G-orbital, B is an isolated subgroup of A (i.e. A/B is torsion free) and all larger orbital subgroups have finite index in A. Thus A/B is a plinth for some subgroup of G of finite index.

LEMMA 5.3   *Let P be a G-orbital prime ideal of KA and let B be a maximal isolated orbital subgroup of A. If $P > (P \cap KB) . KA$, then $\overline{KA}$ is algebraic over $\overline{KB}$.*

Proof.   Suppose that for some $x \in A$ we have $\bar{x}$ not algebraic over $\overline{KB}$. Then the ring generated by $\overline{KB}$ and $\bar{x}$ is the polynomial ring $\overline{KB}[\bar{x}]$ and there is a homomorphism $\overline{KB}[\bar{x}] \to \overline{KB}$ which is the identity on $\overline{KB}$ and which sends $\bar{x}$ to 0. By the Extension Theorem for Places, this homomorphism extends to a place defined on $\overline{KA}$ with associated valuation v. Clearly $\overline{KB}$ is in O_v and is disjoint from M_v. Furthermore, $x \notin W_v$. By the previous two lemmas, W_v is a $\log_B$ subgroup for P and W_v contains a minimal $\log_B$ subgroup $C > B$ which is G-orbital. By assumption, B is a maximal isolated orbital subgroup so A/C is finite. But A/W_v is torsion-free, so $A = W_v$ and $x \in W_v$, a contradiction.

Observe that if A is a plinth for G, then $B = 1$ is maximal isolated orbital. Furthermore, if $P \neq 0$, then $P > (P \cap KB).KA$. Consequently we have

THEOREM 5.4 (G. M. Bergman [3]) *Let A be a finitely generated torsion-free abelian group which is a plinth for G. If $P \neq 0$ is a G-orbital prime ideal of KA, then KA/P is algebraic and hence finite-dimensional over K.*

One aspect of Hilbert's Nullstellensatz is that if R is a finitely generated commutative algebra, then any prime ideal is an intersection of maximal ideals. (This is now known by K. I. Beidar [98] to be false in general for noncommutative algebras.) In particular, if R is a commutative domain, then it is semiprimitive. Theorem 5.4 is merely the starting point in proving the following operator group analog of this Jacobson property. This was obtained for absolute fields by Roseblade [71] and, as a corollary, for all fields by D. L. Harper [35].

THEOREM 5.5 *Let A be a finitely generated torsion-free abelian group which is a plinth for the $\mathfrak{PF}$-group G. Let $\alpha \in KA$ and suppose that for each G-orbital maximal ideal T of KA we have $\alpha^g \in T$ for some $g \in G$. Then $\alpha = 0$.*

We do not prove this result but we will see how it applies. If $H \subseteq G$ then, with respect to the conjugation action, H is orbital if and only if $|G : N_G(H)|$ is finite.

LEMMA 5.6 *Let H be an infinite orbital subgroup of the $\mathfrak{PF}$-group G. Then there exist subgroups $1 \neq A \subseteq H$ and $L \lhd G$ with $A \lhd L$, A a plinth for L, $L/C_L(A)$ abelian and G/L finite.*

Proof. Since H is infinite it has a characteristic torsion-free abelian subgroup $B \neq 1$. Then B is also orbital and we choose $A \neq 1$ to be a subgroup of B of smallest rank which is still orbital. Thus if $N = N_G(A)$, then $|G : N| < \infty$ and A is a plinth for N. In particular, $\bar{N} = N/C_N(A)$ is an irreducible linear group so Mal'cev's theorem implies that $\bar{N}$ is abelian-by-finite. Thus there exists $L \lhd G$ with G/L finite, $L \subseteq N$ and $L/C_L(A)$ abelian. Replacing A by $A \cap L$ completes the proof.

PROPOSITION 5.7 [71]   *Let G be in $\mathfrak{PF}$ and let A be a normal plinth of G. Then every irreducible KG-module has KA-torsion.*

Proof. Let M be an irreducible KG-module and suppose that M is torsion-free as a KA-module. By Theorem 4.3 M is α^G-free for some nonzero α in KA. Moreover, if M_0 is the free KA-submodule given by this property, then $M_0 \neq 0$ since KA acts in a torsion-free manner.

Let T be any G-orbital maximal ideal of KA. If $I = \bigcap_{g \in G} T^g$, then I is a G-invariant ideal of KA contained in T. Furthermore, $I \neq 0$ since T is nonzero orbital and KA is prime. Now MI is easily seen to be a nonzero KG-submodule of M, since $I \neq 0$. Hence $MI = M$ and in particular $MT = M$.

Since M_0 is a nonzero free KA-module, we have $M_0 \neq M_0 T$ and we can choose $v \in M_0 \setminus M_0 T$. But $v \in M = MT$ so there exist $v_1, v_2, \ldots, v_k \in M$ and $\tau_1, \tau_2, \ldots, \tau_k \in T$ with $v = \Sigma v_i \tau_i$. By definition of the α^G-free property, there exists a finite product of G-conjugates of α which annihilates each v_i modulo M_0. Hence since KA is commutative there exists $\beta = \alpha^{g_1} \alpha^{g_2} \ldots \alpha^{g_r}$ with $v_i \beta \in M_0$ for all i and thus $v\beta \in M_0 T$. If $\beta \notin T$ then since T is maximal we could write $1 = \beta \gamma + \tau$ for some $\gamma \in KA$ and $\tau \in T$. But then

$$v = v(\beta\gamma + \tau) = (v\beta)\gamma + (v\tau) \in M_0 T$$

a contradiction. Thus $\beta = \alpha^{g_1} \alpha^{g_2} \ldots \alpha^{g_r} \in T$ and since T is prime we have $\alpha^{g_i} \in T$ for some i.

We have therefore shown that α satisfies the hypothesis of Theorem 5.5 and we conclude that $\alpha = 0$, a contradiction.

The existence of KA-torsion in the irreducible KG-module M implies that M is an induced module. Combining this with Lemma 5.6 leads to a proof, by induction on $h(G)$, of the main result.

THEOREM 5.8 (Roseblade [71])   *Let G be a $\mathfrak{PF}$-group and let K be an absolute field. Then every irreducible KG-module is finite-dimensional over K.*

It is now an immediate consequence of Corollary 4.5 and Theorem 5.8 that, for G in $\mathfrak{PF}$, every irreducible $\mathbb{Z}G$-module is finite.

6. The Artin–Rees Property

It remains to show that the finitely generated monolithic $\mathbb{Z}G$-modules are finite and for this, suitable versions of the Artin–Rees property are required.

Let R be a right Noetherian ring and let I be a two-sided ideal of R. We say that I has the (*weak*) *AR property* if for all finitely generated R-modules M and submodules U

$$MI^n \cap U \subseteq UI$$

for some integer n. If R is commutative, then every ideal has the AR property. For noncommutative rings this is not true, but useful sufficient conditions do exist.

If I is an ideal of R, we let $R(I)$ denote the graded subring of the polynomial ring $R[t]$ given by

$$R(I) = R + It + I^2 t^2 + \ldots .$$

We say that I is *centrally generated* if it is generated as an ideal by central elements. Furthermore, I is *polycentral* if there exists a sequence of ideals $0 = I_0 \subseteq I_1 \subseteq \ldots \subseteq I_n = I$ such that each I_i/I_{i-1} is a centrally generated ideal of R/I_{i-1}. The following is essentially the Artin–Rees lemma.

LEMMA 6.1   *Let I be an ideal of the Noetherian ring R.*

(i) *If I is centrally generated, then $R(I)$ is Noetherian.*
(ii) *If $R(I)$ is Noetherian, then I has the AR property.*

Proof.   (i) If I is generated by the central elements $x_1, x_2, \ldots, x_m$, then $R(I)$ is generated as a ring by R and the central elements $x_1 t, x_2 t, \ldots, x_m t$. The Hilbert basis theorem implies that $R(I)$ is Noetherian.

(ii) Let M be a finitely generated R-module and form the finitely generated graded module

$$M' = M + MIt + MI^2 t^2 + \ldots$$

for $R' = R(I)$. Let U be a submodule of M, then

$$U' = U + (MI \cap U)t + (MI^2 \cap U)t^2 + \ldots$$

is an R'-submodule and hence is also finitely generated by assumption. If the generators are contained in the first n summands then from

$$(MI^n \cap U)t^n \subseteq (U+(MI\cap U)t+\ldots+(MI^{n-1}\cap U)t^{n-1}).R'$$

we deduce immediately that $MI^n \cap U \subseteq UI$.

A more technical formulation of this yields an analogous result for polycentral ideals.

THEOREM 6.2 (Y. Nouazé and P. Gabriel [63])   *If I is a polycentral ideal of the Noetherian ring R, then I has the AR property.*

For group rings of $\mathfrak{P}\mathfrak{F}$-groups more can be done.

THEOREM 6.3 (Roseblade [73])   *Let G be a $\mathfrak{P}\mathfrak{F}$-group, let $H \triangleleft G$ and let R be Noetherian. If I is a polycentral, G-invariant ideal of RH, then the extended ideal $I.RG$ of RG has the AR property.*

Proof.   We sketch the argument for I centrally generated where the goal is to show that $RG(I.RG)$ is Noetherian. First Lemma 6.1 (i) implies that $RH(I)$ is Noetherian. Hence since G normalizes this ring, it follows as in Theorem 2.1 that $S = \langle RH(I), G \rangle$, the subring of $RG[t]$ generated by these two subsets, is also Noetherian. But it is easy to see that S is in fact equal to $RG(I.RG)$, so Lemma 6.1 (ii) yields the result.

Ideals of particular interest here are the augmentation ideals $\mathfrak{g}$, namely the kernels of the natural homomorphisms $RG \to R$. If G is in $\mathfrak{P}\mathfrak{F}$, we say that G is p-nilpotent for a prime p if and only if all finite homomorphic images of G are p-nilpotent, i.e. p'-by-p. As a consequence of Theorem 6.3 and a good deal of additional work we have

THEOREM 6.4 (Roseblade [73], P. F. Smith [82])   *Let G be in $\mathfrak{P}\mathfrak{F}$, let $H \triangleleft G$ and let K be a field. Then $\mathfrak{h}KG$ has the AR property if and only if*

(i) char $K = 0$ *and H is finite-by-nilpotent, or*
(ii) char $K = p > 0$ *and H is p-nilpotent.*

Part (ii) above becomes all the more interesting when one observes that for each prime p every $\mathfrak{P}\mathfrak{F}$-group has a p-nilpotent normal subgroup of finite index. Indeed the centralizer of all the p-chief factors of G is the unique largest such subgroup. Here, by a p-chief factor we mean a chief factor whose order is divisible by p.

A somewhat different approach to the existence of ideals with the AR property appears in Jategaonkar [45]. Using it or Theorem 6.3 we obtain

PROPOSITION 6.5 [45, 73]   *If G is a $\mathfrak{P}\mathfrak{F}$-group, then every finitely generated monolithic $\mathbb{Z}G$-module is finite.*

Proof.   Assume that G acts faithfully on the finitely generated monolithic $\mathbb{Z}G$-module M. If M_0 is the lith, then by Theorem 5.8 M_0 is finite and $M_0 p = 0$ for some prime p. Let H be a normal subgroup of G of finite index which acts trivially on M_0.

Let A be a characteristic abelian subgroup of H and let I be the centrally generated ideal $I = p\mathbb{Z}A + \mathfrak{a}$ of $\mathbb{Z}A$. By Theorem 6.3, the extended ideal $J = I . \mathbb{Z}G$ has the AR property and hence $MJ^n \cap M_0 \subseteq M_0 J$ for some n. Observe that both p and $\mathfrak{a}$ annihilate M_0 so $M_0 J = 0$. But then MJ^n is a $\mathbb{Z}G$-submodule of M not containing the lith. We conclude therefore that $MJ^n = 0$ and hence that $MI^n = 0$.

Let us first take $A = 1$. Then $MI^t = 0$ for some t implies that $Mp^t = 0$ and hence M is a periodic abelian group of period dividing p^t. Now let A be arbitrary and say $MI^n = 0$. If $a \in A$, then since both p^t and $(a-1)^n$ map to zero in $\text{End}(M)$, it is clear that the image of $\mathbb{Z}\langle a \rangle$ is a finite ring. Thus $\langle a \rangle$ is finite and A is periodic. Since A is an arbitrary characteristic abelian subgroup of H, this implies that H and hence G is finite.

The main result is now an immediate consequence of the above, Lemma 4.1 and the residual finiteness of $\mathfrak{P}\mathfrak{F}$-groups.

THEOREM 6.6 (Jategaonkar [45], Roseblade [73])   *If G is a finitely generated $\mathfrak{A}\mathfrak{P}\mathfrak{F}$-group, then G is residually finite.*

Finally we mention

THEOREM 6.7 (D. Segal [77])   *Let G be a finitely generated $\mathfrak{A}\mathfrak{P}\mathfrak{F}$-group. Then G has a normal subgroup of finite index which is residually finite nilpotent.*

This then implies (see Segal [101]) that every periodic group of automorphisms of a finitely generated nilpotent-by-polycyclic group is locally finite.

7. Nilpotence of the Frattini Subgroup

If G is an arbitrary group with maximal subgroups, then its Frattini subgroup $\Phi(G)$ is the intersection of all these. As is well known, if G is finite then $\Phi(G)$ is nilpotent. This is also true for G in $\mathfrak{P}\mathfrak{F}$ by K. A. Hirsch [42]. Here we discuss to what extent this remains true for larger classes of finitely generated groups. We remark that if $\bar{G}$ is a homomorphic image of G, then the image of $\Phi(G)$ is contained in $\Phi(\bar{G})$.

Let G be a finitely generated $\mathfrak{A}\mathfrak{P}\mathfrak{F}$-group. Then G has a normal abelian subgroup A with $\bar{G} = G/A$ polycyclic-by-finite and with A a finitely generated $\mathbb{Z}\bar{G}$-module. If $H = \Phi(G)$, then $\bar{H} = HA/A \subseteq \Phi(\bar{G})$ and hence $\bar{H}$ is a normal nilpotent subgroup of $\bar{G}$. Let B be any maximal submodule of A. Then A/B is an irreducible $\mathbb{Z}\bar{G}$-module, so it is finite and G/B is polycyclic-by-finite. Again this implies that HB/B is a nilpotent normal subgroup of G/B, so it centralizes all chief factors of G/B. In particular, H and hence $\bar{H}$, centralizes A/B.

As an indication of what can be deduced about such $\mathbb{Z}\bar{G}$-modules, we prove the following result from Hall [37].

LEMMA 7.1 *Let G be in $\mathfrak{P}\mathfrak{F}$ and let M be a finitely generated $\mathbb{Z}G$-module. Let α be central in $\mathbb{Z}G$ and assume that α annihilates the quotients M/N for all maximal $\mathbb{Z}G$-submodules N. Then $M\alpha^n = 0$ for some n.*

Proof.   Assume that $M\alpha^n \neq 0$ for all n and let U be a submodule of M maximal with the property that $M\alpha^n \nsubseteq U$ for all n. Replacing M by M/U, we can now assume that every nonzero submodule of M contains some $M\alpha^n$. This implies that $\mathrm{ann}_M\alpha = 0$ since otherwise $\mathrm{ann}_M\alpha \supseteq M\alpha^t$ for some t and hence $M\alpha^{t+1} = 0$.

Write $R = \mathbb{Z}G$. Since α is central in R, we can localize R at α and then form the localized R_α-module $\bar{M} = M \otimes R_\alpha$. Since $\mathrm{ann}_M\alpha = 0$ it follows that $\bar{M} \supseteq M$ and that α is an automorphism on $\bar{M}$. Now let $C = \langle x \rangle$ be an infinite cyclic group and set $\bar{R} = RC = \mathbb{Z}[G \times C]$. Then we make $\bar{M}$ into an $\bar{R}$-module by letting R act by way of its image in R_α and by letting x act as the automorphism α.

Suppose $\bar{V}$ is a nonzero $\bar{R}$-submodule of $\bar{M}$. Then $\bar{V} \cap M$ is a nonzero R-submodule of M so, by assumption, $M\alpha^n \subseteq \bar{V} \cap M \subseteq \bar{V}$ for some n. Thus since x acts like α on $\bar{M}$ we have $M \subseteq \bar{V}x^{-n} = \bar{V}$ and hence clearly $\bar{M} = \bar{V}$. Thus $\bar{M}$ is a simple $\mathbb{Z}[G \times C]$-module, so Corollary 4.5 and Theorem 5.8 imply that $\bar{M}$, and hence M, is finite. But α is one-to-one on M so it must be onto and this contradicts the fact that α annihilates the simple quotients M/N.

The above is a module-theoretic analog of the version of the commutative Nullstellensatz which asserts that if $\alpha \in K[x_1, x_2, \ldots, x_n]$ is contained in all maximal ideals which contain a fixed ideal I, then α is nilpotent modulo I. Furthermore, in the classical proof of this fact, an additional indeterminate is adjoined and the larger polynomial ring is studied.

The preceding lemma is used group theoretically as follows. Let A be a normal abelian subgroup of G with $\bar{G} = G/A$ in $\mathfrak{P}\mathfrak{F}$ and let $\bar{z}$ be a central element of $\bar{G}$. Suppose that $\bar{z}$ centralizes all quotients A/B with B a maximal $\mathbb{Z}\bar{G}$-submodule of A. Then $\alpha = \bar{z} - 1$ is a central element of $\mathbb{Z}\bar{G}$ which annihilates each such factor so the above yields $0 = A\alpha^n = A(\bar{z}-1)^n$ for some n. This translates multiplicatively to the n-fold commutator equation $1 = [A, \bar{z}, \bar{z}, \ldots, \bar{z}]$ and hence the group $\langle A, z \rangle$ is nilpotent. This is a key idea in the proof of the following result.

We say that a group G is *n-step nilpotent* if it has a subnormal series of length n with each quotient nilpotent.

THEOREM 7.2 (Hall [37])   *Let G be a finitely generated $(n+1)$-step nilpotent group. Then $\Phi(G)$ is n-step nilpotent and this bound is sharp.*

In particular, if G is a finitely generated abelian-by-nilpotent group, then $\Phi(G)$ is nilpotent. Indeed, this fact is the crucial step in the proof of Theorem 7.2. To handle finitely generated $\mathfrak{A}\mathfrak{P}\mathfrak{F}$-groups we require

THEOREM 7.3 (Roseblade [71])   *Let M be a finitely generated $\mathbb{Z}G$-module with G in $\mathfrak{P}\mathfrak{F}$ and let H be a normal nilpotent subgroup of G. If $\mathfrak{h}$ annihilates the quotients M/N for all maximal $\mathbb{Z}G$-submodules N, then $M\mathfrak{h}^n = 0$ for some n.*

While we do not prove this, we do prove the following group-theoretic consequence.

COROLLARY 7.4 [72]   *Let G be a finitely generated $\mathfrak{A}\mathfrak{P}\mathfrak{F}$-group. Then $\Phi(G)$ is nilpotent.*

Proof.   Let A be a normal abelian subgroup of G with $\bar{G} = G/A$ in $\mathfrak{P}\mathfrak{F}$ and let $H = \Phi(G)$. Then, by the remarks at the beginning of this section, $\bar{H} = HA/A$ is a normal nilpotent subgroup of $\bar{G}$ which centralizes each quotient A/B with B a maximal $\mathbb{Z}\bar{G}$-submodule of A. Thus $\mathfrak{h}$ annihilates each such quotient. Since A is a finitely generated $\mathbb{Z}\bar{G}$-module, Theorem 7.3 yields $A\mathfrak{h}^n = 0$ for some n. In multiplicative notation this becomes the n-fold commutator equation $[A, H, H, \ldots, H] = 1$. Since A is abelian, this implies that A is contained in the nth term of the upper central series for AH. But $AH/A = \bar{H}$ is nilpotent, so we conclude that AH and hence H is nilpotent.

We close with a module-theoretic result related, to the Frattini question. An R-module M is *residually simple* if the intersection of all its maximal submodules is zero. Furthermore, M is poly-(residually simple) if there exists a sequence of R-submodules

$$0 = M_0 \subseteq M_1 \subseteq \ldots \subseteq M_n = M$$

with each M_i/M_{i-1} residually simple.

THEOREM 7.5 (C. J. B. Brookes [6]) *Let G be a $\mathfrak{P}\mathfrak{F}$-group. Then every finitely generated $\mathbb{Z}G$-module is poly-(residually simple).*

This was first proved for G finitely generated nilpotent by Segal [78] using the Krull dimension of modules as a tool (see also [8]). The more general result uses a new group-ring dimension function called calibre. As a consequence we have

COROLLARY 7.6 [78] *Let G be a finitely generated abelian-by-nilpotent group. If H is any subgroup of G, then $\Phi(H)$ is nilpotent.*

8.   Zalesskii-type Subgroups

For the remainder of this essay we will assume that the coefficient ring is a field K. As we have observed, if G is in $\mathfrak{P}\mathfrak{F}$, then KG is Noetherian. Since prime ideals are the basic building blocks in the Goldie theory of Noetherian rings, it is important to determine the structure of these ideals of KG.

Before we consider this specific problem, we discuss the general question of describing ideals in KG for G arbitrary. There are a number of inductive approaches. For example, let I be an ideal of KG and let H be a normal subgroup of G. Then we say that H *controls* I if $I = (I \cap KH) . KG$. Equivalently this occurs when KH contains generators for I as a two-sided ideal. Note that $I \cap KH$ is a G-invariant ideal of KH and that the controlling condition is also equivalent to $I = L . KG$ for some G-invariant ideal L of KH. When this occurs we must have $L = I \cap KH$. Obviously the smaller H is and the more knowledge we have about $I \cap KH$, the more we understand the structure of I. It can be shown that for any ideal I, there exists a unique normal subgroup $\mathscr{C}(I)$, called the *controller* of I, such that $H \lhd G$ controls I if and only if $H \supseteq \mathscr{C}(I)$ (cf. [68], p. 304). Thus one aspect of the description of I is the determination of $\mathscr{C}(I)$.

Let I be a nonzero ideal of KG and let $H \lhd G$. Then an *intersection theorem* is a result which asserts, under suitable hypotheses on I, H or G/H, that $I \cap KH \neq 0$. A *generalized intersection theorem* asserts that I is controlled by H. In this section we discuss an extremely powerful generalized intersection theorem.

An ideal L of KG is said to be *annihilator free* if $\alpha . \mathfrak{u} \subseteq L$ for $\alpha \in KG$ and some infinite subgroup U of G implies that $\alpha \in L$.

If W acts on G, recall that

$$D_G(W) = \{x \in G \mid |W : C_W(x)| < \infty\}.$$

Recall also that an object is said to be W-orbital if it has only finitely many W-conjugates. Thus $D_G(W)$ is the set of W-orbital elements of G. Observe that in general it is quite possible for $D_G(W)$ not to be normal in G.

LEMMA 8.1   *Let W act on G and assume that $D = D_G(W)$ is normal in G. Let L be an annihilator-free ideal of KD which is both G-invariant and W-invariant. If $\alpha \in KG$ is W-orbital modulo $L . KG$, then $\alpha \in KD + L . KG$.*

Proof.   Let W_0 be a subgroup of finite index in W which centralizes α modulo $L . KG$. Write $\alpha = \Sigma_i \alpha_i x_i$ with $\alpha_i \in KD$ and with the x_i in distinct cosets of D in G.

Suppose some x_i has infinitely many W-conjugates modulo D. Then we can choose $w \in W_0$ with $x_i^w \notin x_j D$ for all j. For this w, the equation $\alpha^w - \alpha \in L . KG$ implies that $\alpha_i^w \in L$ and hence $\alpha_i \in L$. We may now delete all such $\alpha_i x_i$ from α and assume that all the x_i have only finitely many W-conjugates modulo D.

With this assumption, there exists a subgroup W_1 of finite index in W_0 such that W_1 centralizes all α_i and all x_i modulo D. If $w \in W_1$, then $\alpha^w - \alpha \in L . KG$ and

$$\alpha^w - \alpha = \Sigma_i \alpha_i (x_i^w - x_i) = \Sigma_i \alpha_i (x_i^w x_i^{-1} - 1) x_i.$$

Thus $\alpha_i(x_i^w x_i^{-1} - 1) \in L$ for all i and w. Finally if $x_i \notin D$, then x_i has infinitely many W-conjugates and hence infinitely many W_1-conjugates. In particular, if U_i is the subgroup of D generated by the commutators $\{x_i^w x_i^{-1} \mid w \in W_1\}$, then U_i is infinite and $\alpha_i . \mathfrak{u}_i \subseteq L$. Since L is annihilator free, we conclude that all such $\alpha_i \in L$. The only term not covered by this argument is contained in KD so we have $\alpha \in KD + L . KG$.

A group G is said to be *hyper-Δ* (or *FC-hypercentral*) if for every nontrivial homomorphic image $\bar{G} \neq 1$ of G we have $\Delta(\bar{G}) \neq 1$. More generally, suppose W acts on G. Then G is a *hyper-$D(W)$* group if for all W-invariant normal subgroups N of G with $\bar{G} = G/N \neq 1$ we have $1 \neq D_{\bar{G}}(W) \lhd \bar{G}$.

If I is an ideal of KG and $D \lhd G$, we say that I is *above* the ideal $I \cap KD$ of KD.

THEOREM 8.2 (A. E. Zalesskii [96])   *Let W act on G and assume that G is hyper-$D(W)$ with $D = D_G(W)$ normal in G. Let L be an annihilator-free ideal of KD which is both G-invariant and W-invariant. If I is a W-orbital ideal of KG above L, then D controls I.*

Proof.   We consider the set $\mathcal{N}$ of all W-invariant normal subgroups N of G containing D such that J is a W-orbital ideal of KN above L then D controls J. Certainly $D \in \mathcal{N}$. It follows from Zorn's lemma that $\mathcal{N}$ has a maximal member N. Suppose that $N \neq G$. Then we can find $H > N$ so that $H/N = D_{G/N}(W)$. We show that H is in $\mathcal{N}$.

To this end, let I be a W-orbital ideal of KH with $I \cap KD = L$. We show by induction on the number $n+1$ of cosets of N meeting $\operatorname{supp} \alpha$ that if $\alpha \in I$ then $\alpha \in L . KH$. Suppose $\alpha = \Sigma_0^n \alpha_i x_i \in I$ with $\alpha_i \in KN$ and with the x_i in distinct cosets of N in H. Replacing α by αx_0^{-1} if necessary, we may assume that $x_0 = 1$. Define J to be

$$J = \{\beta_0 \mid \Sigma_0^n \beta_i x_i \in I, \quad \beta_i \in KN\}.$$

Then J is an ideal of KN. Since I is W-orbital and $H/N = D_{G/N}(W)$, there exists a subgroup W_0 of finite index in W which normalizes I and centralizes all x_i modulo N. Clearly this group W_0 normalizes J, so J is W-orbital.

Since $I \supseteq L . KH$ we have $J \supseteq L . KN$ and $J \cap KD \supseteq L$. Conversely let $\beta_0 \in J \cap KD$. Then for suitable $\beta_i \in KN$ we have $\beta = \Sigma_0^n \beta_i x_i \in I$. Since $\beta_0 \in KD$, there exists a subgroup W_1 of finite index in W_0 which centralizes β_0. Then for all $w \in W_1$, $\beta^w - \beta \in I$ has support meeting at most n cosets of N. Hence by induction, $\beta^w - \beta \in L . KH$. In other words, β is W-orbital modulo $L . KH$ and Lemma 8.1 implies that $\beta \in KD + L . KH$. But $\beta \in I$ and $I \supseteq L . KH$ so we conclude that $\beta_0 \in I \cap KD = L$.

Thus $J \cap KD = L$ and we see that $J = L . KN$ since N is in $\mathcal{N}$. In particular, $\alpha_0 \in J = L \cdot KN \subseteq I$ and $\alpha - \alpha_0 = \Sigma_1^n \alpha_i x_i \in I$. Induction implies that $\alpha - \alpha_0 \in L \cdot KH$ and therefore $\alpha \in L \cdot KH$. Thus H is in $\mathcal{N}$, contradicting the maximality of N. We conclude that $N = G$.

If $W \lhd G$, then W acts on G by conjugation and centralizes the quotient G/W. In particular, if W is hyper-Δ, then G is hyper-$D(W)$. Thus we obtain the generalized intersection theorem

COROLLARY 8.3 (Zalesskii [96])   *Let W be a normal hyper-Δ subgroup of G. Then $D_G(W)$ controls every ideal above an annihilator-free one.*

Furthermore, since the zero ideal is annihilator free, we have the ordinary intersection theorem

COROLLARY 8.4 (Zalesskii [95])   *Let W be a normal hyper-Δ subgroup of G and set $D = D_G(W)$. If I is a nonzero ideal of KG then $I \cap KD \neq 0$.*

We remark that these methods are related to hypercentrality in group rings (see Roseblade and Smith [75]). These results are of interest if $I \cap KD$ is smaller in some sense, or easier to understand, than I itself. This will happen when W is large in G so that D, or rather the action of G on D, can be expected to be small. It can also happen when W is comparatively small, but contains $D_G(W)$, for then D equals $\Delta(W)$ and is an FC-group. We say that a subgroup D of G is a *Zalesskii-type* subgroup of G if D equals $D_G(W) \subseteq W$ for some normal hyper-Δ subgroup W of G. It was shown by Zalesskii [95] that Zalesskii-type subgroups exist when G is solvable. This is analogous to Fitting's theorem, which asserts that if G is solvable then G has a normal nilpotent subgroup containing its centralizer. We offer a slight generalization of this as Theorem 8.7.

LEMMA 8.5   *Let G, A and B be hyper-Δ groups.*

(i) *If $1 \neq N \lhd G$, then $N \cap \Delta(G) \neq 1$.*

(ii) *If $A, B \lhd H$, then AB is hyper-Δ.*

(iii) *Suppose G acts as permutations on a set Ω with each element of G moving only finitely many points. Then all G-orbits on Ω are finite.*

Proof. Parts (i) and (ii) are standard, having the same proof as the analogous result for hypercentral groups. For (iii), we may assume that $G \neq 1$ is transitive and faithful on Ω. Since $\Delta(G) \neq 1$, G has a finitely generated normal subgroup $N \neq 1$. Since each generator of N moves only finitely many points, N moves only finitely many points. But this set of points is G-invariant so Ω is finite.

LEMMA 8.6   *Let $H \lhd G$ with $G = D_G(H)$. If G/H is a hyper-Δ group, then G is hyper-Δ.*

Proof.   Since any homomorphic image of G has this same structure, we need only show that $G \neq 1$ implies $\Delta(G) \neq 1$. We do this in four steps. Note that $H = D_H(H)$ is an FC-group.

Step 1.   Let $1 \neq B \lhd G$ with $B \subseteq H$. If $\Delta(G/C_G(B)) \neq 1$, then $\Delta(G) \neq 1$.

Proof.   Let $x \in G$ map to a nonidentity element of $\Delta(G/C_G(B))$ and choose W a subgroup of finite index in G with $[x, W] \subseteq C_G(B)$. Since $x \notin C_G(B)$, fix $a \in B$ with $[x, a] \neq 1$. We show that $[x, a] \in \Delta(G)$. If $w \in W$, then $x^w = cx$ for some $c \in C_G(B)$, so $[x, a]^w = [cx, a^w] = [x, a^w]$. In other words, all W-conjugates of $[x, a]$ are of the form $[x, b]$ with $b \in B$. But the latter are finite in number since $B \subseteq H$ implies $|B : C_B(x)| < \infty$. Since $|G : W| < \infty$, this fact is proved.

Step 2.   Let $1 \neq S \lhd G$ with $S \subseteq H$. If S is semisimple, then $S \subseteq \Delta(G)$ and hence $\Delta(G) \neq 1$.

Proof.   By definition, $S = \prod_i S_i$ is the weak direct product of the finite nonabelian simple groups S_i. Let $\Omega = \bigcup_i S_i$. Since G permutes the factors S_i by conjugation, G acts as permutations on Ω. If $x \in G$, then since $S \subseteq H$ we have $|S : C_S(x)| < \infty$, so x centralizes a normal subgroup of S of finite index. But the normal subgroups of S are partial products of the S_i, so x moves only finitely many points of Ω. Since H is an FC-group, the H-orbits on Ω are all finite. Now G/H acts on the set Λ of these orbits and, by Lemma 8.5 (iii), the orbits of G/H on Λ are finite. Thus the orbits of G on Ω are finite, so $\Omega \subseteq \Delta(G)$ and hence $S \subseteq \Delta(G)$.

Step 3.   If H has finite residual type, then $\Delta(G) \neq 1$.

Proof.   We say H has residual type n if it is a subdirect product of finite groups of order $\leqslant n$. Moreover, H has finite residual type if it has residual type n for some n. We suppose H has residual type n and proceed by induction on n. If $n = 1$, then $H = 1$ so $G = G/H$ is hyper-Δ and hence $\Delta(G) \neq 1$. Now suppose $n > 1$ and note that H is an FC-group. If H is torsion free, then H is abelian and the result follows from Step 1 with $B = H$. Thus we assume that H has torsion elements and we can choose $1 \neq A$ to be a finite minimal normal subgroup of H. Then A

is either abelian or semisimple. In the latter case, $S = A^G$ is also semi-simple and Step 2 yields the result.

Finally assume A is abelian and let A^g be any G-conjugate of A. Since H has residual type n, there exists $N \lhd H$ with $|H/N| \leqslant n$ and $A^g \not\subseteq N$. But A^g is also a minimal normal subgroup of H, so $A^g \cap N = 1$ and hence N centralizes A^g. Thus $C_H(A^g) \supseteq A^g \times N > N$ and we have $|H : C_H(A^g)| < n$. Set $B = A^G$ and consider $\bar{G} = G/C_G(B)$. Observe that $\bar{H} = HC_G(B)/C_G(B) \simeq H/C_H(B)$ and $C_H(B) = \cap_g C_H(A^g)$. Thus $\bar{H}$ is of finite residual type less than n. If $\bar{G} = 1$, then B is central and $\Delta(G) \neq 1$. If $\bar{G} \neq 1$ then, by induction, $\Delta(\bar{G}) \neq 1$ and Step 1 yields this result.

Step 4. $\Delta(G) \neq 1$.

Proof.   We may assume $H \neq 1$ and let $1 \neq A$ be a finitely generated normal subgroup of H. Then $|H : C_H(A)| = n$ for some n and we set $B = A^G$. As above, if $\bar{G} = G/C_G(B)$ and if $\bar{H} = HC_G(B)/C_G(B)$, then $\bar{H}$ is of residual type n. Thus if $\bar{G} \neq 1$, then $\Delta(\bar{G}) \neq 1$ by Step 3 and $\Delta(G) \neq 1$ by Step 1. On the other hand, if $\bar{G} = 1$ then B is central and the lemma is proved.

We now follow the argument in the solvable case to obtain

THEOREM 8.7   *Let G be a group with a normal series*

$$1 = G_0 \subseteq G_1 \subseteq \ldots \subseteq G_n = G$$

such that each quotient G_{i+1}/G_i is hyper-Δ. Then G has a normal hyper-Δ subgroup W with $D_G(W) \subseteq W$.

Proof.   We define the subgroups W_i of G_i inductively by $W_0 = 1$ and $W_{i+1} = W_i . D_{G_{i+1}}(W_i)$. Since $G_i \lhd G$ we see by induction that $W_i \lhd G$. Moreover, $W_{i+1} \supseteq W_i$ so $D_{G_{i+1}}(W_{i+1}) \subseteq D_{G_{i+1}}(W_i) \subseteq W_{i+1}$. We now prove by induction that each W_i is hyper-Δ. Set $B = D_{G_{i+1}}(W_i)$ and

$$H = B \cap G_i = D_{G_i}(W_i) \subseteq W_i$$

so $B = D_B(H)$. Furthermore, B/H is contained isomorphically in the hyper-Δ group G_{i+1}/G_i, so B/H is hyper-Δ. Thus B is hyper-Δ by Lemma 8.6, and hence so is $W_{i+1} = W_i . B$ by Lemma 8.5 (ii). The result follows with $W = W_n$.

The above certainly applies when G is $\mathfrak{P}\mathfrak{F}$, so such groups always have Zalesskii-type subgroups. In this case we have a more precise construction. Let F denote the largest finite normal subgroup of G and let $F \subseteq D \subseteq W \subseteq G$ where W/F is the Fitting subgroup of G/F and D/F is the center of W/F. Then W is finite-by-nilpotent, so surely hyper-Δ, and it is easy to see (cf. [68], p. 518) that $D = D_G(W)$. We call D *the Zalesskii subgroup* of the $\mathfrak{PF}$-group G.

9.   Orbitally Sound Groups

Let G be a $\mathfrak{PF}$-group and let D be its Zalesskii subgroup. Then an ideal I of KG will be controlled by D provided $I \cap KD$ is annihilator free. Thus we seek reasonable sufficient conditions for the latter to occur. There are actually two aspects to this problem. First, we cannot allow I to contain u for U an infinite subgroup of D. Second, we require that G satisfy a certain group-theoretic property. We consider the latter first.

We say that the $\mathfrak{PF}$-group G is *orbitally sound* if for every orbital subgroup H of G we have $|H : \mathrm{core}_G H|$ finite. Here $\mathrm{core}_G H$ is the largest normal subgroup of G contained in H and H orbital means as usual that $|G : N_G(H)|$ is finite. It is not true that every $\mathfrak{PF}$-group G is orbitally sound, but we will show that any such group has an orbitally sound normal subgroup of finite index and we will characterize the largest such G_0. It is clear that the orbitally sound property is inherited by homomorphic images and by subgroups of finite index. Furthermore, there are several equivalent formulations of this property which we consider later. The main goal, however, is to prove the existence of G_0.

Let $\mathbb{Q}$ denote the field of rational numbers. A finite-dimensional $\mathbb{Q}G$-module V is said to be a *rational plinth* for G if V is an irreducible $\mathbb{Q}L$-module for all subgroups L of finite index in G. If A is a finitely generated torsion-free abelian group and if G acts on A, then A is a plinth for G if and only if $V = A \otimes \mathbb{Q}$ is a rational plinth. Thus A is a plinth if and only if no proper pure subgroup of A is G-orbital.

A normal series

$$1 = G_0 \subseteq G_1 \subseteq \ldots \subseteq G_n = G$$

is called a *plinth series* for G if each quotient G_i/G_{i-1} is either finite or a plinth. It is not necessarily true that any G has a plinth series, but if it does, certain other properties follow. Assume that G does have a plinth series and suppose we refine it by adding one more term, say $G_{i-1} < T < G_i$. If G_i/G_{i-1} is finite, then we have merely added one more finite quotient. If $A = G_i/G_{i-1}$ is a plinth, then $B = T/G_{i-1}$, being G-invariant, must have finite index in A. Hence again we have added just one more finite quotient, namely G_i/T. Furthermore, since $|A:B|$ is finite, we see that $A \otimes \mathbb{Q} = B \otimes \mathbb{Q}$ as $\mathbb{Q}G$-modules. Thus this new series is also a plinth series. It now follows immediately from the Schreier–Zassenhaus theorem that any normal series for G can be refined to a plinth series. Furthermore, each i with G_i/G_{i-1} a plinth gives rise to a rational plinth $V_i = (G_i/G_{i-1}) \otimes \mathbb{Q}$ and the collection of these rational plinths $V_1, V_2, \ldots, V_t$ is independent of the particular series and hence is an invariant of G.

Suppose L is a subgroup of finite index in G and set $L_i = L \cap G_i$. Then $1 = L_0 \subseteq L_1 \subseteq \ldots \subseteq L_n = L$ is a plinth series for L. Indeed if G_i/G_{i-1} is a plinth for G, then L_i/L_{i-1} is an L-invariant subgroup of G_i/G_{i-1} of finite index and hence the rational plinth for L is precisely the restriction V_{iL}. Finally, if $N \lhd G$, then we can refine the series $1 \subseteq N \subseteq G$ to a plinth series. By considering the part of this series above N, we see that G/N has a plinth series and we can recognize the appropriate rational plinths for G/N.

We say that G has *property* (*) if:

(i) G has a plinth series with rational plinths $V_1, V_2, \ldots, V_t$;
(ii) for all i, $G/C_G(V_i)$ is abelian; and
(iii) if L is any subgroup of G of finite index, then $V_i \cong V_j$ if and only if $V_{iL} \cong V_{jL}$.

In view of our remarks above, property (*) is inherited by subgroups of finite index and by homomorphic images. The goal now is to show that any G has a normal subgroup of finite index with property (*) and then that groups with this property are orbitally sound.

LEMMA 9.1 *Let G be in $\mathfrak{PF}$. Then G has a subgroup of finite index with property* (*), *which can be chosen normal in G.*

Proof. We may assume that G is infinite. Then by Lemma 5.6, G has a subgroup L of finite index and an L-plinth $A \lhd L$ such that $L/C_L(A)$ is abelian. By considering L/A and applying induction on $h(G)$, we conclude that L has a subgroup of finite index satisfying (i) and (ii).

We may now assume that G satisfies (i) and (ii). If $V_{iT} \cong V_{jT}$ for some subgroup T of finite index in G, choose one such T, say T_{ij}. Then setting $T_0 = \bigcap_{ij} T_{ij}$, we see that $|G:T_0|$ is finite and T_0 has property (*).

Thus we have achieved the first goal.

LEMMA 9.2   *Let G have property $(*)$ and suppose $1 < B < C \subseteq L \subseteq G$ is a normal series for G with G/L finite. Furthermore, suppose $C = A \times B$, where A and B are normal plinth subgroups of L. Then A is normal in G.*

Proof.   It is clear that B and C/B are plinths for G and hence $V_1 = B \otimes \mathbb{Q}$ and $V_2 = (C/B) \otimes \mathbb{Q}$ are rational plinths. Let $W = C \otimes \mathbb{Q}$. Then W is a $\mathbb{Q}G$-module with submodule V_1. Furthermore, since $C = B \times A$ with $A \lhd L$, we see that $W_L = V_{1L} + (A \otimes \mathbb{Q})$ and that $A \otimes \mathbb{Q} \simeq V_{2L}$. We show now that $A \otimes \mathbb{Q}$ is a $\mathbb{Q}G$-submodule of W.

Suppose first that $V_{1L} \not\cong V_{2L}$. Then $A \otimes \mathbb{Q}$ and $V_1 = B \otimes \mathbb{Q}$ are the unique irreducible $\mathbb{Q}L$-submodules of W. Since $L \lhd G$, G permutes these submodules, and hence since G fixes V_1 it must also fix $A \otimes \mathbb{Q}$. Thus $A \otimes \mathbb{Q}$ is a $\mathbb{Q}G$-submodule in this case.

Now suppose $V_{1L} \cong V_{2L}$. Then since G has property $(*)$ we know that $V_1 \cong V_2$. Observe that $|G/L|$ is finite and that V_{1L} is a direct summand of W_L. Thus since $\mathbb{Q}$ has characteristic 0, a lemma of D. G. Higman implies that V_1 is a direct summand of W as a $\mathbb{Q}G$-module. Hence $W \cong V_1 \oplus V_2 \cong V_1 \oplus V_1$. Let ρ denote the irreducible representation of $\mathbb{Q}G$ associated with V_1. Then since $\dim_{\mathbb{Q}} V_1$ is finite and $G/C_G(V_1)$ is abelian, $\rho(\mathbb{Q}G) = F$ is a finite field extension of $\mathbb{Q}$ and V_1 is a one-dimensional F-vector space. Hence W is a 2-dimensional F-space. Furthermore, $\rho(\mathbb{Q}L) \subseteq \rho(\mathbb{Q}G) = F$ and since $\mathbb{Q}L$ acts irreducibly on V_1 we see that $\rho(\mathbb{Q}L) = F$ also. But $A \otimes \mathbb{Q}$, being a $\mathbb{Q}L$-submodule of W, is an F-subspace of W and hence we conclude again that $A \otimes \mathbb{Q}$ is G-invariant.

Finally, since A is a pure subgroup of C, we have $A = C \cap (A \otimes \mathbb{Q})$. Thus, since $A \otimes \mathbb{Q}$ is G-invariant we conclude that $A \lhd G$.

LEMMA 9.3   *If G has property $(*)$, then G is orbitally sound.*

Proof.   We proceed by induction on $h(G)$ and let H be an orbital subgroup of G. Since $(*)$ is inherited by homomorphic images, we may divide out by $\mathrm{core}_G H$ and assume that $\mathrm{core}_G H = 1$. Suppose by way of contradiction that H is infinite. Then, by Lemma 5.6, there exist subgroups $A \subseteq H$ and $L \lhd G$ with $A \lhd L$, A a plinth for L and G/L finite. Since $A \subseteq H$ we have $\mathrm{core}_G A = 1$. Now L is infinite, so L has a characteristic torsion-free subgroup T, and, by refining the normal series $1 < T \subseteq L \subseteq G$ to a plinth series, we see that G has a normal subgroup $B \subseteq L$ which is a plinth for G.

If $A \cap B \neq 1$, then since $A \lhd L$ and B is a plinth for L, it follows that $|B : A \cap B|$ is finite, and hence that $A \cap B \supseteq B^m$ for some m. But $B^m \lhd G$, so this contradicts $\mathrm{core}_G A = 1$. Thus we have $A \cap B = 1$.

Finally, $h(G/B) < h(G)$ and AB/B is orbital in G/B. Thus by induction there exists a subgroup $C \lhd G$ with $B \subseteq C \subseteq AB$ and $|AB:C|$ is finite. Thus since $A \cap B = 1$ and $A, B \lhd L$, we see that $C = B \times (A \cap C)$. Moreover, $A \cap C \lhd L$ and $|A: A \cap C|$ is finite. Thus $A \cap C$ is also a plinth of L, so Lemma 9.2 applies. We conclude that $A \cap C \lhd G$ and again we obtain the contradiction that $\mathrm{core}_G A \neq 1$.

Lemmas 9.1 and 9.3 show that any $\mathfrak{P}\mathfrak{F}$-group has an orbitally sound normal subgroup of finite index. We now proceed to sharpen this result. A subgroup H of G is said to be an *isolated orbital subgroup* if H is orbital and has infinite index in every orbital subgroup L properly containing it. If H is any orbital subgroup of G, define its *isolator* $i_G(H)$ to be

$$i_G(H) = \langle L \mid L \supseteq H \text{ is orbital and } |L:H| \text{ is finite} \rangle.$$

LEMMA 9.4   *If H is an orbital subgroup of G, then $i_G(H)$ is the unique isolated orbital subgroup of G containing H and having H as a subgroup of finite index.*

Proof.   We first show that if L_1, L_2 are orbital subgroups of G with each $|L_i: H|$ finite, then $|\langle L_1, L_2 \rangle : H|$ is finite. For this it suffices to assume that $G = \langle L_1, L_2 \rangle$ and that $\mathrm{core}_G H = 1$. Since these subgroups are all orbital and $|L_i: H|$ is finite, we can find a normal subgroup A of G of finite index such that A normalizes H and $[A, L_i] \subseteq H$ for $i = 1, 2$. The commutator identity $[a, xy] = [a, y][a, x][a, x, y]$ now implies that if $[A, x] \subseteq H$ and $[A, y] \subseteq H$, then $[A, xy] \subseteq H$. Indeed $A \lhd G$ implies that $[a, x] \in A$ and hence $[a, x, y] \in [A, y] \subseteq H$. Thus since $[A, L_i] \subseteq H$ we have $[A, G] \subseteq H$ and hence, since $A \lhd G$, $[A, G] \subseteq \mathrm{core}_G H = 1$. In other words, A is a central subgroup of finite index, and since $\mathrm{core}_G H = 1$ we have $A \cap H = 1$ and H is finite. Therefore, L_1 and L_2 are finite orbital subgroups of G and hence, by Dietzmann's lemma, $G = \langle L_1, L_2 \rangle$ is finite.

Finally, since G satisfies max, we choose L maximal subject to L orbital and $|L:H|$ finite. By the above, any orbital subgroup L_i with $|L_i: H|$ finite must be contained in L. Thus $i_G(H) = L$ has the appropriate properties.

Define $\mathrm{nio}(G)$ to be the intersection of the normalizers of all the isolated orbital subgroups of G. Then we have

THEOREM 9.5 (Roseblade [74])   *Let G be a $\mathfrak{P}\mathfrak{F}$-group. Then G is orbitally sound if and only if all its isolated orbital subgroups are normal.*

Furthermore, nio(G) *is the largest normal orbitally sound subgroup of G of finite index.*

Proof. Suppose G is orbitally sound and H is isolated orbital. If $H_0 = \text{core}_G H$, then $|H:H_0|$ is finite so $H = i_G(H_0)$. But G normalizes H_0 so it normalizes $i_G(H_0) = H$. Conversely, assume all such isolated orbitals are normal and let H be orbital. Then $|i_G(H):H|$ finite implies that H contains a characteristic subgroup of finite index in $i_G(H)$. Hence since $i_G(H) \lhd G$ we have $|H:\text{core}_G H|$ finite.

Now let G_0 be any orbitally sound normal subgroup of finite index in G. We already know that such a subgroup exists. If H is an isolated orbital subgroup of G, then $H_0 = H \cap G_0$ is G_0-orbital and of finite index in H. By Lemma 9.4, H must be $i_G(H_0)$ and so H_0 is an isolated orbital subgroup of G_0. By the previous paragraph, $H_0 \lhd G_0$. But then G_0 normalizes $H = i_G(H_0)$ so $G_0 \subseteq \text{nio}(G)$ and $|G:\text{nio}(G)|$ is finite.

Conversely let H_0 be an isolated orbital subgroup of nio(G). Then nio(G) normalizes $H = i_G(H_0)$ and hence also $H_0 = H \cap \text{nio}(G)$. Thus nio($G$) is orbitally sound.

We now see how this applies. If I is an ideal of KG, we let $I^\dagger = \{x \in G \mid x - 1 \in I\} = (I+1) \cap G$. Since $I^\dagger$ is the kernel of the natural homomorphism $G \to KG/I$, we have $I^\dagger \lhd G$. We say that I is *faithful* if $I^\dagger = 1$ and that I is *almost faithful* if $I^\dagger$ is finite. Since I is the complete inverse image of a faithful ideal of $K[G/I^\dagger]$, it usually suffices to study faithful ideals. The following lemma is an immediate consequence of the Noetherian property of KG (cf. [74], Lemma 5).

LEMMA 9.6 *Let P be a prime ideal of KG and let $H \lhd G$. Then there exists a G-orbital prime ideal Q of KH with $P \cap KH = \bigcap_{x \in G} Q^x$.*

Finally we have

THEOREM 9.7 (Roseblade [74]) *Let G be an orbitally sound $\mathfrak{PF}$-group and let D be its Zalesskii subgroup. If P is a faithful prime ideal of KG, then D controls P. Furthermore, there exists a G-orbital almost faithful prime ideal Q of KD with $P \cap KD = \bigcap_{x \in G} Q^x$.*

Proof. By Lemma 9.6 there exists a G-orbital prime ideal Q of KD with $P \cap KD = \bigcap_{x \in G} Q^x$. Since $(Q^x)^\dagger = (Q^\dagger)^x$, it is clear that $Q^\dagger$ is a G-orbital subgroup of G. The orbitally sound property now implies that there exists $N \lhd G$ with $|Q^\dagger:N|$ finite. But then $\mathfrak{n} \subseteq Q^x$ for all $x \in G$, so $\mathfrak{n} \subseteq P$ and hence $N \subseteq P^\dagger = 1$. Thus $Q^\dagger$ is finite and Q is almost faithful.

We can now show that $P \cap KD$ is annihilator free. Suppose $\alpha \cdot \mathfrak{u} \subseteq$

$P \cap KD$ for $\alpha \in KD$ and U some infinite subgroup of D. Since D is an FC-group, it has a center of finite index. Hence, by replacing U by a subgroup of finite index, we can assume that U is an infinite central subgroup. But then, since Q^x is prime, $\alpha . \mathfrak{u} \subseteq Q^x$ and $\mathfrak{u} \not\subseteq Q^x$, we conclude that $\alpha \in Q^x$ for each $x \in G$. Thus $\alpha \in P \cap KD$ and Corollary 8.3 yields the result.

10.   Orbital Primes in Abelian Group Rings

Let G be a polycyclic-by-finite group. Then, by Theorem 9.7, G has an orbitally sound normal subgroup $G_0 = \mathrm{nio}(G)$ of finite index. Furthermore, if P is a faithful prime ideal of KG_0 and if D is the Zalesskii subgroup of G_0, then P is controlled by D. It remains to study $P \cap KD$. We do know that D is an FC-group and that $P \cap KD = \bigcap_{x \in G_0} Q^x$, where Q is a G_0-orbital almost faithful prime of KD. Thus the structure of D is nice, but we still know very little about the action of G_0 on D. It would certainly be better if we could replace D by the smaller group $\Delta(G_0)$ which is centralized by some subgroup of finite index in G_0. It is the goal of this section to relate the prime ideals of the group rings of D and $\Delta(G_0)$. For this we prove a far-reaching generalization of Theorem 5.4.

In Lemmas 10.1–10.3, A is a finitely generated abelian group acted upon by the arbitrary group G. Moreover, P is a G-orbital prime ideal of KA and $\overline{}: KA \to KA/P$ denotes the natural epimorphism onto the commutative integral domain $\overline{KA}$. We recall from section 5 that if $P > (P \cap KU) . KA$, then $W \supseteq U$ is a $\log_U$ subgroup for P if for all $\alpha \in P \backslash (P \cap KU) . KA$ there exist $x, y \in \mathrm{supp}\, \alpha$ with $xy^{-1} \in W \backslash U$.

LEMMA 10.1   *Let $A > B > C$ with B a maximal isolated orbital subgroup, with C orbital and with B/C torsion-free. Suppose $P > (P \cap KB) . KA$ and $P \cap KB = (P \cap KC) . KB$. Then there exists an isolated orbital subgroup $W > C$ such that $B \cap W = C$ and $|A : BW|$ is finite. Furthermore, $\overline{KA}$ is not integral over $\overline{KW}$.*

Proof.   Since B/C is torsion-free, we have $B = C \times D$ and D can be given a linear ordering. Since $P \cap KB = (P \cap KC) . KB$, it follows easily that $\overline{KB} \cong \overline{KC}[D]$, the group ring of D over $\overline{KC}$. Hence if $D^+ = \{d \in D \mid d \geqslant 1\}$ under some fixed ordering, then $\overline{KC}[D^+]$ is a ring and there is a homomorphism $\overline{KC}[D^+] \to \overline{KC}$ which is the identity on $\overline{KC}$ and which sends each $d > 1$ to zero. By the Extension Theorem for Places, this homomorphism extends to a place on $\overline{KA}$ with corresponding valuation v.

Observe that $P > (P \cap KB) \cdot KA = (P \cap KC) \cdot KA$ and that $\overline{KC}$ is contained in the valuation ring O_v but meets its maximal ideal M_v in zero. Thus, by Lemma 5.2, $W_v = \{a \in A \mid v(\bar{a}) = 1\}$ is a $\log_C$ subgroup for P and A/W_v is torsion-free. Furthermore, $W_v > C$ and, by construction, $W_v \cap D = 1$ so $W_v \cap B = C$. By Lemma 5.1, W_v contains a minimal $\log_C$ subgroup $E > C$ which is orbital. Then $B \cap E = C$ so $BE > B$ is orbital and hence $|A:BE|$ is finite, since B is maximal isolated orbital. Now $E/C \cong BE/B$ and $W_v/C \cong BW_v/B$ so $|W_v:E|$ is finite. Hence W_v/E is the torsion subgroup of A/E and W_v is therefore also orbital.

Finally, since $\overline{KW_v} \subseteq O_v$ and $\bar{D} \nsubseteq O_v$, we see that $\overline{KA}$ is not integral over $\overline{KW_v}$. The result follows with $W = W_v$.

The next lemma uses results of commutative ring theory to guarantee that only finitely many valuations of interest occur. The necessary background material can be found in [48] (especially p. 76).

LEMMA 10.2 *Let B be a maximal isolated orbital subgroup of A. If $P > (P \cap KB) \cdot KA$ and if A/B is not infinite cyclic, then $\overline{KA}$ is integral over $\overline{KB}$.*

Proof. By Lemma 5.3, $\overline{KA}$ is algebraic over $\overline{KB}$. Thus if S is the field of fractions of $\overline{KA}$ and if T is the field of fractions of $\overline{KB}$, then $|S:T|$ is finite. Furthermore, if I is the integral closure of $\overline{KB}$ in S, then since $\overline{KB}$ is a finitely generated K-algebra, the same is true of I and thus I is Noetherian. Suppose by way of contradiction that $\overline{KA}$ is not integral over $\overline{KB}$. Then $\overline{KA} \nsubseteq I$ and hence $\overline{KA} \nsubseteq \cap I_p$. Here p ranges over the height-1 primes of I and I_p denotes the localization at p. But KA is finitely generated over K so the set Λ of height-1 primes with $\overline{KA} \nsubseteq I_p$ is finite.

Since P and B are G-orbital, there is a subgroup G_0 of finite index in G which normalizes P and B. Thus G_0 acts on $\overline{KA}$ and $\overline{KB}$, so it acts on I and hence permutes the finitely many primes in Λ. Observe that each I_p is a discrete valuation subring of S with valuation v_p. Thus if $W_p = \{a \in A \mid v_p(\bar{a}) = 1\}$, then it is clear from all of the above that W_p is orbital and that $W_p \supseteq B$. Now v_p is discrete, so A/W_p is a subgroup of an infinite cyclic group. Since B is maximal isolated orbital with A/B not infinite cyclic, we conclude that $W_p = A$ for all $p \in \Lambda$. But then $\overline{KA} \subseteq I_p$, a contradiction. Thus $\overline{KA}$ is integral over $\overline{KB}$.

We remark that the conclusion of Lemma 10.2 is not true if A/B is infinite cyclic. We now assume in addition that P is an almost faithful prime and we set $F = D_A(G) = \{a \in A \mid |G:C_G(a)| < \infty\}$. Then A/F is torsion-free. Indeed, suppose $b \in A$ with $b^n \in F$. Then $|G: C_G(b^n)|$ is finite and $C_G(b^n)$ permutes the finitely many nth roots of b^n. Thus b is orbital and $b \in F$. The goal is to show that F controls P. We start with

LEMMA 10.3 *The subgroup F controls P if either*

(i) *F is a maximal orbital subgroup of A, or*
(ii) *every element of A/F is orbital.*

Proof. (i) Since F is finitely generated, there exists a subgroup G_0 of finite index in G which centralizes F and normalizes P. Suppose $P > (P \cap KF) . KG$. Then, by Lemma 5.3, $\overline{KA}$ is algebraic over $\overline{KF}$. Since G_0 acts on $\overline{KA}$ and centralizes $\overline{KF}$, we conclude that G_0 induces a finite group of automorphisms on $\overline{KA}$. But $A/P^{\dagger}$ is contained isomorphically in $\overline{KA}$ so every element of $A/P^{\dagger}$ is also G_0-orbital. Finally, since P is almost faithful, $P^{\dagger}$ is finite and this yields $A = F$, a contradiction. Thus $P = (P \cap KF) . KA$.

(ii) Since P is almost faithful, $P \cap KF$ is annihilator free. Furthermore, since each element of A/F is orbital we see that A is a hyper-$D(G)$ group. Theorem 8.2 yields the result.

We can now prove the appropriate extension of Theorem 5.4.

THEOREM 10.4 (Roseblade [74]) *Let A be a finitely generated abelian group with a group of operators G and set $F = D_A(G)$. If P is a G-orbital almost faithful prime ideal of KA, then F controls P.*

Proof. We proceed by induction on $h(A)$. We may assume that $A \neq F$ and in particular that $h(A) > 0$. By Lemma 10.3(i) we may further assume that F is not maximal isolated orbital in A. Suppose by way of contradiction that $P > (P \cap KF) . KA$ and recall that A/F is torsion-free.

Let B be any maximal isolated orbital subgroup of A containing F. By induction, $P \cap KB = (P \cap KF) . KB$, and therefore $P > (P \cap KB) . KA$. Hence, by Lemma 5.3, $\overline{KA}$ is algebraic over $\overline{KB}$. Since B/F is torsion-free, we have $B = F \times D$ and then $P \cap KB = (P \cap KF) . KB$ implies that $\overline{KB} \cong \overline{KF}[D]$, the group ring of D over $\overline{KF}$. Thus computing transcendence degrees over $\overline{KF}$ we have

$$\text{t.d. } \overline{KA} = \text{t.d. } \overline{KB} = h(D).$$

Therefore $h(B/F) = h(D)$ is a constant for all such B and hence $r = h(A/B)$ is also invariant.

There are two cases to consider according as $r = 1$ or $r > 1$.

Case $r = 1$. We show that the elements of A/F are orbital. Set $U/F = D_{A/F}(G)$, so that A/U is torsion free and U is G-invariant. If $A \neq U$, then there exists a maximal isolated orbital subgroup B with $A > B \supseteq U$. Since $P \cap KB = (P \cap KF) . KB$, Lemma 10.1 applies with $C = F$ and there exists an orbital subgroup $W > F$ with $B \cap W = F$ and $|A : BW|$ finite. But $r = 1$ so A/B is infinite cyclic and hence W/F is infinite cyclic. It follows that the elements of W/F are orbital, so $W \subseteq U$. But then $B \cap W \supseteq W > F$, a contradiction. Thus $A = U$ and this contradicts Lemma 10.3(ii).

Case $r > 1$. Since $A/F \neq 1$ is torsion-free and F is not maximal isolated orbital, we can find subgroups B and C with $A > B > C \supseteq F$, with B a maximal isolated orbital subgroup of A and with C a maximal isolated orbital subgroup of B. Since $P \cap KB = (P \cap KF) . KB$ and $P \cap KC = (P \cap KF) . KC$ by induction, we have $P \cap KB = (P \cap KC) . KB$ and Lemma 10.1 applies. We conclude that there exists an orbital subgroup $W > C$ with A/W torsion free, $W \cap B = C$ and $|A : BW|$ finite. Note that G has a subgroup G_0 of finite index which normalizes B, C and W and that $BW/W \cong B/C$ as G_0-modules. Thus W is maximal isolated orbital in BW and since $|A : BW|$ is finite and A/W is torsion-free, we conclude that W is a maximal isolated orbital subgroup of A. By Lemma 10.1, $\overline{KA}$ is not integral over $\overline{KW}$. On the other hand, since $r > 1$, A/W is not infinite cyclic. Thus Lemma 10.2 asserts that $\overline{KA}$ is integral over $\overline{KW}$, and we have the required contradiction.

A consequence of this is an affirmative solution to the ergodic conjecture of Farkas and Passman [22].

COROLLARY 10.5 [74] *Let A be a finitely generated torsion-free abelian group and let G be a group of operators on A with $D_A(G) = 1$. If Q is a faithful prime ideal of KA, then $\bigcap_{x \in G} Q^x = 0$.*

Finally, since finitely generated FC-groups are close to abelian, the following corollary comes as no surprise.

COROLLARY 10.6 [74] *Let H be a finitely generated FC-group with a group of operators G and set $F = D_H(G)$. If P is a G-orbital almost faithful prime ideal of KH, then $F \lhd H$ and F controls P.*

11. Prime Ideals in Polycyclic Group Rings

Let G be a $\mathfrak{P}\mathfrak{F}$-group and let P be a prime ideal of KG. We say that P is a *standard prime* if:

(i) $\Delta(G)$ controls P;
(ii) $P \cap K[\Delta(G)] = \bigcap_{x \in G} Q^x$ with Q an almost faithful prime of $K\Delta(G)$.

Furthermore, we say P is *image standard* if the image of P in $K[G/P^\dagger]$ is a standard prime. The key result is then

THEOREM 11.1 (Roseblade [74]) *Let G be a $\mathfrak{P}\mathfrak{F}$-group. If G is orbitally sound, then all prime ideals of KG are image standard.*

Proof. Let P be a prime ideal of KG. Since the orbitally sound property is inherited by homomorphic images, we may assume that P is faithful. If D denotes the Zalesskii subgroup of G, then by Theorem 9.7 $P = (P \cap KD) . KG$ and there is a G-orbital almost faithful prime ideal Q of KD with $P \cap KD = \bigcap_{x \in G} Q^x$.

We study the action of G on the FC-group D. If F denotes the G-orbital elements of D, then $F = D_D(G) = \Delta(G)$. Thus since Q is a G-orbital almost faithful prime of KD, we conclude from Corollary 10.6 that $Q = (Q \cap KF) . KD$. It then follows easily that $P = (P \cap KF) . KG$ with $P \cap KF = \bigcap_{x \in G} (Q \cap KF)^x$. Finally, by Lemma 9.6, $Q \cap KF = \bigcap_{y \in D} T^y$ for some prime ideal T of KF. Furthermore, since Q is almost faithful and $|D: Z(D)|$ is finite, we see that T is almost faithful. Thus since $P \cap KF = \bigcap_{x \in G} T^x$ and $F = \Delta(G)$, we conclude that P is standard.

Thus the structure of the primes P for G orbitally sound depend only upon the groups $P^\dagger$ and $\Delta(G/P^\dagger)$. Since the latter group is essentially a central factor of G (its centralizer in G has finite index), the above theorem reduces a highly noncommutative question to a commutative one. The first result along this line was proved in Zalesskii [93]. It concerned primitive ideals in group rings of nilpotent groups and it influenced some of the later developments.

In addition, by Theorem 9.6, $|G: \text{nio } G|$ is finite and nio G is orbitally sound. This allows certain ring-theoretic parameters to be computed. We mention one example.

The prime length of a ring R is the largest n for which there exists a chain of primes $P_0 > P_1 > \ldots > P_n$. If such chains are of unbounded length, then the prime length of R is infinite. If G is in $\mathfrak{P}\mathfrak{F}$, then G has a normal subgroup H of finite index with a plinth series. The observations of section 9 imply that the number of infinite factors in the series is an invariant for G called its *plinth length*.

THEOREM 11.2 [74]   *If G is in $\mathfrak{PF}$, then the prime length of KG is equal to the plinth length of G.*

This was known previously for G finitely generated nilpotent and in the general case it was known that the prime length of KG was bounded above by the Hirsch number $h(G)$ (see Smith [83]).

It is not known whether for all primes $P \supseteq Q$ of KG every saturated chain of primes joining P to Q has the same length. It is true however for orbitally sound groups G.

It remains to describe the prime ideals of KG for G an arbitrary polycyclic-by-finite group. The argument here contains three ingredients. The first is Theorem 11.1, which leaves us a mere finite index away from G. The second is the general theory of prime ideals in crossed products of finite groups, as developed by Lorenz and Passman [56]. This applies since, if H is a normal subgroup of G of finite index, then KG is a crossed product of the finite group G/H over the ring KH. The final ingredient is the following Δ-method result, which generalizes Lemma 3.1. Recall that $\theta: KG \to K[\Delta(G)]$ is the natural projection.

PROPOSITION 11.3 (Lorenz and Passman [55])   *Let G be a $\mathfrak{PF}$-group and let I be an ideal of KG satisfying*

 (i) *I is controlled by $\Delta = \Delta(G)$, and*
 (ii) *$I \cap K\Delta = \bigcap_i Q_i$ where each Q_i is an almost faithful prime ideal of $K\Delta$.*

If J_1 and J_2 are ideals of KG with $J_1 J_2 \subseteq I$, then $\theta(J_1)\theta(J_2) \subseteq \theta(I)$.

We deduce from this that I is a semiprime ideal. Furthermore, if the Q_i are all G-conjugate, then I is prime.

Let H be a subgroup of G and let L be an ideal of KH. Then we define L^G by
$$L^G = \bigcap_{x \in G} L^x . KG.$$

It is a simple matter to see that L^G is the largest ideal contained in $L.KG$. Observe that if $H \lhd G$ then $L^G = (\bigcap_{x \in G} L^x) . KG$ and therefore H controls L^G.

Again if $H \subseteq G$ we let $\nabla_G(H)$ denote the complete inverse image in $N_G(H)$ of $\Delta(N_G(H)/H)$. Finally, suppose $N \lhd G$ and I is an ideal of KG. Then we say that I is almost N-faithful if $I^\dagger \subseteq N$ and $|N: I^\dagger| < \infty$. We now state the three results of Lorenz and Passman [58].

THEOREM 11.4 (Existence)   *Let G be in $\mathfrak{PF}$ and let P be a prime ideal of KG. Then there exists an isolated orbital subgroup N of G and an almost N-faithful prime ideal L of $K[\nabla_G(N)]$ with $P = L^G$.*

We call any such N above a *vertex* for P. Moreover, for this N, any such L is a *source* of P.

THEOREM 11.5 (Uniqueness)   *If P is a prime ideal of KG, then the vertices of P are unique up to conjugation in G. Furthermore, if N is any such vertex, then the sources for this N are unique up to conjugation by $N_G(N)$.*

THEOREM 11.6 (Converse)   *If N is an isolated orbital subgroup of G and if L is an almost N-faithful prime ideal of $K[\nabla_G(N)]$, then L^G is a prime ideal of KG.*

To see how these results relate to Theorem 11.1, let G be orbitally sound and let P be a prime ideal of KG. By Theorem 11.4, let N be a vertex for P and L a corresponding source. Since N is an isolated orbital subgroup, it follows from Theorem 9.5 that $N \lhd G$. Hence $D = \nabla_G(N) \lhd G$ and by Theorem 11.4 $P = (P \cap KD) . KG$ with $P \cap KD = \bigcap_{x \in G} L^x$. Furthermore, $|N : L^\dagger|$ is finite and hence, since $N \lhd G$, we see that $|N : P^\dagger|$ is finite. In particular, if P is faithful, then N is finite, $D = \Delta(G)$ and P is a standard prime. Thus in the orbitally sound case, Theorem 11.4 recaptures Theorem 11.1 and the vertex of an arbitrary prime P is $i_G(P^\dagger)$. Moreover, we have

COROLLARY 11.7   *Let G be a $\mathfrak{P}\mathfrak{F}$-group. If all prime ideals of KG are image standard, then G is orbitally sound.*

Proof.   Let N be an isolated orbital subgroup of G. If $L = \mathfrak{n} . K[\nabla_G(N)]$, then since $\nabla_G(N)/N$ is torsion-free abelian and $L^\dagger = N$, it follows that L is an almost N-faithful prime ideal of $K[\nabla_G(N)]$. Thus, by Theorems 11.5 and 11.6, if $P = L^G$ then P is a prime ideal with vertex N unique up to G-conjugation. By hypothesis N is normal in G, so Theorem 9.5 yields the result.

We close this section with an important extension of Theorem 11.6.

THEOREM 11.8 (K. A. Brown [9])   *Let N be an isolated orbital subgroup of G and let L be a G-orbital prime ideal of KN. Then L^G is a prime ideal of KG.*

12.   Irreducible Modules

In this section we discuss the irreducible representations of group algebras of $\mathfrak{P}\mathfrak{F}$-groups G. We start by considering the primitivity of KG

and then, more generally, we describe all primitive ideals. Finally we study certain types of irreducible KG-modules.

A field K is *absolute* if it is algebraic over a finite field. If K is absolute, then it follows from Theorem 5.8 that:

   (i) every irreducible KG-module is finite-dimensional over K;
   (ii) all primitive ideals of KG are maximal;
   (iii) KG is primitive if and only if $G = 1$.

Thus the representation theory over such fields effectively reduces to that of finite groups. Therefore we will assume throughout the remainder of this section that K is nonabsolute.

The question of the primitivity of KG was settled for fields of characteristic 0 by Domanov [16] and Farkas and Passman [22]. Roseblade [74] obtained the result for arbitrary fields.

THEOREM 12.1   *KG is primitive if and only if $\Delta(G) = 1$.*

The more general problem of interest is to describe the primitive ideals of KG. Since these ideals are necessarily prime, the results of the previous section are relevant. The following formulation is from [58].

THEOREM 12.2 (Roseblade [74])   *Let P be a prime ideal of KG with vertex N and corresponding source L. Then P is primitive if and only if $\dim_K K[\nabla_G(N)]/L$ is finite.*

If R is a ring, then the *primitive length* of R is the largest n for which there exists a chain of primitive ideals $P_0 > P_1 > \ldots > P_n$. If such chains are of unbounded length, then we say that the primitive length of R is infinite. Let G_0 be a normal subgroup of finite index in G having a plinth series (cf. section 9). We call those infinite factors in this series which are of rank greater than 1 the *eccentric factors*. The number of such factors is easily seen to be an invariant of G, its *eccentric length*.

THEOREM 12.3 (Roseblade [74])   *The primitive length of KG is equal to the eccentric length of G.*

Observe that G is nilpotent-by-finite if and only if its plinth length is $h(G)$ or equivalently if and only if its eccentric length is 0. Moreover, a ring R has primitive length 0 if and only if all its primitive ideals are maximal. Thus the above implies that G is nilpotent-by-finite if and only if all primitive ideals of KG are maximal. This is an earlier result due to Zalesskii [93] and Snider [84]. Furthermore, it was shown by

Snider [87] that all primitive ideals of KG are intersections of maximal ideals.

If R is a ring then the set Priv R of all primitive ideals of R can be viewed as a topological space with the Jacobson topology. Thus a collection of primitive ideals is closed if it is the set of all primitive ideals containing some ideal of R. A topological space is said to be a *Baire space* if a countable intersection of dense open sets is always dense. The next result (which we do not prove) was proved for polycyclic groups by Farkas [19] and extended to the general case by Lorenz and Passman [57].

THEOREM 12.4 *Suppose K is nondenumerable. Then* Priv R *is a Baire space for every homomorphic image R of KG.*

Let H be a subgroup of G and let V be a KH-module. Then $V^G = V \otimes_{KH} KG$ is the KG-module induced from V. Note that if $\{x_i\}$ is a right transversal for H in G, then V^G is the K-vector space direct sum $\Sigma_i V \otimes x_i$. Furthermore, if L is the annihilator of V in KH, then the induced ideal L^G is the annihilator of V^G in KG.

A module M is said to be *finite induced* if there exists a subgroup $H \subseteq G$ and a finite K-dimensional KH-module V with $M = V^G$. It is not true that every irreducible KG-module is of this form. Indeed we have

THEOREM 12.5 (I. M. Musson [61])   *Every irreducible KG-module is finite induced if and only if every nilpotent subgroup of G is abelian-by-finite.*

In particular, if G is nilpotent then all irreducible KG-modules are finite induced if and only if G is abelian-by-finite. This result is due to S. D. Berman and V. V. Sharaya [4] and D. Segal [79] and a slightly stronger formulation is given by D. L. Harper [38]. In spite of Theorem 12.5, the finite induced irreducible modules are ample in the following sense.

THEOREM 12.6 (Farkas and Snider [26])   *Let P be a primitive ideal of KG. Then P is the annihilator of an irreducible finite induced KG-module.*

For nilpotent groups, this result can be found in Segal [79].

We now mention two results about individual irreducible KG-modules which are based on Proposition 5.7.

THEOREM 12.7 (Harper [39]) *Let M be an irreducible KG-module, acted upon faithfully by G. Suppose there exists an orbital subgroup $B \neq 1$ of G with $B \cap \Delta(G) = 1$. Then there exists a subgroup H of G with $h(H) < h(G)$ and an irreducible KH-module V with $M = V^G$.*

For the second, we define the *plinth socle* of G, to be the subgroup of G generated by all abelian subgroups A with $|G: N_G(A)|$ finite and with A a plinth in its normalizer.

THEOREM 12.8 (Musson [61]) *Let B be the plinth socle of G. If M is an irreducible KG-module, then M_B, the restriction of M to B, is a direct sum of finite K-dimensional irreducible KB-modules.*

Finally, we discuss $J(KG)$, the Jacobson radical of KG. If G is poly-(infinite cyclic), then it follows by induction on $h(G)$ that KG is semi-primitive or equivalently that $J(KG) = 0$. But any $\mathfrak{P}\mathfrak{F}$-group has such a poly-(infinite cyclic) normal subgroup of finite index. This then implies that $J(KG)$ is nilpotent. More precisely we have

PROPOSITION 12.9 *If char $K = 0$ then KG is semiprimitive. If char $K = p > 0$, and if Δ^p denotes the largest finite normal subgroup of G generated by p-elements, then $J(KG) = J(K\Delta^p) \cdot KG$. In particular in the latter case, KG is semiprimitive if and only if $\Delta^p = 1$.*

13. Concluding Remarks

This completes our detailed discussion of group rings of polycyclic-by-finite groups. As indicated in the introduction, we have emphasized those aspects of the subject which stem most directly from Philip Hall's papers [35], [36] and [37]. However, other, perhaps more ring-theoretic, topics have also been extensively studied. We briefly consider some of these below.

Idempotents

If $\alpha = \Sigma a_x x \in KG$, then the *trace* tr($\alpha$) of α is defined to be a_1, the identity coefficient. This simple function carries a good deal of information. If $e \in KG$ is an idempotent and char $K = 0$, then it was proved by I. Kaplansky (cf. [47]) that tr(e) is real and satisfies $0 \leqslant$ tr(e) $\leqslant 1$. Later, it was shown by Zalesskii [94] that tr(e) is in fact rational. Furthermore, if char $K = p > 0$ then tr(e) $\in \mathbb{F}_p$. Now let G be a $\mathfrak{P}\mathfrak{F}$-group and let char $K = 0$. If $\text{tr}(e) = r/s$ with r and s relatively prime integers, then, by Cliff and Sehgal [12], G must contain an element of prime order q for each prime divisor of s. Numerical information of this sort is extremely useful in studying projective modules. As a first step towards the zero divisor problem, it was shown by Formanek [31] that KG has only trivial idempotents if G is torsion-free. Furthermore, Cliff [11] showed if char $K = p > 0$ and G has only p-torsion, then again KG has only trivial idempotents. Some of this work has been extended by H. Bass [2] and A. Weiss [92] to more general trace functions.

Zero divisors

If G contains a nonidentity element of finite order then it is easy to see that KG contains nontrivial zero divisors. If G is torsion-free, no such easy construction exists. The zero divisor conjecture, which has eluded proof for over 30 years, asserts that if G is torsion-free then KG is a domain. For G polycyclic-by-finite the conjecture was settled by means of some of the most exciting mathematics in this whole subject. Formanek [29] dealt with supersolvable groups. In Brown [7] the abelian-by-finite case in characteristic 0 was settled along with some special cases in characteristic $p > 0$. Then Farkas and Snider [25] completed the $\mathfrak{P}\mathfrak{F}$-group case in characteristic 0 and again obtained results in characteristic p. Linnell [52] handled abelian-by-finite groups in characteristic p and then finally G. H. Cliff [11] obtained the complete characteristic p result. Some progress on more general groups appears in Snider [86].

A trivial unit in KG is an element of the form $u = kg$, with $g \in G$ and $k \in K^{\times}$. If G is not torsion-free then it is easy to see that KG has non-trivial units, unless both $|K|$ and $|G|$ are finite and quite small. The unit conjecture asserts that if G is torsion-free then KG has only trivial units. This problem is known to be more difficult than the zero divisor question. Indeed if G is torsion-free and KG has only trivial units, then KG is a domain. Moreover, the zero divisor question is a ring-theoretic statement; the unit question is not. While the zero divisor problem for group rings of $\mathfrak{P}\mathfrak{F}$-groups has been settled, the unit problem for these groups remains untouched. Even so, it is possible that both problems are really equivalent. If G is a unique product group, then KG is a domain. Conversely, if KG is a domain for all fields K, it may be possible, by considering generic elements in $K[\zeta_1, \zeta_2, \ldots]G$, to show that G is a unique product group. If this can be done, then G will also be a two unique product group by Strojnowski [105], and hence have only trivial units.

Goldie rank

If G is polycyclic-by-finite with no nonidentity finite normal subgroup, then KG is a prime Noetherian ring and therefore has a classical ring of quotients which is an $n \times n$ matrix ring over a division ring. The Goldie rank of KG is this number n and it is of interest to compute it. Since $n = 1$ if and only if KG has no zero divisors, this problem is related to but more difficult than the zero divisor question. The conjecture of Farkas [20] is that the Goldie rank of KG is equal to the least common multiple of the orders of the finite subgroups of G. This has been verified for supersolvable groups by Lorenz [53] and Rosset [76] and for some abelian-by-cyclic groups by Gabber and Rosset [33]. It is also known to be true if char $K = p > 0$ and G is the extension of a poly-(infinite cyclic) group by an elementary abelian p-group.

Projective modules

The study of projective modules over group rings of $\mathfrak{PF}$-groups is analogous to the study over polynomial rings. We ask whether the finitely generated modules are stably free and then whether the stably free modules are in fact free. The first question is related to the zero divisor problem since KG is a domain if all its finitely generated projective modules are stably free. An affirmative answer to this question is known for KG with G poly-(infinite cyclic) (J.-P. Serre [81]) and, via topological methods, for $\mathbb{Z}G$ with G a torsion-free $\mathfrak{PF}$-group (F. T. Farrell and W. C. Hsiang [27]). In addition, if R is a Noetherian ring of finite Krull dimension, then so is RG by Smith [83]. It therefore follows from the work of Stafford [104] that if M is a finitely generated RG-module, sufficiently large in a certain sense, then $M = M' \oplus RG$ for some submodule M'. Moreover, there is a cancellation theorem which asserts that, for these large modules, M' is unique up to isomorphism. A special case of these facts was proved by Brown, Lenagan and Stafford [10] by showing that, for suitable R, RG is weakly ideal invariant, a technical condition on the Krull dimension of modules, and then invoking the earlier work of Stafford [90]. The second question, as to whether stably free modules are necessarily free, is the analog of the Serre conjecture. It has a positive solution for KG with G a finitely generated free abelian group (R. G. Swan [91]). On the other hand, J. Lewin [49] proved that if G is a torsion-free $\mathfrak{PF}$-group and if all stably free modules for the rational group ring $\mathbb{Q}G$ are free, then G must be nilpotent. Indeed G must be abelian, by a result announced by Artamonov [97]. Using this, Farkas and Marciniak [21] construct a group ring KG for G in $\mathfrak{PF}$ with an idempotent e such that

e is not conjugate, by a unit of KG, to an idempotent with support in a finite subgroup of G. Furthermore, Artamonov [1] showed that if G is a non-abelian poly-(torsion-free abelian) group, then KG has a finitely generated projective module which is not free. Finally, Artamonov [97] has announced a characterization of the infinite, finitely generated nilpotent groups G with the property that all projective $\mathbb{Z}G$-modules are free. These are precisely the groups $G = A \times B$, where A is free abelian of finite rank and where $B = 1$ or a cyclic group of prime order $p \leqslant 19$.

Injective modules

The injective modules of commutative Noetherian rings have been characterized via the well known Matlis theory (see Sharpe and Vámos [102]). Certain injective KG-modules for $\mathfrak{P}\mathfrak{F}$-groups G enjoy similar properties. We say that the KG-module V is *locally finite* if vKG is finite dimensional over K for all $v \in V$. Such modules are equal to the ascending union of their socle series. Furthermore, it is known from Donkin [17] that if W is an essential extension of a locally finite KG-module V, then W is also locally finite. This applies to the injective hull $E(V)$ of a finite-dimensional irreducible KG-module V. In this case, $E(V)$ is known to be an Artinian KG-module with a Noetherian endomorphism ring. Furthermore, if char $K = p > 0$, then $E(V)$ has only finitely many distinct isomorphism types of KG-composition factors. In characteristic 0, this occurs if and only if G is nilpotent-by-finite. This study was begun by Snider [85]. The above results in characteristic p are due to Musson [60] and Jategaonkar [46], while in characteristic 0 they were obtained by Donkin [17] using certain Hopf algebra techniques. Analogous results also exist in Musson [60] for the integral group ring $\mathbb{Z}G$. In [60] it is also shown that for K a nonabsolute field, $E(V)$ is Artinian for all irreducible KG-modules V (not just the finite-dimensional ones) if and only if G is abelian-by-finite. Additional results on $E(V)$ and certain specific constructions for G nilpotent are in Musson [61, 62]. Finally, Brown [9] has begun the study of more general indecomposable injectives.

Division ring of quotients

If G is a torsion-free $\mathfrak{P}\mathfrak{F}$-group, then KG is a Noetherian domain. Hence by Goldie's theorem, this group algebra has a classical ring of quotients $Q(KG)$ which is a division ring. Farkas, Schofield, Snider and Stafford [23] consider the isomorphism question and show that if G_1 and G_2 are nilpotent, then $Q(KG_1) \cong Q(KG_2)$ if and only if $G_1 \cong G_2$. Additional results appear in Lorenz [54]. If G is poly-(infinite cyclic), then $Q(KG)$ is a universal field of fractions for KG so that an analog of the Extension Theorem for Places holds in these division rings. This was proved by Passman [69] using localization techniques from Smith [82]. Finally there is the study of linear groups over division rings D which are generated over their center by polycyclic-by-finite groups. For example, A. I. Lichtman [50] showed that any noncentral normal subgroup of the multiplicative group $D^\times$ contains a nonabelian free group. Furthermore, Lichtman [51] proved, in analogy with Burnside's theorem, that any periodic subgroup of $GL_n(D)$ is locally finite. A survey of some of this material can be found in Snider [89].

References

[1] Artamonov, V. A., Projective non-free modules over group rings of solvable groups (Russian), *Mat. Sb.* **116** (1981), 232–244.
[2] Bass, H., Euler characteristics and characters of discrete groups, *Invent. Math.* **35** (1976), 155–196.
[3] Bergman, G. M., The logarithmic limit-set of an algebraic variety, *Trans. Amer. Math. Soc.* **157** (1971), 459–469.
[4] Berman, S. D., Sharaya, V. V., On irreducible complex representations of finitely generated nilpotent groups (Russian), *Ukrainian Math. J.* **29** (1977), 435–442.
[5] Bovdi, A. A., 'Group Rings' (Russian), Uzgorod (1974).
[6] Brookes, C. J. B., The residual simplicity of modules over polycyclic groups, Ph.D. Dissertation, Univ. of Cambridge (1981).
[7] Brown, K. A., On zero divisors in group rings, *Bull. London Math. Soc.* **8** (1976), 251–256.
[8] Brown, K. A., Modules over polycyclic groups have many irreducible images, *Glasgow Math. J.* **22** (1981), 141–150.
[9] Brown, K. A., The structure of modules over polycyclic groups, *Math. Proc. Cambridge Phil. Soc.* **89** (1981), 257–283.
[10] Brown, K. A., Lenagan, T. H., Stafford, J. T., K-theory and stable structure of some Noetherian group rings, *Proc. London Math. Soc.* (3) **42** (1981), 193–230.
[11] Cliff, G. H., Zero divisors and idempotents in group rings, *Canad. J. Math.* **32** (1980), 596–602.
[12] Cliff, G. H., Sehgal, S. K., On the trace of an idempotent in a group ring, *Proc. Amer. Math. Soc.* **62** (1977), 11–14.
[13] Connell, I. G., On the group ring, *Canad. J. Math.* **15** (1963), 650–685.
[14] Dixmier, J., 'Enveloping Algebras', North-Holland, Amsterdam (1977).
[15] Domanov, O. I., A prime but not primitive regular ring (Russian), *Uspehi Mat. Nauk* **32** (1977), 219–220.
[16] Domanov, O. I., Primitive group algebras of polycyclic groups (Russian), *Sibirsk. Mat. Zurn.* **19** (1978), 37–43.
[17] Donkin, S., Locally finite representations of polycyclic-by-finite groups, *Proc. London Math. Soc.* (3) **44** (1982), 333–349.

[18] Duflo, M., Certaines algèbres de type fini sont des algèbres de Jacobson, *J. Algebra* **27** (1973), 358–365.

[19] Farkas, D. R., Baire category and Laurent extensions, *Canad. J. Math.* **31** (1979), 824–830.

[20] Farkas, D. R., Group rings: an annotated questionnaire, *Comm. in Algebra* **8** (1980), 585–602.

[21] Farkas, D. R., Marciniak, Z. S., Idempotents in group rings: a surprise, *J. Algebra* **81** (1983), 266–267.

[22] Farkas, D. R., Passman, D. S., Primitive Noetherian group rings, *Comm. in Algebra* **6** (1978), 301–315.

[23] Farkas, D. R., Schofield, A. H., Snider, R. L., Stafford, J. T., The isomorphism question for division rings of group rings, *Proc. Amer. Math. Soc.* **85** (1982), 327–330.

[24] Farkas, D. R., Snider, R. L., On group algebras whose simple modules are injective, *Trans. Amer. Math. Soc.* **194** (1974), 241–248.

[25] Farkas, D. R., Snider, R. L., K_0 and Noetherian group rings, *J. Algebra* **42** (1976), 192–198.

[26] Farkas, D. R., Snider, R. L., Induced representations of polycyclic groups, *Proc. London Math. Soc.* (3) **39** (1979), 193–207.

[27] Farrell, F. T., Hsiang, W. C., The Whitehead group of poly-(finite or cyclic) groups, *J. London Math. Soc.* (2) **24** (1981), 308–324.

[28] Fisher, J. W., Snider, R. L., Prime von Neumann regular rings and primitive group rings, *Proc. Amer. Math. Soc.* **44** (1974), 244–250.

[29] Formanek, E., The zero divisor question for supersolvable groups, *Bull. Austral. Math. Soc.* **9** (1973), 67–71.

[30] Formanek, E., Group rings of free products are primitive, *J. Algebra* **26** (1973), 508–511.

[31] Formanek, E., Idempotents in Noetherian group rings, *Canad. J. Math.* **15** (1973), 366–369.

[32] Formanek, E., Snider, R. L., Primitive group rings, *Proc. Amer. Math. Soc.* **36** (1972), 357–360.

[33] Gabber, O., Rosset, S., On the Goldie rank conjecture in group rings (to appear).

[34] Goldie, A. W., 'The Structure of Noetherian Rings', *Springer Lecture Notes in Math.* **246** (1972), 213–321.

[35] Hall, P., Finiteness conditions for soluble groups, *Proc. London Math. Soc.* (3) **4** (1954), 419–436.

[36] Hall, P., On the finiteness of certain soluble groups, *Proc. London Math. Soc.* (3) **9** (1959), 595–622.

[37] Hall, P., The Frattini subgroups of finitely generated groups, *Proc. London Math. Soc.* (3) **11** (1961), 327–352.

[38] Harper, D. L., Primitive irreducible representations of nilpotent groups, *Math. Proc. Cambridge Phil. Soc.* **82** (1977), 241–247.

[39] Harper, D. L., Primitivity in representations of polycyclic groups, *Math. Proc. Cambridge Phil. Soc.* **88** (1980), 15–31.

[40] Hartley, B., Injective modules over group rings, *Quart. J. Math.* (2) **28** (1977), 1–29.

[41] Hartley, B., Locally finite groups whose irreducible modules are finite dimensional, *Rocky Mt. J. Math.* **13** (1983), 255–263.

[42] Hirsch, K. A., On infinite soluble groups V, *J. London Math. Soc.* **29** (1954), 250–251.

[43] Irving, R. S., Some primitive group rings, *J. Algebra* **56** (1979), 274–281.

[44] Irving, R. S., Generic flatness and the Nullstellensatz for Ore extensions, *Comm. in Algebra* **7** (1979), 259–278.

[45] Jategaonkar, A. V., Integral group rings of polycyclic-by-finite groups, *J. Pure Appl. Algebra* **4** (1974), 337–343.

[46] Jategaonkar, A. V., Morita duality and Noetherian rings, *J. Algebra* **69** (1981), 358–371.

[47] Kaplansky, I., 'Fields and Rings', Univ. of Chicago Press, Chicago (1969).

[48] Kaplansky, I., 'Commutative Rings', Univ. of Chicago Press, Chicago (1974).

[49] Lewin, J., Projective modules over group-algebras of torsion-free groups, *Michigan Math. J.* **29** (1982), 59–64.

[50] Lichtman, A. I., On normal subgroups of the multiplicative group of skew fields generated by a polycyclic-by-finite group, *J. Algebra* **78** (1982), 548–577.

[51] Lichtman, A. I., On linear groups over a field of fractions of a polycyclic group ring, *Israel J. Math.* **42** (1982), 318–326.

[52] Linnell, P. A., Zero divisors and idempotents in group rings, *Math. Proc. Cambridge Phil. Soc.* **81** (1977), 365–368.

[53] Lorenz, M., The Goldie rank of prime supersolvable group algebras, *Mitteilungen aus dem Math. Sem. Giessen* **149** (1981), 115–129.

[54] Lorenz, M., Division algebras generated by finitely generated nilpotent groups, *J. Algebra* **85** (1983), 368–381.

[55] Lorenz, M., Passman, D. S., Centers and prime ideals in group algebras of polycyclic-by-finite groups, *J. Algebra* **57** (1979), 355–386.

[56] Lorenz, M., Passman, D. S., Prime ideals in crossed products of finite groups, *Israel J. Math.* **33** (1979), 89–132. (Addendum, ibid. **35** (1980), 311–322.)

[57] Lorenz, M., Passman, D. S., Integrality and normalizing extensions of rings, *J. Algebra* **61** (1979), 289–297.

[58] Lorenz, M., Passman, D. S., Prime ideals in group algebras of polycyclic-by-finite groups, *Proc. London Math. Soc.* (3) **43** (1981), 520–543.

[59] Mihalev, A. V., Zalesskii, A. E., 'Group Rings' (Russian), Modern Problems of Mathematics, Vol. 2, VINITI, Moscow (1973).

[60] Musson, I. M., Injective modules for group rings of polycyclic groups I, *Quart. J. Math. Oxford* **31** (1980), 429–448. (Part II, ibid. 449–466.)

[61] Musson, I. M., Irreducible modules for polycyclic group algebras, *Canad. J. Math.* **33** (1981), 901–914.

[62] Musson, I. M., On the structure of certain injective modules over group algebras of solvable groups of finite rank, *J. Algebra* **85** (1983), 51–75.

[63] Nouazé, Y., Gabriel, P., Idéaux premiers de l'algèbre enveloppante d'une algèbre de Lie nilpotente, *J. Algebra* **6** (1967), 77–99.

[64] Passi, I. B. S., 'Group Rings and their Augmentation Ideals', *Springer Lecture Notes in Math.* **715** (1979).

[65] Passman, D. S., Nil ideals in group rings, *Michigan Math. J.* **9** (1962), 375–384.

[66] Passman, D. S., 'Infinite Group Rings', Marcel Dekker, New York (1971).

[67] Passman, D. S., Primitive group rings, *Pacific J. Math.* **47** (1973), 499–506.

[68] Passman, D. S., 'The Algebraic Structure of Group Rings', Wiley-Interscience, New York (1977).

[69] Passman, D. S., Universal fields of fractions for polycyclic group algebras, *Glasgow Math. J.* **23** (1982), 103–113.

[70] Quillen, D., On the endomorphism ring of a simple module over an enveloping algebra, *Proc. Amer. Math. Soc.* **21** (1969), 171–172.

[71] Roseblade, J. E., Group rings of polycyclic groups, *J. Pure Appl. Algebra* **3** (1973), 307–328.

[72] Roseblade, J. E., The Frattini subgroup in infinite soluble groups, *in* 'Three Lectures on Polycyclic Groups', *Queen Mary College Mathematics Notes* (1973).

[73] Roseblade, J. E., Applications of the Artin–Rees lemma to group rings, *Symp. Math.* **17** (1976), 471–478.

[74] Roseblade, J. E., Prime ideals in group rings of polycyclic groups, *Proc. London Math. Soc.* (3) **36** (1978), 385–447. (Corrigenda, ibid. **38** (1979), 216–218.)

[75] Roseblade, J. E., Smith, P. F., A note on hypercentral group rings, *J. London Math. Soc.* (2) **13** (1976), 183–190.

[76] Rosset, S., Miscellaneous results on the Goldie rank conjecture (to appear).

[77] Segal, D., On abelian-by-polycyclic groups, *J. London Math. Soc.* (2) **11** (1975), 445–452.

[78] Segal, D., On the residual simplicity of certain modules, *Proc. London Math. Soc.* (3) **34** (1977), 327–353.

[79] Segal, D., Irreducible representations of finitely generated nilpotent groups, *Math. Proc. Cambridge Phil. Soc.* **81** (1977), 201–208.

[80] Sehgal, S. K., 'Topics in Group Rings', Marcel Dekker, New York (1978).

[81] Serre, J-P., Cohomologie des groupes discrets, *in* 'Prospects in Mathematics', *Annals of Math. Studies No. 70*, Princeton (1971), 77–169.

[82] Smith, P. F., Localization in group rings, *Proc. London Math. Soc.* (3) **22** (1971), 69–90.

[83] Smith, P. F., On the dimension of group rings, *Proc. London Math. Soc.* (3) **25** (1972), 288–302. (Corrigenda, ibid. **27** (1973), 766–768.)

[84] Snider, R. L., Primitive ideals in group rings of polycyclic groups, *Proc. Amer. Math. Soc.* **57** (1976), 8–10.

[85] Snider, R. L., Injective hulls of simple modules over group rings, *in* 'Ring Theory', Marcel Dekker, New York (1977), 223–226.

[86] Snider, R. L., The zero divisor conjecture for some solvable groups, *Pacific J. Math.* **90** (1980), 191–196.

[87] Snider, R. L., On primitive ideals in group rings, *Arch. Math.* **38** (1982), 423–425.

[88] Snider, R. L., Solvable groups whose irreducible modules are finite dimensional, *Comm. in Algebra* **10** (1982), 1477–1485.

[89] Snider, R. L., The division ring of fractions of a group ring, *Springer Lecture Notes in Math.* **1029** (1983), 325–339.

[90] Stafford, J. T., Stable structure of non-commutative Noetherian rings, *J. Algebra* **47** (1977), 244–267. (Part II, ibid. **52** (1978), 218–235.)

[91] Swan, R. G., Projective modules over Laurent polynomial rings, *Trans. Amer. Math. Soc.* **237** (1978), 111–120.

[92] Weiss, A., Idempotents in group rings, *J. Pure Appl. Algebra* **16** (1980), 207–213.

[93] Zalesskii, A. E., Irreducible representations of finitely generated nilpotent torsion-free groups, *Math. Notes* **9** (1971), 117–123.

[94] Zalesskii, A. E., On a problem of Kaplansky, *Soviet Math.* **13** (1972), 449–452.

[95] Zalesskii, A. E., On the semisimplicity of a modular group algebra of a solvable group, *Soviet Math.* **14** (1973), 101–105.

[96] Zalesskii, A. E., The Jacobson radical of the group algebra of a solvable group is locally nilpotent (Russian), *Izv. Akad. Nauk SSSR, Ser. Mat.* **38** (1974), 983–994.

Additional references

[97] Artamonov, V. A., Projective modules over group rings of polycyclic-by-finite groups, *Fifth Symp. on Ring Theory, Algebras and Modules*, Novosibirsk (1982).

[98] Beidar, K. I., On radicals of finitely generated algebras (Russian), *Uspehi Mat. Nauk* **36** (1981), 203–204.

[99] Brown, K. A., Primitive group rings of soluble groups, *Arch. Math.* **36** (1981), 404–413.

[100] Farkas, D. R., Noetherian group rings: An exercise in creating folklore and intuition, Proceedings of an Oberwolfach Conference (to appear).

[101] Segal, D., Unipotent groups of module automorphisms over polycyclic group rings, *Bull. London Math. Soc.* **8** (1976), 174–178.

[102] Sharpe, D. W., Vámos, P., 'Injective Modules', Cambridge Univ. Press, London (1972).

[103] Snider, R. L., Group rings with finite endomorphism dimension, *Arch. Math.* **41** (1983), 219–225.

[104] Stafford, J. T., Generating modules efficiently, *J. Algebra* **69** (1981), 312–346.

[105] Strojnowski, A., A note on u.p. groups, *Comm. in Algebra* **8** (1980), 231–234.

[106] Wehrfritz, B. A. F., Groups whose irreducible representations have finite degree, *Math. Proc. Cambridge Phil. Soc.* **90** (1981), 411–421.

[107] Wehrfritz, B. A. F., Groups whose irreducible representations have finite degree II, *Proc. Edinburgh Math. Soc.* **25** (1982), 237–243.

[108] Wehrfritz, B. A. F., Groups whose irreducible representations have finite degree III, *Math. Proc. Cambridge Phil. Soc.* **91** (1982), 397–406.

Finitely Presented Soluble Groups

RALPH STREBEL

1. Background 258
2. Examples of infinitely related groups 264
3. Examples of finitely presented soluble groups 270
4. Finitely presented nilpotent-by-cyclic groups 278
5. Finitely presented centre-by-metabelian groups I 280
6. General theory of the geometric invariant 287
7. Finitely presented centre-by-metabelian groups II 304
8. Some open problems 310
References 312

It was realized early in the 1970's that finitely presented soluble groups, even metabelian ones, could be far more complex than had previously been thought, and yet their structure has to be rather restricted. Our aim in this essay is to give an account of the progress which has been made since then. Much of this depends on an important geometric invariant, closely related to G. M. Bergman's logarithmic limit set ([13], cf. D. S. Passman's essay), which was defined by R. Bieri and R. Strebel [20]. This is introduced in section 1, which also contains a sketch of the background and a statement of the main results, Theorems A, A*, B, B1, B2, C, D and Corollaries BD and BD1. In the rest of the essay (apart from the last section, which deals with some open questions) these theorems and further results are discussed, and some of them proved. Specifically:

Subsection 3.2 has a proof of Theorem A and a sketch of the proof of Theorem A*. Generalizations of Theorem A* are stated as Theorems 13 and 14. The analysis of the proof of A* is taken up in Example 42.

Theorem B and its generalization B* are established in subsection 2.2. In subsection 5.2 Theorem B1 is derived from Theorem B; the proof of Theorem B2 is outlined in subsection 5.3. Theorem B2 is also deduced from B1 and Theorem 40 in subsection 6.5.

GROUP THEORY: essays for Philip Hall
ISBN 0-12-304880-X

Theorem C is proved in subsection 5.4, and in section 7 we explain the method of proof of Theorem D and carry it through in an important special case.

I am very grateful to Robert Bieri for his constructive criticism of this survey and to the Deutsche Forschungsgemeinschaft for its financial support.

1. Background

1.1 The earliest examples of presentations of infinite groups are connected with geometry or with the theory of functions of one complex variable. They yield groups like the fundamental groups of the closed orientable surfaces, the inhomogeneous modular group $PSL_2(\mathbb{Z})$ and, more generally, the finitely generated *Fuchsian groups* of the first kind; *knot groups* (W. Wirtinger, 1905 [59]; M. Dehn, 1910 [27]) and *braid groups* (E. Artin, 1925 [2]). These presentations are all finite, i.e. they involve finitely many generators and finitely many defining relations. This was also true of certain groups arising within algebra, such as the automorphism groups of finitely generated free groups and the automorphism groups of finitely generated free abelian groups (J. Nielsen, 1924 [44, 45]; W. Magnus, 1934 [40]).

It was B. H. Neumann [43] who first made the point that finite presentability was a special property enjoyed by rather few groups. He proved that there are continuously many 2-generator groups, of which only countably many can be finitely presented; and he exhibited explicit examples of finitely generated, infinitely related groups. Is there an *a priori* reason why the groups previously studied are finitely presented?

For surface, knot or braid groups one can argue as follows: such a group is the fundamental group of a path-connected, topological space which can be built up from simpler spaces in finitely many steps, starting with a space whose fundamental group is trivial. At each step the fundamental group G of the new space is related to the fundamental groups G_i of spaces previously obtained, in one of the following ways:

(I) G is the pushout of the diagram $G_1 \xleftarrow{\varphi_1} G_0 \xrightarrow{\varphi_2} G_2$; or equivalently, $G \cong D/R$, where $D = G_1 * G_2$ and $R = \langle \varphi_1(g) \cdot \varphi_2(g)^{-1} \mid g \in G_0 \rangle^D$.

(II) There are two homomorphisms $\varphi_1, \varphi_2 : G_0 \to G_1$ and an infinite cyclic group $\langle t \rangle$ such that $G \cong E/S$, where $E = G_1 * \langle t \rangle$ and

$$S = \langle t^{-1}\varphi_1(g)t \cdot \varphi_2(g)^{-1} \mid g \in G_0 \rangle^E.$$

(III) G is an extension of G_1 by G_2.

(IV) G contains G_1 as a subgroup of finite index.

Inductively, the groups G_0, G_1 and G_2 are known to be finitely pre-sented, and general arguments then show that the new group G can also be finitely presented. In case (II), for example, it suffices to choose a finite set of generators Y_0 of G_0 and to notice that the normal closure in $G_1 * \langle t \rangle$ of the finite set $\{t^{-1}\varphi_1(g)t . \varphi_2(g)^{-1} \mid g \in Y_0\}$ coincides with the normal closure of $\{t^{-1}\varphi_1(g)t . \varphi_2(g)^{-1} \mid g \in G_0\}$.

If the homomorphisms φ_1, φ_2 in (I) and (II) are injective, (I) and (II) become:

(I') G is the *generalized free product* $G_1 *_{G_0} G_2$, with factors G_1, G_2 and amalgamated subgroup G_0, and

(II') G is the *HNN-extension* $\langle G_1, t; t^{-1}st = \sigma(s) \text{ for } s \in S \rangle$, having base group G_1, stable letter t, associated subgroups $S = \varphi_1(G_0)$ and $T = \varphi_2(G_0)$, and isomorphism $\sigma: S \xrightarrow{\sim} T$ given by $\sigma(\varphi_1(g_0)) = \varphi_2(g_0)$.

Let $\mathfrak{C}_0 = (1)$ be the trivial class and $\mathfrak{C}_n$ $(n \geqslant 1)$ be the class of all groups G which satisfy (I'), (II'), (III) or (IV) with G_0, G_1 and G_2 in $\mathfrak{C}_{n-1}$. The union $\mathfrak{C} = \bigcup_n \mathfrak{C}_n$ is the class of all *constructible* groups studied by Baumslag and Bieri in [9]. All constructible groups are finitely presentable. (Incidentally, (III) may be omitted without loss.)

1.2   When is a constructible group *soluble*? Suppose the generalized free product $G = G_1 *_{G_0} G_2$ contains no non-abelian free subgroup. If G_0 equals one of the factors, G coincides with the other factor; if G_0 has index 2 in both G_1 and G_2, then G is an extension of G_0 by an infinite dihedral group. Since an extension of G_0 by an infinite cyclic group $\langle t \rangle$ is an HNN-extension of G_0, where the base group and associated subgroups are all equal to G_0, it follows that an application of (I') leading to a soluble group can be replaced by an application of (II') followed by an application of (IV).

An HNN-extension contains non-abelian free subgroups unless one of the associated subgroups coincides with the base group. Assuming that the first associated subgroup coincides with the base group B, the HNN-extension can be written as

$$G = \langle B, t; t^{-1}bt = \sigma(b) \text{ for all } b \in B \rangle.$$

The obvious homomorphism $B \to G$ is injective and will be treated as an inclusion. The normal closure of B in G is the union of the ascending chain

$$\ldots \subseteq t^{-1}Bt \subseteq B \subseteq tBt^{-1} \subseteq t^2Bt^{-2} \subseteq \ldots.$$

In view of this, one calls an HNN-extension *ascending* if the first associated subgroup coincides with the base group. (The dependence of this on an order for the associated subgroups is clearly inessential.) Thus ascending HNN-extensions with soluble base groups are again soluble. One should view ascending HNN-extensions as a kind of localization in group theory: cf. for example P. Hall in Lemma 9 of [34]; also [50], p. 457.

The above considerations show that a constructible, soluble group can be built up from the trivial group by iterated use of either

(II'') an ascending HNN-extension over a base group already obtained, or

(IV'') passage to a soluble supergroup containing a previously obtained group with finite index.

The class of constructible soluble groups turns out to be closed with respect to taking homomorphic images and subgroups of finite index (cf. p. 271). Since the class is also closed under extensions, it may lay claim to being the largest class of soluble groups which are finitely presented for natural reasons. It contains all polycyclic groups (these were proved to be finitely related by P. Hall in [33]), and also the one-relator group $\langle a, t; t^{-1}at = a^m \rangle$, which is not polycyclic if $|m| > 1$. We return to this in subsection 3.1.

1.3 In the early 1970's G. Baumslag and V. N. Remeslennikov independently discovered a finitely presented metabelian group which differs radically from any finitely presented soluble group seen before.

THEOREM A [5, 49] *The 3-generator 3-relator group*

$$G = \langle a, s, t; a^t = aa^s, [s, t] = 1 = [a, a^s] \rangle$$

is metabelian and its derived group is free abelian of rank $\aleph_0$.

Baumslag and Remeslennikov succeeded in generalizing this example, again independently, to the following astonishing embedding result.

THEOREM A* [6, 49] *Every finitely generated metabelian group can be embedded in a finitely presented metabelian group.*

A natural question posed by Baumslag in his survey [8] is

Problem 1.   Is there any way of discerning finitely presented metabelian groups from the other finitely generated metabelian groups?

The first results about structural restrictions imposed on a finitely generated, infinite soluble group by the requirement that it be finitely related, were published in 1978 by Bieri and Strebel [18]. Their main result, when applied to soluble groups, implies

THEOREM B [18]   *Assume G is a soluble group and N is a normal subgroup with infinite cyclic quotient. If G can be finitely presented, then G is an ascending HNN-extension over a finitely generated base group B contained in N.*

Notice that every finitely presented infinite soluble group $\tilde{G}$ has a subgroup G of finite index to which Theorem B can be applied. From this point of view Theorem B describes a structural peculiarity common to all finitely presented infinite soluble groups.

Also in 1978, J. R. J. Groves published an application of Theorem B to centre-by-metabelian groups. To gain the proper perspective here, we recall that Hall [33, 34] showed that there is a marked difference between the classes of finitely generated metabelian and finitely generated centre-by-metabelian groups: groups of the first class satisfy max-n, the maximal condition on normal subgroups, and are residually finite, whereas groups of the second class need not have either of these properties. Hall proved indeed that every countable abelian group occurs as the centre of a suitable 2-generator centre-by-metabelian group. These particular differences disappear for finitely presented groups:

THEOREM C [31]   *If G is a finitely presented centre-by-metabelian group, then $G'/Z(G')$ is finitely generated. In particular, G is abelian-by-polycyclic, satisfies max-n and is residually finite.*

1.4   Suppose now that G is a finitely presented soluble group whose abelianization $G^{ab} = G/G'$ has torsion-free rank $r_0(G^{ab})$ greater than 1. Then G contains infinitely many normal subgroups N to which Theorem B can be applied and the impact of all these applications on the abelianized derived group $A = (G')^{ab}$ can be expressed in terms of a new geometric invariant Σ_A. To explain this we need a few definitions. Let Q be a finitely generated abelian group. The homomorphisms

$v: Q \to \mathbb{R}$ of Q into the additive group of the reals form a real vector space $\mathrm{Hom}(Q, \mathbb{R})$ of dimension $n = r_0(Q)$. We make this into a topological space via an isomorphism to the n-dimensional real vector space $\mathbb{R}^n$ and the usual norm on $\mathbb{R}^n$. The topology is independent of the chosen isomorphism. (For more information on this point and the ones immediately below, see subsection 5.1.) Call two homomorphisms v, v' equivalent if one is a positive real multiple of the other. The equivalence classes $[v]$ represented by non-zero homomorphisms will then form a compact topological space $S(Q)$, which is homeomorphic to the unit sphere $\mathbb{S}^{n-1}$ in $\mathbb{R}^n$. (If Q is finite, the set $S(Q)$ is empty and, by convention, has dimension -1.) There is a particularly important subset of $S(Q)$ consisting of the points $[v]$ such that v has the form $Q \twoheadrightarrow \mathbb{Z} \hookrightarrow \mathbb{R}$. Such points are called *discrete*, or of rank 1, and the set of all these will be denoted by $\mathrm{dis}\, S(Q)$. Clearly $\mathrm{dis}\, S(Q)$ is a totally disconnected space and is not compact if $n > 1$; whereas the sphere $S(Q)$ is a compact manifold and is connected when $n > 1$.

For each v in $\mathrm{Hom}(Q, \mathbb{R})$ define the set

$$Q_v = \{q \in Q \,|\, v(q) \geq 0\}.$$

Note that Q_v depends only on the class $[v]$ of v and that it is a submonoid of Q. Let $\mathbb{Z}Q_v$ denote the monoid ring of Q_v. It is a subring of $\mathbb{Z}Q$.

Finally, we attach to each finitely generated $\mathbb{Z}Q$-module A the geometric invariant Σ_A, defined by

$$\Sigma_A = \{[v] \in S(Q) \,|\, A \text{ is finitely generated over } \mathbb{Z}Q_v\}.$$

Thus Σ_A records the subrings $\mathbb{Z}Q_v$ of $\mathbb{Z}Q$ over which A is finitely generated.

THEOREM B1 *Suppose G is a finitely presented soluble group and M a normal subgroup with infinite, abelian factor group $Q = G/M$. Let A denote the abelian group M^{ab} equipped with the right Q-action induced by conjugation. If $[v]$ is discrete then either $[v]$ or $[-v]$ belongs to Σ_A.*

In other words, Theorem B1 asserts that $\mathrm{dis}\, S(Q)$ is a subset of $\Sigma_A \cup (-\Sigma_A)$, where $-\Sigma_A$ is the antipodal set of Σ_A.

1.5 It turns out that Σ_A is always an open subset of $S(Q)$. Also, Σ_A has a certain open cover $\{O_\lambda \,|\, \lambda \in C(A)\}$, defined on p. 284, where $C(A)$ denotes the centralizer of A in $\mathbb{Z}Q$. When $\Sigma_A \cup (-\Sigma_A)$ is the whole sphere $S(Q)$, in which case we say A is a *tame* module, the compactness of $S(Q)$ implies that there exists a finite subset Λ of $C(A)$ such that

$$S(Q) = \bigcup \{ O_\lambda \cup (-O_\lambda) \mid \lambda \in \Lambda \}.$$

The existence of such a finite cover of $S(Q)$ is a key ingredient in the proof of the following partial converse of Theorem B1.

THEOREM D (Bieri and Strebel [20], 3.1) *Every extension of a finitely generated, tame $\mathbb{Z}Q$-module by a finitely generated abelian group Q can be finitely presented.*

Consider now a short exact sequence $A \lhd G \twoheadrightarrow Q$, where A is a finitely generated $\mathbb{Z}Q$-module and Q is a finitely generated abelian group. A comparison of Theorems B1 and D brings to light an unsatisfactory gap: if G is finitely presented Theorem B1 implies only that $\Sigma_A \cup (-\Sigma_A)$ contains dis $S(Q)$, but in order to obtain a finite presentation for G by means of Theorem D, one needs that $\Sigma_A \cup (-\Sigma_A)$ equals $S(Q)$. It is possible to close this gap by strengthening Theorem B1 to

THEOREM B2 (Bieri and Strebel [20], 4.1) *If G is a finitely presented soluble group and M is a normal subgroup with abelian factor group Q, then the finitely generated $\mathbb{Z}Q$-module M^{ab} is tame.*

1.6 Theorem B2 and Theorem D have an immediate corollary that is worth stating explicitly.

COROLLARY BD *Assume G is a finitely generated metabelian group and let A denote the finitely generated $\mathbb{Z}G^{ab}$-module obtained by equipping the abelian group G' with the G^{ab}-action induced by conjugation. Then the following three statements are equivalent:*

(i) *G can be finitely presented;*
(ii) *the semi-direct product $G^{ab} \ltimes A$ can be finitely presented;*
(iii) *A is a tame $\mathbb{Z}G^{ab}$-module.*

Corollary BD characterizes the finitely related metabelian groups among the finitely generated metabelian groups, thus giving an answer to Baumslag's Problem 1 in terms of the geometric invariant. It turns out that the computation of this invariant has numerous interesting facets. We take up this problem in section 5.

Another consequence of Theorems B2 and D is the very striking result

COROLLARY BD1 *If G is a finitely presented soluble group, then any metabelian image of G is finitely presented.*

2. Examples of Infinitely Related Groups

We begin with some remarks on the notion of finite presentation, followed by a discussion of wreath products, a construction that generally yields infinitely related groups. In subsection 2.2 we prove Theorem B and in 2.3 we use Theorem B to obtain a characterization of those relatively free soluble groups which can be finitely presented.

2.1 Preliminaries

The following is an important observation of B. H. Neumann ([43], (8); cf. [50], p. 52):

LEMMA 1 *If G is a finitely presented group and $\pi: F \twoheadrightarrow G$ is a homomorphism of a finitely generated free group F onto G, then $\ker \pi$ is the normal closure of a finite set.*

We record two consequences. Let G be a finitely generated group and suppose we wish to show that G cannot be finitely presented. According to Lemma 1 we can prove this as follows: first choose a finite generating set $\mathscr{X}$ of G which is easy to work with; next let $\pi: F(\mathscr{X}) \to G$ be the obvious epimorphism and choose a sequence $r_1, r_2, r_3, \ldots$ of normal generators of $\ker \pi$; finally, verify that none of the obvious surjections

$$\langle \mathscr{X}; r_1, \ldots, r_n \rangle \twoheadrightarrow G \quad (n = 1, 2, \ldots)$$

is an isomorphism.

The second consequence concerns extensions

$$(1) \qquad\qquad\qquad N \lhd E \twoheadrightarrow G,$$

where E is finitely generated. The lemma implies that N is the normal closure of finitely many elements whenever G is finitely related. For example, if N is abelian and G polycyclic, then, as P. Hall pointed out in [33], N must be a finitely generated $\mathbb{Z}G$-module (via conjugation by E). If, on the other hand, G is a finitely generated group and N is an infinitely generated central subgroup of E, we can conclude that G cannot be finitely related.

Example 2 (P. Hall [33]) *The wreath product* $\langle b \rangle \wr \langle a \rangle$ *of two infinite cyclic groups is an infinitely related 2-generator group.*

Indeed, by the proof of Theorem 7 of [33] there is a central extension $Z \lhd E \twoheadrightarrow (\langle b \rangle \wr \langle a \rangle)$ with Z free abelian of infinite rank and E generated by 2 elements.

Multiplicator and central extensions. The problem of constructing central extensions with infinitely generated kernel is closely related to the question of whether the multiplicator of the quotient group is infinitely generated. Quite generally, for any group G, its multiplicator $M(G)$ can be computed by the Schur–Hopf formula

$$M(G) \cong (R \cap [F, F])/[R, F],$$

where $R \lhd F \xrightarrow{\pi} G$ is a group extension with F free. The right-hand side of this formula for $M(G)$ is independent of π (to within natural isomorphism) (see, e.g. [50], pp. 334–336 for some basic facts about the multiplicator).

Now π determines the exact sequence of abelian groups

$$0 \rightarrow M(G) \rightarrow R/[R, F] \rightarrow F/[F, F] \rightarrow F/R \cdot [F, F] \rightarrow 0$$

and the central extension

(2) $R/[R, F] \lhd F/[R, F] \twoheadrightarrow F/R \cong G.$

Thus to prove that the multiplicator of the finitely generated group G is infinitely generated is equivalent to constructing a central extension (2) with $F/[R, F]$ finitely generated but $R/[R, F]$ infinitely generated. Nevertheless, it may be advantageous to bring $M(G)$ into play: for $M(G)$ is, by definition, $H_2(G, \mathbb{Z})$, the second homology group of G with integral coefficients, and thus the whole machinery of homological algebra is available.

As a sample of results about the multiplicator that bear on the question of finite presentability, we quote

THEOREM 3 (N. Blackburn [24]) *The multiplicator of the wreath product* $B \wr A$ *of two finitely generated groups B and A is finitely generated if, and only if, the multiplicators of both B and A are finitely generated and either A is finite or B is perfect.*

Theorem 3, when applied with B and A both infinite cyclic, confirms Example 2; but it fails to tell us whether, for instance, $A_5 \wr \langle a \rangle$, is finitely or infinitely related—for the alternating group A_5 has multiplicator of order two and the multiplicator of an infinite cyclic group is trivial. This and many similar examples suggest that the multiplicator is not always a sufficiently fine indicator of whether a finitely generated group is finitely or infinitely related. This suspicion is confirmed by the following counterpart to Theorem 3.

THEOREM 4 (G. Baumslag [3])   *The wreath product $B \wr A$ of two finitely generated groups is finitely related if, and only if, either B is trivial and A is finitely related, or A is finite and B is finitely related.*

Proof.   Suppose B is non-trivial and A is infinite. Let $F = A * B$ and $H = B^F = *\{B^a \mid a \in A\}$. For any inverse closed subset T of $\hat{A} = A \setminus \{1\}$, let $\bar{T}$ be the normal closure in H of $\langle [B^{a_1}, B^{a_2}] \mid a_1 a_2^{-1} \in T \rangle$. Evidently $\bar{T}$ is normal in F. If $T_1 < T_2$, then $\bar{T}_1 < \bar{T}_2$. To see this, let $a_0 \in T_2 \setminus T_1$. All of $\bar{T}_1$ is in the kernel of the natural projection $H \twoheadrightarrow B * B^{a_0}$, but $\bar{T}_2$ is not because $[B, B^{a_0}]$ is not. In particular, $\hat{A}$ is not of the form $\bar{T}$ for any finite subset T. Hence $F/\hat{\bar{A}}$ cannot be finitely presented (Lemma 1). Since $F/\hat{\bar{A}}$ is $B \wr A$ in disguise, we are done. The converse is easy.

2.2   Proof of Theorem B

The theorem will be an easy consequence of

THEOREM B* ([18], p. 259, Theorem A)   *Let G be a finitely presented group, N a normal subgroup of G with infinite cyclic factor group G/N and let g generate a complement of N in G. Then there exist finitely generated subgroups S, T, B of N, with S and T in B, so that conjugation by g induces an isomorphism $\tau: S \xrightarrow{\sim} T$ and that the inclusion of B in G and the assignment $t \mapsto g$ induce an isomorphism*

$$\psi: \langle B, t; t^{-1}st = \tau(s) \text{ for all } s \in S \rangle \xrightarrow{\sim} G.$$

This isomorphism exhibits G as an HNN-extension with finitely generated base group B, stable letter t and finitely generated associated subgroups S and T.

Proof.   Choose a finite number of elements $b_1, \ldots, b_f$ in N so that $G = \langle b_1, \ldots, b_f, g \rangle$. Let F be the free group on $\{a_1, \ldots, a_f, x\}$ and π the homomorphism $F \to G$ determined by $a_i \mapsto b_i$ $(1 \leqslant i \leqslant f)$, $x \mapsto g$. The kernel R of π is the normal closure of, say, $r_1, \ldots, r_k$. Each r_j is a product of powers of conjugates $a_i^{x^l}$. As k is finite, there is a natural number m such that every r_j is contained in

$$U = \langle a_i^{x^l} \mid 1 \leqslant i \leqslant f \text{ and } -m \leqslant l \leqslant m \rangle.$$

Define now

$$S = \langle b_i^{g^l} \mid 1 \leqslant i \leqslant f \text{ and } -m \leqslant l \leqslant m-1 \rangle$$

and $T = S^g$. If $B = \langle S, b_i^{g^m} \mid 1 \leqslant i \leqslant f \rangle$, then S, T are subgroups of B. The inclusion of B in G and the assignment $t \mapsto g$ induce a homomorphism

$$\psi : \tilde{G} = \langle B, t; t^{-1}st = \tau(s) \text{ for } s \in S \rangle \to G,$$

where $\tau(s) = g^{-1}sg \in T$. Since $\{b_1, \ldots, b_f\} \subseteq B$, this map ψ is surjective. Next the assignments $a_i \mapsto b_i$, where $1 \leqslant i \leqslant f$ and $x \mapsto t$ induce a homomorphism χ of F onto the HNN-extension $\tilde{G}$. The defining relations of $\tilde{G}$ state that $(b_i^{g^l})^t = b_i^{g^{l+1}}$ for $-m \leqslant l \leqslant m-1$ and $1 \leqslant i \leqslant f$. This implies that $b_i^{t^l} = b_i^{g^l}$ in $\tilde{G}$ for all $-m \leqslant l \leqslant m$ and all i. Consequently χ maps $r_j = r_j(a_i^{x^l})$ to

$$r_j(b_i^{t^l}) = r_j(b_i^{g^l}) = 1 \in B.$$

So χ induces an epimorphism $F/R \twoheadrightarrow \tilde{G}$. As the composite $\psi \circ \chi$ is π, it follows that ψ is an isomorphism.

Deduction of Theorem B from Theorem B*. Consider an HNN-extension

$$(3) \qquad \langle B, t; t^{-1}st = \tau(s) \text{ for all } s \in S \rangle$$

with associated subgroups S and $T = \sigma(S)$. If $S \neq B \neq T$, pick a in $B \setminus S$ and b in $B \setminus T$, and set $x = at^{-1}$ and $y = bt$. Then x and y generate a subgroup of G which is free on $\{x, y\}$, as can readily be seen with the help of the Normal Form Theorem for HNN-extensions (see, e.g., [39], p. 181).

Now apply Theorem B* to a finitely presented soluble group G. This does not possess non-abelian free subgroups, and hence G must be isomorphic to an HNN-extension (3) with either $S = B$ or $T = B$. Defining $\varepsilon = 1$ in the first and $\varepsilon = -1$ in the second case, setting $\sigma = \tau^\varepsilon$ and replacing t by t^ε, one obtains an isomorphism

$$\psi : \langle B, t; t^{-1}bt = \sigma(b) \text{ for all } b \in B \rangle \xrightarrow{\sim} G$$

with $\psi(t) = g^{\varepsilon}$ and $\sigma(b) = g^{-\varepsilon} b g^{\varepsilon}$. This isomorphism exhibits G as an ascending HNN-extension over the finitely generated base group B, as is asserted by Theorem B.

ADDENDUM 5   *The assertion of Theorem B* remains true if the hypothesis that G be finitely presented is replaced by the weaker hypothesis that G be of type FP_2 over some commutative ring k with $1 \neq 0$* (see the proof of Theorem A in [18]).

Recall that a group G is said to be of type FP_2 over a commutative ring k (all rings are assumed non-zero throughout this essay) if there exists a free presentation $F/R \overset{\sim}{\to} G$ with F free of finite rank and such that the scalar extension $R^{ab} \otimes_{\mathbb{Z}} k$ of the relation module R^{ab} is finitely generated over the group algebra kG; see [14], p. 20, and [18], p. 261, for more information.

A finitely presented group G is of course of type FP_2 over k, no matter what k is; but there exist groups which are of type FP_2 over some k without being finitely related (see [20], p. 464, or [56], §4.2).

Various reasons make it advisable to use the assumption FP_2 over some k rather than the assumption of finite presentability. First, one may hope to gain some insight into when the two assumptions are equivalent. For example, Theorem B2 remains true if G is merely of type FP_2 over some k. In conjunction with Theorem D this fact yields the result that *every metabelian group which is of type FP_2 over some k can be finitely presented* ([20], Theorem 5.4). Incidentally, the analogous statement for groups G, where G' is nilpotent of class 2, is false ([20], p. 464).

Secondly, the cohomology theory of groups often enables one to establish an FP_2 property for groups which are not known to admit a finite presentation. A striking example is provided by the recent classification of B. Eckmann and H. Müller [28] of Poincaré-duality groups with an infinite cyclic image: their first step is to use Addendum 5 to Theorem B* to get an HNN-decomposition with finitely generated base group and associated subgroups.

2.3   Further examples

We now use Theorem B as a tool for proving that certain finitely generated soluble groups cannot be finitely presented. This enterprise relies heavily upon

OBSERVATION 6   *The homomorphic image of an ascending HNN-extension by a normal subgroup contained in the normal closure of the base group is again an ascending HNN-extension.*

Example 7.   Consider the 1-relator group

$$\tilde{G} = \langle a, b;\ b^{-1}a^2 b = a^3 \rangle.$$

It is an HNN-group with infinite cyclic base group $\langle a \rangle$, stable letter b, and associated subgroups $S = \langle a^2 \rangle$ and $T = \langle a^3 \rangle$, but it is not ascending and it contains non-abelian free subgroups: indeed, $\tilde{G}''$ is free of rank $\aleph_0$ (this can be deduced from the subgroup theorem for HNN-extensions and the fact that $\tilde{G}/\tilde{G}'''$ is non-hopfian; cf. [10], p. 50). Now put $G = \tilde{G}/\tilde{G}''$. The group G' is isomorphic to the additive group of the subring $\mathbb{Z}[\frac{1}{6}]$ of $\mathbb{Q}$ and conjugation by $b \cdot \tilde{G}''$ induces on it multiplication by $\frac{3}{2}$. Every non-trivial finitely generated subgroup $B \leqslant \mathbb{Z}[\frac{1}{6}]$ is cyclic and so neither multiplication by $\frac{3}{2}$ nor by $\frac{2}{3}$ maps B into itself. Consequently, G is not an ascending HNN-extension

$$\langle B, t;\ t^{-1}bt = \sigma(b) \rangle$$

with $t = (b\tilde{G}'')^{\varepsilon}$ and $B \leqslant G'$ a *finitely generated* base group. However, G is an ascending HNN-extension $\langle B, t;\ t^{-1}bt = \sigma(b) \rangle$ with *infinitely generated* base group $B = G'$. Theorem B and Observation 6 allow us to deduce that no soluble quotient $\tilde{G}/M$, where $M \leqslant \tilde{G}''$, can be finitely presented. In particular,

$$G = \langle b \rangle \wr \mathbb{Z}[\tfrac{1}{6}]$$

is infinitely related, despite the fact that its multiplicator is trivial (see, e.g. [10], §1.8).

A good source of finitely generated but infinitely related soluble groups are *relatively free groups*. A. L. Šmel'kin [52] proved that the groups $F/\mathfrak{A}^l(F) = F/F^{(l)}$ are infinitely related provided F is a free non-cyclic group and $l \geqslant 2$; here $\mathfrak{A}^l$ denotes the variety of all soluble groups of derived length at most l. A far-reaching generalization of Šmel'kin's result is

THEOREM 8 ([18], Theorem B; [53], Theorem A)   *Let* $\mathfrak{B}$ *be a soluble variety and* $\tilde{G} = \langle x_1, \ldots, x_m;\ r_1, r_2, \ldots, r_n \rangle$ *a finitely presented group with* $m \geqslant n+2$. *Then the canonical image* $G = \tilde{G}/\mathfrak{B}(\tilde{G})$ *of* $\tilde{G}$ *in* $\mathfrak{B}$ *is infinitely related, unless* $\mathfrak{B}$ *is nilpotent-by-(locally finite). In particular, a finitely generated relatively free group* $F/\mathfrak{B}(F)$ *of rank* $\geqslant 2$ *of a soluble variety can be finitely presented if, and only if,* $F/\mathfrak{B}(F)$ *is nilpotent-by-finite.*

Proof.   According to a theorem by J. R. J. Groves [29], Theorem A(ii), a soluble variety $\mathfrak{V}$ is either contained in some variety $\mathfrak{N}_c \mathfrak{B}_e$, or it contains a variety $\mathfrak{A}_p \mathfrak{A}$ for some prime p. Here $\mathfrak{N}_c$ denotes the variety of all nilpotent groups of class at most c, while $\mathfrak{B}_e$ is the Burnside variety of exponent e, and $\mathfrak{A}$, $\mathfrak{A}_p$ are the varieties respectively, of abelian groups and elementary abelian p-groups. We may suppose $\mathfrak{V} \subseteq \mathfrak{A}_p \mathfrak{A}$. Since $m > n$, $\tilde{G}^{ab}$ is infinite and so there is a basis $\{z_1, \ldots, z_{m-1}, y\}$ of the free group F on $\{x_1, \ldots, x_m\}$ such that all relators $r_1, \ldots, r_n$ have exponent sum zero with respect to y. Write R for the normal closure in F of $r_1, \ldots, r_n$. Let U denote the normal closure in F of $\{z_1, \ldots, z_{m-1}\}$, and let N be the canonical image of U in $\tilde{G}$. From the presentation of N as U/R one derives easily an $\mathbb{F}_p \langle y \rangle$-module presentation of the $\mathbb{F}_p \langle y \rangle$-module $\bar{N} = N/N' . N^p$ of the form

$$(\mathbb{F}_p \langle y \rangle)^n \to (\mathbb{F}_p \langle y \rangle)^{m-1} \to \bar{N} \to 0.$$

As $n < m - 1$, the quotient $\bar{N}$ is not an $\mathbb{F}_p \langle y \rangle$-torsion module and consequently $\bar{N}$ is an *infinite* elementary abelian p-group. On the other hand, every finitely generated subgroup of $\bar{N}$ is finite and so $\bar{G} = G/N' . N^p = \langle y \rangle \ltimes \bar{N}$ cannot be an ascending HNN-extension of the form

$$\langle B, t; \ t^{-1} b t = \sigma(b) \text{ for } b \in B \rangle,$$

with $t = y^\varepsilon$ and B finitely generated. Consequently we can conclude from Theorem B and Observation 6 that $G = \tilde{G}/\mathfrak{V}(\tilde{G})$ is infinitely related.

3.   Examples of Finitely Presented Soluble Groups

The examples fall into the several classes: of constructible soluble, of metabelian, of nilpotent-by-abelian and of abelian-by-nilpotent groups. We shall discuss them in order.

3.1   Constructible soluble groups

We begin with a lemma summarizing three easy ways of constructing finitely presented soluble groups from given finitely presented soluble groups.

LEMMA 9

(i) *Every extension of a finitely presented group by another finitely presented group is finitely presented.*

(ii) *If $H \leqslant G$ is a subgroup of finite index in G, and if one of H or G is finitely presented, so is the other.*

(iii) *If B is a finitely presented group and $\sigma: B \rightarrowtail B$ is an injective endomorphism, the ascending HNN-extension*

$$\langle B, t; t^{-1}bt = \sigma(b) \text{ for } b \in B \rangle$$

is finitely presented.

Proof.

(i) We sketch the argument (cf. P. Hall [33], p. 426, or [50], p. 52). Let $N \lhd G \twoheadrightarrow G/N$ be the given extension and $\langle \mathcal{X}; \mathcal{R} \rangle \backsimeq N$ and $\langle \bar{\mathcal{Y}}; \mathcal{S} \rangle \backsimeq G/N$ finite presentations. Then G has a finite presentation with generating set $\mathcal{X} \cup \mathcal{Y}$, where $\mathcal{Y} \subseteq G$ lifts $\bar{\mathcal{Y}} \subseteq G/N$, and three types of relators: those in $\mathcal{R}$, lifts $s \cdot w_s(\mathcal{X})^{-1}$ of the relators $\bar{s} \in \mathcal{S}$ and relators determined by the action of $\mathcal{Y}$ on N induced by conjugation. (This third set is finite since $\mathcal{Y}$ is finite and N is finitely generated.)

(ii) If G is finitely presented, apply the Reidemeister–Schreier procedure to a finite presentation of G to obtain a finite presentation of H. Conversely, if H is finitely presented, pass to $H_G = \bigcap_{g \in G} H^g \leqslant H$. Then H_G has finite index in H hence is finitely presented by the argument just given and G, an extension of H_G by the finite group G/H_G, will be finitely presented by (i).

(iii) This is clear.

Basic properties of constructible soluble groups.   Recall from p. 259 that a soluble group is constructible if, and only if, it can be built up from the trivial group in finitely many steps by passing from a group H already obtained to a group G containing H as a subgroup of finite index, or passing to an ascending HNN-extension with base H.

The properties of constructible soluble groups which are of interest in our context, are summarized by

THEOREM 10

(i) *The class of constructible soluble groups is closed under the passage to subgroups of finite index, to extensions and to homomorphic images.*

(ii) *Every constructible soluble group is finitely presented, satisfies the maximal condition on normal subgroups, has finite Hirsch length, finite Prüfer rank and is (torsion-free)-by-finite. In particular it admits a faithful $\mathbb{Q}$-linear representation and is minimax and residually finite.*

(iii) *Every constructible soluble group contains a torsion-free subgroup of finite index that can be built up from the trivial group by merely using HNN-extensions.*

(iv) *A soluble group G can be embedded in a constructible soluble group if, and only if, G is a (torsion-free)-by-finite minimax group.*

Proof. (i) and (ii) are due to G. Baumslag and R. Bieri [9]: Claim (i) is proved there in **2.3** and **3.4**; the parts of (ii) dealing with finite presentation, finite Hirsch length, finite Prüfer rank and the property of being (torsion-free)-by-finite are proved by induction on the length of the group-building process (cf. [9], Prop. 1 and **3.3**, Remark 2). The property max-n follows from the fact that quotients of constructible soluble groups are again constructible and hence finitely related, and from Lemma 1. The last sentence of (ii) is standard: cf. [58], pp. 25, 45, 51. Part (iii) is a by-product of a characterization of constructible soluble groups in terms of a geometric invariant, due to Bieri and Strebel [23], p. 18: see Theorem 17 below for more details. Part (iv) is due to D. Kilsch [38].

The final part of Theorem 10 shows that finitely generated subgroups of constructible soluble groups need not be finitely presented. Here is an explicit instance:

Example 11.   The constructible group $B = \langle a, s; a^s = a^2 \rangle \cong \langle s \rangle \ltimes \mathbb{Z}[\frac{1}{2}]$ admits an injective endomorphism σ taking a to a^3 and fixing s. The corresponding HNN-extension is

$$G = \langle B, t; t^{-1}bt = \sigma(b) \text{ for all } b \text{ in } B \rangle$$

$$\cong \langle a, s, t; a^s = a^2, a^t = a^3, s^t = s \rangle$$

$$\cong \langle s, t; [s, t] = 1 \rangle \ltimes \mathbb{Z}[\frac{1}{6}].$$

The subgroup $N = \langle a, ts^{-1} \rangle$ is a normal subgroup with infinite cyclic quotient; N is isomorphic to the metabelian top $\tilde{G}/\tilde{G}''$ of the group $\tilde{G}$ discussed in Example 7, and is therefore infinitely related.

A group G is termed *coherent* if all its finitely generated subgroups are finitely related. Example 11 is a constructible soluble group which is not coherent. That coherent soluble groups are rather special is the content of

THEOREM 12 (Bieri and Strebel [19, 23], Groves [30])   *The following statements about the finitely generated soluble group G are equivalent:*

(i) *G is coherent;*
(ii) *G is polycyclic or an ascending HNN-extension over a polycyclic base group;*
(iii) *G and all its finitely generated subgroups are constructible.*

Further characterizations of finitely generated coherent soluble groups are given in the cited papers.

3.2 Theorems A and A*

The striking example which we stated as Theorem A is a finitely presented metabelian group which is not constructible (since, for example, its Hirsch length is infinite—cf. Theorem 10(ii)). To prove Theorem A we proceed as follows. Let B denote the wreath product of two infinite cyclic groups so that B has a presentation

$$B \cong \langle a, s; [a, a^{s^j}] = 1 \text{ for all } j > 0 \rangle.$$

The group B also has a linear representation

$$B \cong \left\langle \bar{a} = \begin{pmatrix} 1 & 1 \\ 0 & 1 \end{pmatrix}, \bar{s} = \begin{pmatrix} 1 & 0 \\ 0 & X \end{pmatrix} \right\rangle \leqslant GL_2(\mathbb{Q}(X)),$$

where X denotes an indeterminate. Conjugation by the diagonal matrix $\bar{t} = \begin{pmatrix} 1 & 0 \\ 0 & 1+X \end{pmatrix}$ induces an injective endomorphism of B, leading to the ascending HNN-extension

$$G \cong \langle B, t; a^t = aa^s, s^t = s \rangle$$
$$\cong \langle a, s, t; a^t = aa^s, [s, t] = 1 = [a, a^{s^j}] \text{ for } j > 0 \rangle.$$

Quite unexpectedly, all the commutator relations $[a, a^{s^j}] = 1, j > 0$, are now redundant except for $j = 1$. For suppose a, s, t are group elements with

$$a^t = aa^s \quad \text{and} \quad [s, t] = 1 = [a, a^s].$$

Assume a commutes with $a^s, \ldots, a^{s^n}$ for some $n \geqslant 1$; then

$$1 = [a, a^{s^n}]^t = [a^t, a^{ts^n}] = [aa^s, a^{s^n} a^{s^{n+1}}] = [a, a^{s^{n+1}}].$$

It is easy enough to see that G' is free abelian of infinite rank (cf. also Remark (ii) on p. 294).

This example and its proof prompts one to ask for what type of finitely generated metabelian groups a similar trick works. To see where the problem lies, let G be a finitely generated metabelian group. It gives rise to an extension

(4) $A \lhd G \twoheadrightarrow G/A = Q,$

where Q is finitely generated abelian and A is a finitely generated $\mathbb{Z}Q$-module. We assume Q is free abelian, say of rank n. Since $\mathbb{Z}Q$ is noetherian by Hilbert's Basis Theorem, A has a finite free presentation

$$\mathbb{Z}Q^h \rightarrow \mathbb{Z}Q^m \rightarrow A \rightarrow 0.$$

If we make the simplifying assumption that the extension (4) splits, G will have a presentation of the form

(5) $\langle a_1,\ldots,a_m,y_1,\ldots,y_n; r_1 = \ldots = r_h = [y_i,y_{i_1}] = [a_j,a_{j_1}^w] = 1 \rangle,$

where $1 \leqslant i < i_1 \leqslant n$, $1 \leqslant j \leqslant j_1 \leqslant m$ and $w = y_1^{l_1}\ldots y_n^{l_n}$ with $(l_1,\ldots,l_n) \in \mathbb{Z}^n$. The relators $r_1, r_2, \ldots, r_h$ define the module A. The presentation (5) has infinitely many defining relations, but this is solely due to the infinite set of commutators $[a_j, a_{j_1}^w]$. On closer analysis it turns out that the trick in the proof of Theorem A can be adapted to the more general group (4), provided Q admits a basis $\{s_1, t_1, \ldots, s_g, t_g\}$ (with $2g = n$) such that the annihilator ideal $\text{Ann}_{\mathbb{Z}Q}(A)$ of A contains for each $j \in \{1, 2, \ldots, g\}$ an element of the form

(6) $t_j - p_j(s_j),$

where $p_j(X)$ is a non-constant integral polynomial with constant term and leading coefficient both 1. It will be convenient to call such a polynomial *special*. (See [6], §6, for more details.)

Sketch of the proof of Theorem A*.   Let G be any finitely generated metabelian group. Theorem A* asserts that G may be embedded in a finitely presented metabelian group. We assume G^{ab} is infinite, as we obviously may. The proof given by Baumslag [6] has three stages. In the first ([6], §5) G is embedded into a split extension $B = G^{ab} \ltimes A_0$, where A_0 is a finitely generated $\mathbb{Z}G^{ab}$-module containing G'. The embedding is achieved by a generalization of the well known Magnus embedding [41]. In the second stage, the $\mathbb{Z}G^{ab}$-module A_0 is embedded in a module A (of fractions) over a larger ring $\mathbb{Z}Q$. Specifically, let

$\{s_1, s_2, \ldots, s_g\} \subseteq G^{ab}$ be a maximal $\mathbb{Z}$-linearly independent set. Using the fact that A_0 is a noetherian $\mathbb{Z}G^{ab}$-module, one can find g monic, integral special polynomials $p_j(X)$, $1 \leqslant j \leqslant g$, such that every $p_j(s_j) \in \mathbb{Z}G^{ab}$ induces an *injective* endomorphism $\tau_j \colon A_0 \rightarrowtail A_0$ (cf. [6], Lemma 7). These endomorphisms are commuting $\mathbb{Z}G^{ab}$-endomorphisms and can be used to define a chain of ascending HNN-extensions $G_1, G_2, \ldots, G_g$ as follows: Extend $\tau_1 \colon A_0 \rightarrowtail A_0$ to an injective endomorphism $\sigma_1 \colon B \rightarrowtail B$ which restricts to the identity on G^{ab}. Then G_1 is defined to be the ascending HNN-extension

$$G_1 = \langle B, t_1; t_1^{-1} b t_1 = \sigma_1(b) \text{ for all } b \text{ in } B \rangle.$$

Next τ_2 can be extended to an injective endomorphism σ_2 of G_1 with σ_2 the identity on G^{ab} and $\sigma_2(t_1) = t_1$. Now use τ_2 to define a second ascending HNN-extension

$$G_2 = \langle G_1, t_2; t_2^{-1} g_1 t_2 = \sigma_2(g_1) \text{ for all } g_1 \text{ in } G_1 \rangle.$$

Proceeding thus one ends up with a group $\tilde{G} = G_g$. This is a split extension $Q \ltimes A$, where $Q \cong G^{ab} \times \langle t_1, \ldots, t_g \rangle$ is abelian and A is obtained from A_0 by inverting $\tau_1, \ldots, \tau_g$, i.e. by localizing. The point of the whole construction is that $\mathbb{Z}Q$ will contain elements

$$\lambda_j = t_j - p_j(s_j)$$

having the form (6) and annihilating A. In the final step these elements λ_j are then used to prove that $\tilde{G} = Q \ltimes A$ has a finite presentation (cf. [6], §6).

The above procedure can be refined in two ways. First, one can do away with the preliminary embedding $G \hookrightarrow G^{ab} \ltimes A_0$ and perform suitable ascending HNN-extensions right from the beginning, ending up with a group $\tilde{G}$, which is shown to be finitely presented using Theorem D.

Secondly, one can adapt the argument to finitely generated centre-by-metabelian groups which are abelian-by-polycyclic. These refinements yield

THEOREM 13 (Strebel [55]) *A finitely generated centre-by-metabelian group G can be embedded in a finitely presented centre-by-metabelian group if G is abelian-by-polycyclic.*

Of course, Theorem C shows that the condition is necessary.

3.3   M. W. Thomson's embedding theorem

The simple trick proving Theorem A can be generalized in another direction. Let F be a field, k its prime ring $\mathbb{Z}$ or $\mathbb{F}_p$ and let Q be a finitely generated subgroup of $F^\times$. For any ring R, let $\mathrm{Tr}(l, R)$ be the group of lower triangular matrices with entries in R. Define

(7) $$G(l, Q) = \{g \in \mathrm{Tr}(l, k[Q]) \mid g_{ii} \in Q, 1 \leqslant i \leqslant l\}.$$

Clearly $G(l, Q)$ is finitely generated. We claim that there exists a finitely generated group P in $F^\times$ containing Q such that $G(l, P)$ is finitely related. Indeed, choose a maximal $\mathbb{Z}$-linearly independent subset $\{s_1, s_2, \ldots, s_g\}$ of Q and take

$$P = \langle Q, 1 + s_1, \ldots, 1 + s_g \rangle.$$

We now prove by induction on l that $G(l, P)$ is finitely related. It is clear for $l = 1$. So let $l > 1$ and consider the set N_l of all matrices in $G(l, P)$ of the form

$$\begin{pmatrix} p_1 & & & & \\ * & 1 & & & \\ \vdots & \vdots & \ddots & & \\ * & 0 & \cdots & 1 & \\ * & 0 & \cdots & 0 & 1 \end{pmatrix}$$

where $p_1 \in P$ and $* \in k[P] \leqslant F$. Clearly N_l is the kernel of a map from $G(l, P)$ to $G(l-1, P)$. Moreover, as $N_l \cong P \ltimes (k[P])^{l-1}$, there are calculations similar to those used in establishing Theorem A that imply N_l is finitely presented. The claim that $G(l, P)$ itself is finitely presented now follows from the inductive assumption that $G(l-1, P)$ is so and Lemma 9(i).

The sketched argument implies that $\mathrm{Tr}(l, F)$ *is locally finitely presentable*, i.e. that every finitely generated group of lower triangular matrices with entries in a field can be embedded in a finitely presented group of lower triangular matrices of the same size and with entries in the same field.

Since every relatively free $\mathfrak{N}_c\mathfrak{A}$-group can be represented by lower triangular $(c+1) \times (c+1)$-matrices with entries in a suitable field [32], one sees that every finitely generated relatively free $\mathfrak{N}_c\mathfrak{A}$-group can be embedded in a finitely presented $\mathfrak{N}_c\mathfrak{A}$-group.

The general method of Baumslag [6] and Remeslennikov [49], together with results from commutative algebra, make it possible to improve the statement above to

THEOREM 14 (Thomson [57], Theorem 1) *Let R be a commutative ring. For every finitely generated subgroup G of $\mathrm{Tr}(l, R)$ there exists a commutative ring S containing R and a finitely presented subgroup $\tilde{G}$ of $\mathrm{Tr}(l, S)$ which contains the canonical image of G in $\mathrm{Tr}(l, S)$.*

3.4 More examples

The linear groups $G(l, Q)$ defined by (7) contain a finite chain of normal subgroups whose factors are finitely generated metabelian and so satisfy max-n. Similarly, the finitely presented groups $\tilde{G}$ of Theorem 14 satisfy max-n. These groups do not therefore answer a question P. Hall had left open in 1954: 'Do there exist soluble groups satisfying FR but not max-n?' ([33], p. 425); but examples constructed by H. Abels [1] in 1978 do.

Let R be a commutative ring generated by a multiplicative group of units Q. Let $l \geqslant 3$ and consider the matrix group

$$(8) \quad H = H(l, R, Q) = \left\{ \begin{pmatrix} 1 & & & & \\ * & q_2 & & & \\ \vdots & \vdots & \ddots & & \\ * & * & \cdots & q_{l-1} & \\ * & * & \cdots & * & 1 \end{pmatrix} \middle| q_i \in Q \text{ and } * \in R \right\}.$$

This group is centre-by-$\mathfrak{N}_{l-2}\mathfrak{A}$ and its centre Z is isomorphic with the additive group of R. If Q is finitely generated, H is also finitely generated and so is R. Hence R is noetherian and H/Z satisfies max-n.

If $l = 3$, H is finitely presented only if H is polycyclic. Indeed, since H is centre-by-metabelian, it follows from Theorem C that H can only be finitely presented if Z is finitely generated; but if the additive group of R is finitely generated, H itself will obviously be polycyclic.

If $l > 3$, H is finitely presented if, and only if, its homomorphic image $Q \upharpoonright R$ is finitely presented. The necessity of this condition follows from Corollary BD1; the sufficiency was proved by H. Abels [1] for $l = 4$, $R = \mathbb{Z}[1/p]$ and $Q = \langle p \rangle$, where p is a prime, and by Strebel (unpublished) in general.

Examples are also known of finitely presented soluble groups with insoluble word problem. These were first constructed by O. G. Kharlampovič [37]. Her work has since been analysed and extended by Baumslag *et al.* [11].

3.5 Examples of finitely presented abelian-by-nilpotent groups
The examples discussed so far are all nilpotent-by-abelian-by-finite. But finitely presented soluble groups need not be like this:

THEOREM 15 (D. J. S. Robinson and R. Strebel [51], Theorem 1)   *Let l be an integer not equal to $-1, 0$ or 1. Then the group G given by*

$$\langle a, s, t; a^t = aa^s, a^{[s,\,t]} = a^l, [a, a^s] = 1 = [[s, t], s] = [[s, t], t] \rangle$$

has the following properties:
(i) *$A = \langle a \rangle^G$ is a free $\mathbb{Z}[1/l]$-module of countably infinite rank, while G/A is free nilpotent of rank 2 and class 2.*
(ii) *A is the Fitting subgroup of G.*
(iii) *G is not nilpotent-by-abelian-by-finite, but every proper quotient of it is metabelian-by-finite cyclic.*

The methods that prove this theorem have been extended in [54] to produce far more general finitely presented abelian-by-nilpotent groups; but despite this, it is not known whether every finitely generated abelian-by-nilpotent group embeds in a finitely presented one.

4.   Finitely Presented Nilpotent-by-Cyclic Groups

4.1   Characterization
The class of all finitely generated nilpotent-by-cyclic groups contains the centre-by-metabelian 3-generator subgroup

$$G = \left\langle \begin{pmatrix} 1 & & \\ 1 & 1 & \\ 0 & 0 & 1 \end{pmatrix}, \begin{pmatrix} 1 & & \\ 0 & 1 & \\ 0 & 1 & 1 \end{pmatrix}, \begin{pmatrix} 1 & & \\ 0 & X & \\ 0 & 0 & 1 \end{pmatrix} \right\rangle$$

of $GL_3(\mathbb{Q}(X))$ (X is an indeterminate), and the class contains all homomorphic images of G. It follows that groups in the class are in general neither residually finite, nor do they satisfy max-n. This was known already from results of Hall [33]. The situation is entirely different for finitely *presented* nilpotent-by-cyclic groups.

As every finitely generated nilpotent-by-(finite cyclic) group is poly-cyclic and thus finitely related, we restrict our attention to nilpotent-by-(infinite cyclic) groups, to which Theorem B applies.

THEOREM 16 (cf. [18], p. 260)   *If G is a finitely generated group having a nilpotent normal subgroup N with infinite cyclic factor group G/N, the following statements are equivalent:*

(i)   *G is coherent.*
(ii)   *G is constructible.*
(iii)   *G is finitely related.*
(iv)   *$G/[N, N]$ is finitely related.*
(v)   *$N/[N, N]$ has finite torsion subgroup and finite torsion-free rank, and G/N has a generator that induces by conjugation an automorphism of $N/[N, N] \otimes_{\mathbb{Z}} \mathbb{Q}$ with integral characteristic polynomial.*

Proof.   Every coherent soluble group is constructible (Theorem 12) and every constructible soluble group is finitely related (Theorem 10, (ii)). Next, if G is finitely related, it is isomorphic by Theorem B to an ascending HNN-extension

$$\langle B, t; t^{-1}bt = \sigma(b) \text{ for all } b \in B \rangle,$$

where B is a finitely generated subgroup of the nilpotent subgroup N, and so polycyclic. By Theorem 12 this implies that G is coherent. This proves the equivalence of (i), (ii), (iii).

Continuing to assume (iii), we have further that, by Observation 6, G/N' is an ascending HNN-extension of the form

(9) $$\langle \bar{B}, t; t^{-1}\bar{b}t = \bar{\sigma}(\bar{b}) \text{ for all } b \in \bar{B} \rangle,$$

where $\bar{B} = B/B \cap N' \overset{\sim}{\rightarrow} BN'/N'$ is finitely generated abelian. Hence G/N' is finitely related by Lemma 9(iii).

Next, if G/N' is finitely related, it is an ascending HNN-extension of the form (9) and so there is a generator gN of G/N and a *finitely generated* subgroup $\bar{B}$ of N^{ab} such that

(10) $(gN')^{-1} . \bar{B} . (gN') \leqslant \bar{B}$ and $\cup \{(gN')^{j} . \bar{B} . (gN)^{-j} \mid j \geqslant 0\} = N^{ab}$.

This gives (v). To deduce (ii) from (10) one aims to find a finitely generated subgroup B of N with properties analogous to (10). This is done by induction on the nilpotency class of N and using the maps

$$N^{ab} \otimes_{\mathbb{Z}} \gamma_{j-1} N/\gamma_j N \longrightarrow \gamma_j N/\gamma_{j+1} N \quad (j \geqslant 2)$$

which send $x . \gamma_2 N \otimes y . \gamma_j N$ to $[x, y] . \gamma_{j+1} N$ (see [18], p. 265, for details). Once B is found, G is seen to be an ascending HNN-extension over the finitely generated nilpotent base group B and hence constructible. This has proved (iii) $\Rightarrow$ (iv) $\Rightarrow$ (ii) and (iv) $\Rightarrow$ (v). The remaining implication (v) $\Rightarrow$ (iv) is clear.

4.2   Comparison with constructible groups

For every prime p, Abel's matrix group $H(4, \mathbb{Z}[1/p], \langle p \rangle)$ is a finitely presented nilpotent-by-abelian group (cf. p. 277). The quotient H/Z is infinitely related because $\mathbb{Z}[1/p]$ is not finitely generated, but H/H'' is finitely related by Corollary BD1. These examples show that the implication

$$G/G'' \text{ finitely related} \Rightarrow G \text{ finitely related}$$

does not hold for finitely generated nilpotent-by-abelian groups. If 'finitely related' is replaced by 'constructible', the implication is true:

THEOREM 17 (Bieri and Strebel [23], Theorem 5.2)   *Let G be a finitely generated soluble group, containing normal subgroups $N \leqslant G_1$ such that $|G: G_1|$ is finite, G_1/N is abelian and N is nilpotent. Then the following statements are equivalent:*

  (i)   *G is constructible.*
 (ii)   *$G/[N, N]$ is constructible.*
(iii)   *$\Sigma^c_{N^{ab}} = S(G_1/N) \backslash \Sigma_{N^{ab}}$ is contained in an open hemisphere of $S(G_1/N)$.*

5.   Finitely Presented Centre-by-Metabelian Groups I

In sections 5, 6 and 7 we shall present a characterization in terms of a geometric invariant of the finitely related groups in the class of all finitely generated centre-by-metabelian groups. In the present section we discuss Theorems B1, B2 and C. These results describe conditions which follow from finite presentability.

5.1 Basic properties of the geometric invariant Σ_A

Let Q be a finitely generated, multiplicative abelian group, k a commutative ring and kQ the associated group algebra. If $v\colon Q \to \mathbb{R}$ is a homomorphism into the additive group of the real numbers, we can extend v to a mapping

$$(11) \qquad v\colon kQ \to \mathbb{R}_\infty = \mathbb{R} \cup \{\infty\},$$

by setting $v(0) = \infty$ and

$$(12) \qquad v(\lambda) = \min \{v(q)\,|\,q \in \operatorname{supp}\lambda\}$$

for $\lambda \neq 0$. Here if $\lambda = \sum\lambda(q)q$, then $\operatorname{supp}\lambda$ denotes the support $\{q \in Q \,|\, \lambda(q) \neq 0\}$ of λ. If the ordering on $\mathbb{R}$ is extended to one on $\mathbb{R}_\infty$ by making ∞ the largest element, and the addition in $\mathbb{R}$ is extended to $\mathbb{R}_\infty$ by $\alpha + \infty = \infty$, then v has the following two properties:

$$(13) \qquad v(\lambda + \mu) \geqslant \min(v(\lambda), v(\mu))$$

and

$$(14) \qquad v(\lambda . \mu) \geqslant v(\lambda) + v(\mu).$$

Moreover, if kQ is a domain, (14) can be strengthened to an equality and v is then a valuation on kQ with values in $\mathbb{R}_\infty$. If, in addition, v is discrete as in subsection 1.4, then v is a discrete valuation (cf. subsection 6.5 below).

Every homomorphism $v\colon Q \to \mathbb{R}$ gives rise to the submonoid

$$(15) \qquad Q_v = \{q \in Q \,|\, v(q) \geqslant 0\}$$

of Q. Clearly Q_v is unaffected if v is replaced by a positive real multiple ρv. Thus each point $[v]$ of the space $S(Q)$ (cf. p. 262) determines a unique monoid Q_v.

Coordinate isomorphisms. Let $\theta\colon Q \twoheadrightarrow \mathbb{Z}^n$ be an epimorphism of Q onto the standard lattice of $\mathbb{R}^n$; here as elsewhere, n denotes the torsion-free rank $r_0(Q)$ of Q. Then θ induces an isomorphism of $\mathbb{R}$-vector spaces

$$(16) \qquad \theta^*\colon \mathbb{R}^n \xrightarrow{\sim} \operatorname{Hom}(\mathbb{Z}^n, \mathbb{R}) \xrightarrow{\sim} \operatorname{Hom}(Q, \mathbb{R})$$

taking x in $\mathbb{R}^n$ to the homomorphism $(q \mapsto \langle x, \theta(q) \rangle)$. We refer to θ^* as a *coordinate isomorphism*.

If $\theta_1 : Q \twoheadrightarrow \mathbb{Z}^n$ is another epimorphism, the composite $\theta_1 \circ \theta^{-1}$ is an automorphism of $\mathbb{Z}^n$ and it induces an $\mathbb{R}$-linear automorphism of $\mathbb{R}^n$ given by an integral matrix with respect to the standard basis of $\mathbb{R}^n$. *Coordinate isomorphisms can therefore be used to transfer to* $\mathrm{Hom}(Q, \mathbb{R})$ *all concepts defined on* $\mathbb{R}^n$ *which are invariant under* $GL_n(\mathbb{Z})$.

The first example of such a concept is the usual topology on $\mathbb{R}^n$. Thus $\mathrm{Hom}(Q, \mathbb{R})$ is a topological space. Next consider elements in $\mathrm{Hom}(Q, \mathbb{R})$ of the form $Q \twoheadrightarrow \mathbb{Z} \hookrightarrow \mathbb{R}$. These give the discrete points on $S(Q)$ and they correspond (under any coordinate isomorphism) to the primitive lattice points of $\mathbb{Z}^n$ in $\mathbb{R}^n$.

The unit sphere $\mathbb{S}^{n-1}$ in $\mathbb{R}^n$ intersects each ray $\{ \rho x \mid 0 < \rho \in \mathbb{R} \}$ ($x \neq 0$) exactly once. The continuous function

$$\theta^* : \mathbb{S}^{n-1} \to S(Q), \quad u \mapsto (q \mapsto \langle u, \theta(q) \rangle)$$

is therefore bijective. Since $\mathbb{S}^{n-1}$ is compact and $S(Q)$ is Hausdorff, θ^* is actually a homeomorphism. Hence $S(Q)$ is homeomorphic to an $(n-1)$-dimensional sphere and we shall call $S(Q)$ the *valuation sphere* of Q. The subspace $\{ x/\|x\| \mid x \in \mathbb{Z}^n \setminus \{0\} \}$ of $\mathbb{S}^{n-1}$ is dense and totally disconnected, and every coordinate isomorphism maps it homeomorphically onto the subspace $\mathrm{dis}\, S(Q)$ of $S(Q)$ consisting of all discrete points.

The notion of a *rationally defined* great subsphere is invariant under $GL_n(\mathbb{Z})$. Such a great subsphere corresponds under a coordinate isomorphism to a subset of $S(Q)$ having the following intrinsic definition:

$$S(Q, Q_1) = \{ [v] \in S(Q) \mid v(Q_1) = \{0\} \},$$

where Q_1 is a subgroup of Q. Note that $S(Q, Q_1)$ has codimension equal to $r_0(Q_1)$. Similarly, the rationally defined open, and closed, hemispheres correspond, respectively to

$$H_q = \{ [v] \in S(Q) \mid v(q) > 0 \}$$

and

$$\bar{H}_q = \{ [v] \in S(Q) \mid v(q) \geqslant 0 \},$$

where q is an element of infinite order in Q. Note that 0-dimensional great subspheres of $S(Q)$, i.e. pairs of antipodal points, can be defined intrinsically on $S(Q)$, irrespective of whether the subsphere is rationally defined or not.

Finally, one may speak of *convex subsets* and *rationally defined convex polyhedra* of $S(Q)$. A subset of $\mathbb{S}^{n-1}$ is convex by definition if it is the intersection of $\mathbb{S}^{n-1}$ with a convex cone; it is called a rationally defined convex polyhedron if it is the intersection of $\mathbb{S}^{n-1}$ with finitely many rationally defined closed halfspaces. The image of such a polyhedron under a coordinate isomorphism is a finite intersection of the form

$$\bar{H}_{q_1} \cap \bar{H}_{q_2} \cap \ldots \cap \bar{H}_{q_f}.$$

One verifies easily that such an intersection is either empty or contains a discrete point.

Remark 18. The evaluation map $E\colon \operatorname{Hom}(Q,\mathbb{R}) \times kQ \to \mathbb{R}_\infty$ sending (v,λ) to $v(\lambda)$ will be used in the important equation (18) below. Note that E does not induce a map $S(Q) \times kQ \to \mathbb{R}_\infty$. A substitute for E defined on $S(Q) \times kQ$ is sometimes needed and is obtained by choosing an epimorphism $\theta\colon Q \twoheadrightarrow \mathbb{Z}^n$ and defining

$$E_\theta\colon S(Q) \times kQ \to \mathbb{R}_\infty$$

by

$$E_\theta([v], 0) = \infty,$$

$$E_\theta([v], \lambda) = \min\{\langle (\theta^*)^{-1}[v], \theta(q)\rangle \mid q \in \operatorname{supp}\lambda\} \quad \text{if } \lambda \neq 0.$$

Then E_θ is continuous with respect to the first variable and it plays an important role in the proof of Theorem 26 and of Theorem D.

The object of central interest for us is the geometric invariant Σ_A, defined for any finitely generated kQ-module A. Recall that

(17) $\Sigma_A = \{[v] \in S(Q) \mid A \text{ is finitely generated over } kQ_v\}$,

where Q_v is given by equation (15). The set Σ_A is hard to work with directly; we shall therefore be interested in alternative expressions for it. Our first one relates it to the centralizer

$$C(A) = C_{kQ}(A) = \{\lambda \in kQ \mid a\lambda = a \text{ for all } a \text{ in } A\}$$

of A in kQ. Notice that $C(A) = 1 + \operatorname{Ann}_{kQ}(A)$.

LEMMA 19 (cf. [20], Proposition 2.1)   *Let A be a finitely generated kQ-module and $v: Q \to \mathbb{R}$ a non-zero homomorphism. Then A is finitely generated over kQ_v if, and only if, there exists λ in $C(A)$ with $v(\lambda) > 0$. Moreover, if A is finitely generated over kQ_v, every set generating A as a kQ-module also generates A as a kQ_v-module.*

Proof.   Suppose $\lambda \in C(A)$ is an element with $v(\lambda) > 0$ and μ is a non-zero element in kQ. Then $v(\mu\lambda^m) \geqslant v(\mu) + mv(\lambda)$ and is non-negative for large enough m, whence $\mu\lambda^m \in kQ_v$. It follows that $a\mu = a\mu\lambda^m$ lies in akQ_v for every a in A, which implies that $akQ = akQ_v$ for all a in A. Consequently, each kQ-generating set of A generates A also as a kQ_v-module.

For the converse, suppose $\{a_1, \ldots, a_h\}$ is a finite subset of A generating A as kQ_v-module, and that q is an element of Q with $v(q) > 0$. Then

$$a_i q^{-1} = \sum_j a_j \lambda_{ij} \quad \text{for } i = 1, \ldots, h,$$

where the λ_{ij} are elements of kQ_v. Thus

$$\sum_j a_j (\delta_{ij} - q\lambda_{ij}) = 0 \quad \text{for } i = 1, \ldots, h.$$

It follows that the determinant $\det(\delta_{ij} - q\lambda_{ij})$ annihilates A. Since $\det(\delta_{ij} - q\lambda_{ij}) = 1 - q\mu$ for some μ in kQ_v, we conclude that $q\mu$ is an element of $C(A)$. We may take $\lambda = q\mu$.

An immediate consequence of Lemma 19 is the useful expression

(18)
$$\Sigma_A = \bigcup_{\lambda \in C(A)} \{[v] \in S(Q) \mid v(\lambda) > 0\}$$

for Σ_A. Now if $\lambda \in C(A)$, the set $O_\lambda = \{[v] \mid v(\lambda) > 0\}$ is open, since it is the intersection of the finitely many subsets $\{[v] \mid v(q) > 0\}$ with $q \in \text{supp}\,\lambda$ and as every such subset is either empty or an open hemisphere. Formula (18) therefore provides a canonical cover of Σ_A by open subsets and implies that Σ_A *is an open subset of the valuation sphere* $S(Q)$. Moreover, (18) shows that Σ_A depends only on the annihilator ideal $\text{Ann}_{kQ}(A)$ of A; indeed one has

(19)
$$\Sigma_A = \Sigma_{kQ/\text{Ann}(A)}.$$

Passage to subgroups of finite index.   Let $\tilde{Q}$ be a second finitely generated abelian group, $\varphi: \tilde{Q} \to Q$ a homomorphism and let $\tilde{A}$ denote the $k\tilde{Q}$-module obtained from A by pulling back the action along φ. If φ has finite cokernel, every non-zero homomorphism $v\colon Q\to\mathbb{R}$ gives rise to a non-zero homomorphism $\tilde{v}=v\circ\varphi\colon \tilde{Q}\to\mathbb{R}$; there is therefore a continuous embedding

$$\varphi^*\colon S(Q)\to S(\tilde{Q}).$$

If also φ has finite kernel, φ^* is a homeomorphism. The relation between the invariants Σ_A and $\Sigma_{\tilde{A}}$ themselves is described by the next lemma; it often allows one to reduce to the case where Q is a free abelian group.

LEMMA 20 (cf. [20], Proposition 2.3)   *If $\varphi\colon \tilde{Q}\to Q$ has finite kernel and cokernel, then $\varphi^*\colon S(Q)\stackrel{\sim}{\to} S(\tilde{Q})$ is a homeomorphism sending Σ_A onto $\Sigma_{\tilde{A}}$.*

5.2   Deduction of Theorem B1 from Theorem B

Let G be a finitely presented soluble group, M a normal subgroup with $Q=G/M$ infinite abelian, and $A=M^{ab}$ the corresponding $\mathbb{Z}Q$-module. Let $v\colon Q\to\mathbb{Z}\hookrightarrow\mathbb{R}$ be discrete. Theorem B1 asserts that $[v]$ or $[-v]$ is in Σ_A.

Let $N/M=\ker v$. Theorem B gives G as an ascending HNN-extension

$$\langle B,t;\ t^{-1}bt=\sigma(b)\ \text{for all } b \text{ in } B\rangle$$

over some finitely generated subgroup B of N. It follows that

(20) $g^{-1}Bg\leqslant B$ and $\cup\{g^{j}Bg^{-j}\,|\,j\geqslant 0\}=N$

for some g in G corresponding to t. Since $\langle N,g\rangle=G$, we have $v(Mg)=\varepsilon=\pm 1$. We claim $[-\varepsilon v]$ is in Σ_A.

For let $Q_0=BM'/M'$ and $A_1=Q_0\cap(M/M')$. Then $Q_1=Q_0/A_1$ is naturally isomorphic to a subgroup of Q and Q_0 is finitely generated metabelian. Hence A_1 is a finitely generated $\mathbb{Z}Q_1$-module. By (20), conjugation by g induces an endomorphism Mg on A_1 with

$$A=\cup\{A_1(Mg)^{j}\,|\,j\leqslant 0\}.$$

We conclude that A is finitely generated over the ring $\mathbb{Z}Q_1[(Mg)^{-1}]\leqslant\mathbb{Z}Q_{(-\varepsilon v)}$. This establishes our claim and so Theorem B1.

5.3 Outline of the proof of Theorem B2

Let G be a finitely presented soluble group, $Q = G/M$ an abelian homomorphic image and let A denote the finitely generated $\mathbb{Z}Q$-module M^{ab}. Theorem B2 asserts that A is tame, i.e. that for every non-zero homomorphism $v: Q \to \mathbb{R}$ either $[v]$ or $[-v]$ belongs to Σ_A.

The proof of Theorem B2 begins with a reduction. Let $\tilde{G}$ be a subgroup of finite index in G, containing M and such that $\tilde{Q} = \tilde{G}/M$ is free abelian. Let $\tilde{A}$ denote the $\mathbb{Z}\tilde{Q}$-module on A by letting $\tilde{Q}$ act via the inclusion $\iota: \tilde{Q} \hookrightarrow Q$. Then $\tilde{A}$ is a finitely generated $\mathbb{Z}\tilde{Q}$-module, and by Lemma 20, ι induces a homeomorphism $\iota^*: S(Q) \to S(\tilde{Q})$ mapping Σ_A onto $\Sigma_{\tilde{A}}$. Since ι^* preserves the antipodal map and $\tilde{G}$ is also finitely presented (by Lemma 9(ii)), we see that we may assume Q is free abelian.

Let $v: Q \to \mathbb{R}$ be a non-zero homomorphism. The plan is to derive from a suitable finite presentation of G an infinite presentation of M, using the Reidemeister–Schreier procedure. The Schreier transversal involved is chosen depending on v in such a way that M can be proved to be a generalized free product $M = M_- *_{M_0} M_+$. Since M is soluble, this allows us to assume after possible readjustment that M is either M_+ or M_-. An analysis of the generating systems of M_+ and M_- leads to the conclusion that A is finitely generated either over $\mathbb{Z}Q_v$ or over $\mathbb{Z}Q_{(-v)}$.

The proof contains two critical points: the choice of a Schreier transversal and the finding of suitable systems of generators of M_+ and M_-. If v has infinite cyclic image, these points can be dealt with explicitly. In general it is best to perform geometric constructions on associated 2-dimensional complexes, rather than work directly with presentations. The details are given in [20], pp. 454–460.

5.4 Proof of Theorem C

Recall that Theorem C says that if G is a finitely presented centre-by-metabelian group and Z is the centre of G', then $B = G'/Z$ is finitely generated.

Assume first that G is a finitely generated centre-by-metabelian group. Set $Q = G^{ab}$ and let A be $(G')^{ab}$ as $\mathbb{Z}Q$-module, so that B is a quotient module of A. Commutation $(x, y) \mapsto [x, y]$ induces an alternating, bimultiplicative, non-degenerate form

$$\beta: B \times B \to G''.$$

Since G'' is central in G, one has for all x, y in G' and all g in G,

(21) $[gxg^{-1}, y] = [x, g^{-1}yg]$.

If the $\mathbb{Z}$-linear extension of $q \mapsto q^{-1}$ to $\mathbb{Z}Q$ is denoted by τ, then (21) shows at once that

$$\beta(b_1\lambda, b_2) = \beta(b_1, b_2\tau(\lambda)) \quad (\lambda \in \mathbb{Z}Q).$$

This relation and the fact that β is non-degenerate imply that τ maps $\mathrm{Ann}_{\mathbb{Z}Q}(B)$ onto itself. It follows from Lemma 19 that Σ_B is invariant under the antipodal map $[v] \mapsto [-v]$, i.e. that $\Sigma_B = -\Sigma_B$. On the other hand, B is a quotient of A, whence $\Sigma_B \supseteq \Sigma_A$ and

(22) $\Sigma_B = \Sigma_B \cup (-\Sigma_B) \supseteq \Sigma_A \cup (-\Sigma_A)$.

Assume now that G is, in addition, a homomorphic image of a finitely presented soluble group. Then $\Sigma_B = S(Q)$ by Theorem B2 and (22) and also $\Sigma_B \supseteq \mathrm{dis}\, S(Q)$ (the discrete points on $S(Q)$) by Theorem B1 and (22). By Proposition 21 below, either of these properties of B entail that B is a finitely generated abelian group.

The following result is a consequence of Theorem 36.

PROPOSITION 21   *Let k be a noetherian commutative ring, Q a finitely generated abelian group and A a finitely generated kQ-module. Then the following three statements are equivalent:*

(i) *The k-module underlying A is finitely generated;*
(ii) $\Sigma_A = S(Q)$;
(iii) $\Sigma_A \supseteq \mathrm{dis}\, S(Q)$.

Note that (i) $\Rightarrow$ (ii) $\Rightarrow$ (iii) is clear. We shall prove (iii) $\Rightarrow$ (i) on p. 301.

6.   General Theory of the Geometric Invariant

The geometric invariant can be studied by two methods. The first is elementary and uses the connection with the centralizer described in Lemma 19 and the geometric arguments invented in [20]. The second exploits the link with valuations and leads to good qualitative results about the geometry of the complements Σ_A^c of Σ_A in $S(Q)$. We continue with the notation of subsection 5.1.

6.1   An immediate consequence of Lemma 19 is that if A and B are finitely generated kQ-modules,

$$(23) \qquad \mathrm{Ann}(A) \subseteq \mathrm{Ann}(B) \quad \text{implies} \quad \Sigma_A \subseteq \Sigma_B.$$

In particular, (23) holds if B is a finitely generated submodule or quotient module of A. On the other hand, if $A' \rightarrowtail A \twoheadrightarrow A''$ is an exact sequence of finitely generated kQ-modules and A', A'' are finitely generated kQ_v-modules, then so is A. Consequently,

$$(24) \qquad A' \rightarrowtail A \twoheadrightarrow A'' \quad \text{implies} \quad \Sigma_A = \Sigma_{A'} \cap \Sigma_{A''}.$$

Details of what follows may be found in [21], pp. 173–4. If I and J are ideals of kQ, then

$$(25) \qquad \Sigma_{kQ/IJ} = \Sigma_{kQ/(I \cap J)} = \Sigma_{kQ/I} \cap \Sigma_{kQ/J}$$

and

$$(26) \qquad \Sigma_{kQ/I} = \Sigma_{kQ/\sqrt{I}},$$

where $\sqrt{I}$ denotes the radical $\{\lambda \in kQ \mid \lambda^m \in I \text{ for some } m \geqslant 1\}$ of I. If kQ is noetherian, the Lasker–Noether Decomposition Theorem and (25), (26) yield:

LEMMA 22 (cf. [21], Theorem 1.1)   *If k is noetherian and $P_1, P_2, \ldots, P_m$ are the prime ideals which are minimal over $\mathrm{Ann}_{kQ}(A)$, then*

$$\Sigma_A = \Sigma_{kQ/P_1} \cap \Sigma_{kQ/P_2} \cap \ldots \cap \Sigma_{kQ/P_m}.$$

This lemma is the starting point for several applications (subsection 6.5, Corollary 35, Theorem 36 and Theorem 40).

6.2   Modules with a principal annihilator ideal

Assume Q is a finitely generated free abelian group, k is an integral domain and A is a finitely generated kQ-module with principal annihilator ideal, say $\mathrm{Ann}_{kQ}(A) = \mu kQ$. *Our aim is to describe $\Sigma_A = \Sigma_{kQ/\mu kQ}$ explicitly.*

It is clear from the definition or from equation (18) that Σ_{kQ} is empty. So we assume μ is non-zero. Call an element q in $\mathrm{supp}\,\mu$ a *corner* if there exists a non-zero homomorphism $v \colon Q \to \mathbb{R}$ for which $v(q)$ is strictly smaller than $v(p)$ for p in $\mathrm{supp}\,\mu \setminus \{q\}$. Using an identification $\theta \colon Q \xrightarrow{\sim} \mathbb{Z}^n$, this definition can be reformulated by saying that the corners of μ correspond under θ to the corners (in the sense of elementary geometry) of the convex hull of the set $\theta(\mathrm{supp}\,\mu)$.

Each corner gives rise to a non-empty open convex subset

$$(27) \qquad C_q = \{[v] \in S(Q) \,|\, v(q) < v(p) \text{ for all } p \text{ in } \mathrm{supp}\,\mu \setminus \{q\}\}$$

$$= \bigcap_p \{H_{pq^{-1}} \,|\, p \in \mathrm{supp}\,\mu \setminus \{q\}\},$$

where H_x denotes the open hemisphere $\{[v] \,|\, v(x) > 0\}$. In addition, every pair q_1, q_2 of distinct elements of $\mathrm{supp}\,\mu$ defines a great hyper-subsphere (cf. subsection 5.1)

$$S(Q, \langle q_1 q_2^{-1} \rangle) = \{[v] \in S(Q) \,|\, v(q_1) = v(q_2)\}.$$

Obviously the sphere $S(Q)$ is the union of the open set

$$(28) \qquad \qquad \bigcup \{C_q \,|\, q \text{ is a corner of } \mu\}$$

and the closed set

$$(29) \quad \bigcup \{S(Q, \langle q_1 q_2^{-1} \rangle) \,|\, q_1 \text{ and } q_2 \text{ are distinct elements of } \mathrm{supp}\,\mu\}.$$

Note that this union need not be disjoint.

The promised description of Σ_A is now

LEMMA 23 (cf. [13], §7; [21], §5) *Let k be a domain, Q a finitely generated free abelian group and $\mathrm{Ann}_{kQ}(A) = \mu kQ \neq 0$. Then*

(i) $\Sigma_A = \bigcup \{C_q \,|\, q \text{ is a corner of } \mu \text{ with } \mu(q) \text{ a unit of } k\}$;

(ii) *if the coefficients of all corners of μ are units of k then $\Sigma_A^c = S(Q) \setminus \Sigma_A$ is contained in a finite union of rationally defined great hyperspheres.*

Proof.
(i) If $v: Q \to \mathbb{R}$ is any homomorphism and λ a non-zero element of kQ, we set

$$\lambda_v = \sum_{v(q) = v(\lambda)} \lambda(q) q.$$

Note that $(\lambda \lambda')_v = \lambda_v \lambda'_v$ for any pair of non-zero elements.

Now assume $[v]$ lies in Σ_A. By Lemma 19 the centralizer

$C(A) = 1 + \mu k Q$ contains an element $1 + \mu \lambda$ say, with $v(1 + \mu \lambda) > 0$. Since $v(1) = 0$ and $v \colon kQ \to \mathbb{R}_\infty$ is a valuation, $v(\mu \lambda)$ must be 0; furthermore, because v is given by the minimum formula (equation (12)) one has $-1 = (\mu \lambda)_v = \mu_v \lambda_v$. So μ_v is a unit, which must be of the form $\mu(q_0) q_0$ because Q is free abelian and k is a domain. Putting all this together, we see that q_0 is a corner of μ with coefficient a unit and that $[v]$ lies in C_{q_0}.

Conversely, if $[v]$ is in C_{q_1} and $\mu(q_1)$ is a unit, then

$$\lambda = 1 + \mu(-\mu(q_1)q_1)^{-1}$$

lies in $C(A) = 1 + \mu k Q$ and $v(\lambda) > 0$.

(ii) The hypothesis on the coefficients of the corners means that Σ_A coincides with the union (28), whence Σ_A^c is contained in the union (29).

Illustration 24.   Let Q be free abelian on $\{s, t\}$ and choose

$$\mu = ms^{-1}t^{-1} + s^{-1} + t^{-1} + n + st \in \mathbb{Z}Q,$$

where $|m| > 1$. Let $\theta \colon Q \xrightarrow{\sim} \mathbb{Z}^2$ be the isomorphism taking s to $(1, 0)$ and t to $(0, 1)$. Visualize μ as the set $\theta(\operatorname{supp}\mu) \subset \mathbb{Z}^2 \subset \mathbb{R}^2$, with each point of $\theta(\operatorname{supp}\mu)$ being assigned the corresponding coefficient (Fig. 1). The corners of μ are $s^{-1}t^{-1}$, s^{-1}, t^{-1} and st, but only the last three have a unit as coefficient. For $A = \mathbb{Z}Q/\mu\mathbb{Z}Q$ therefore, Lemma 23 (i) gives

$$\Sigma_A = C_{s^{-1}} \cup C_{t^{-1}} \cup C_{st}.$$

To calculate the displayed components it is best to work with their pre-images under $\theta^* \colon \mathbb{S}^1 \xrightarrow{\sim} S(Q)$.

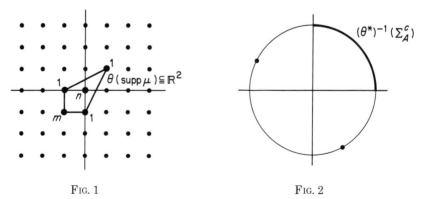

FIG. 1   FIG. 2

Quite generally, if a component C_q has the form (27) and $\theta^*: \mathbb{S}^{n-1} \to S(Q)$ is a coordinate isomorphism,

$$(\theta^*)^{-1}(C_q) = \{u \in \mathbb{S}^{n-1} \mid \langle u, \theta p - \theta q \rangle > 0 \text{ for all } p \in \operatorname{supp}\mu \setminus \{q\}\}.$$

Phrased more intuitively, $(\theta^*)^{-1}(C_q)$ consists of all unit vectors u for which the associated open half-space $\{x \in \mathbb{R}^n \mid \langle u, x \rangle > 0\}$ contains all difference vectors $\theta p - \theta q$, where p ranges over $\operatorname{supp}\mu \setminus \{q\}$.

Using these remarks the reader can easily confirm that $(\theta^*)^{-1}(\Sigma_A^c)$ is the set shown in Fig. 2.

6.3 Results relying on geometric methods

The proof of Theorem D, which tells us that every finitely generated metabelian group with $\Sigma_{G'} \cup -\Sigma_{G'} = S(G^{ab})$ is finitely presented, relies on a geometric induction procedure. This procedure can also be adapted so as to yield module-theoretic results. It is based on a compactness argument incorporated in the next lemma. To state this, let $\mathscr{F}$ denote a finite collection of finite subsets L of $\mathbb{Z}^n$. If x is a lattice point with $\langle x, x \rangle = m+1$, we shall say x *can be taken from the ball* $\mathbb{B}_m = \{x \in \mathbb{Z}^n \mid \langle x, x \rangle \leqslant m\}$ *by means of* $\mathscr{F}$, if there exists L in $\mathscr{F}$ with

$$x + L = \{x + y \mid y \in L\} \subseteq \mathbb{B}_m.$$

This means that $\|x + y\|^2 < \|x\|^2$ for all y in L.

LEMMA 25 (cf. [20], Lemma 1.1)   *Assume that for every u in the sphere $\mathbb{S}^{n-1}$ there exists L in $\mathscr{F}$ such that $\langle u, y \rangle > 0$ for all $y \in L$. Then there exists a natural number m_0 such that every lattice point x with $\langle x, x \rangle = m + 1 > m_0$ can be taken from $\mathbb{B}_m$ by means of $\mathscr{F}$.*

Proof.   Let $f: \mathbb{S}^{n-1} \to \mathbb{R}$ be defined by

$$f(u) = \max_{L} \min_{y} \{\langle u, y \rangle \mid y \in L \in \mathscr{F}\}.$$

By assumption f is everywhere positive. As f is continuous and $\mathbb{S}^{n-1}$ is compact, $C = \inf\{f(u) \mid u \in \mathbb{S}^{n-1}\}$ is positive. Next let D be the radius of the smallest ball containing all L of $\mathscr{F}$, i.e.

$$D = \max_{L} \max_{y} \{\|y\| \mid y \in L \in \mathscr{F}\}.$$

Take m_0 to be the integral part of $(D^2/2C)^2$, so that $(D^2/2C)^2 \in [m_0, m_0 + 1)$.

If x is a lattice point with $\langle x, x \rangle \geqslant m_0 + 1$, there is by the definition of C an L_x in $\mathscr{F}$ such that

$$\min_y \{\langle -x/\|x\|, y \rangle \mid y \in L_x\} \geqslant C,$$

or, equivalently, such that $\max_y \{\langle x, y \rangle \mid y \in L_x\} \leqslant -C\|x\|$. But then every lattice point $x + y$ in $x + L_x$ satisfies

$$\|x + y\|^2 = \|x\|^2 + 2\langle x, y \rangle + \|y\|^2 \leqslant \|x\|^2 - 2C\|x\| + D^2 < \|x\|^2.$$

So x can indeed be taken from $\mathbb{B}_{\langle x, x \rangle - 1}$ by means of $\mathscr{F}$.

Modules finitely generated over k. Our first application of Lemma 25 deals with modules A for which $\Sigma_A = S(Q)$.

THEOREM 26 (cf. [20], Theorem 2.4) *Let Q be a finitely generated abelian group, k a commutative ring and A a finitely generated kQ-module. Then $\Sigma_A = S(Q)$ if, and only if, A is finitely generated over k.*

Proof. If A is finitely generated as k-module, it is obvious that $\Sigma_A = S(Q)$. Conversely, suppose $\Sigma_A = S(Q)$. Equation (18) says that $S(Q)$ is covered by open subsets $\{[v] \mid v(\lambda) > 0\}$, where λ ranges over $C(A)$. Since $S(Q)$ is compact, there exists a finite subset $\Lambda \subset C(A) \setminus \{0\}$ with

$$S(Q) = \bigcup_{\lambda \in \Lambda} \{[v] \mid v(\lambda) > 0\}.$$

Choose an epimorphism $\theta \colon Q \twoheadrightarrow \mathbb{Z}^n$ with finite kernel and define $\mathscr{F}$ to consist of all finite sets $L_\lambda = \theta(\operatorname{supp} \lambda)$, where $\lambda \in \Lambda$. Then $\mathscr{F}$ satisfies the assumptions of Lemma 25 (use the mapping E_θ defined in Remark 18) and provides us with a natural number m_0 such that every lattice point x with $\|x\|^2 = m + 1 > m_0$ can be taken from the ball $\mathbb{B}_m$ by means of $\mathscr{F}$. Set $B_m = \theta^{-1}(\mathbb{B}_m)$ in what follows. We claim that

$$(30) \qquad\qquad akQ = akB_{m_0}$$

for every a in A. Indeed, if $m \geqslant m_0$ and $q \in Q$ is an element with $\|\theta q\|^2 = m + 1$, there exists $\lambda_q \in \Lambda$ with $\theta q + L_{\lambda_q} \subseteq \mathbb{B}_m$. Since λ_q centralizes A and $\theta(\operatorname{supp} \lambda_q) = L_{\lambda_q}$, this implies that

$$aq = a\lambda_q q \in akB_m.$$

Consequently $akB_{m+1} = akB_m$. The desired equality (30) follows. Finally, if $\mathscr{A} \subset A$ is a finite kQ-generating set of A, the finite set $\{aB_{m_0} \mid a \in \mathscr{A}\}$ generates A over k.

Modules finitely generated over some kQ_1. Theorem 26 can be generalized in several ways (cf. [21], §4). One such is

COROLLARY 27 ([21], Corollary 4.5)   *Let Q, k and A be as in Theorem 26 and let Q_1 be a subgroup of Q. Then $S(Q, Q_1) \subseteq \Sigma_A$ if, and only if, A is finitely generated over kQ_1.*

Proof.   We shall exploit the fact that Theorem 26 holds also with kQ_1 as base ring. Choose a subgroup Q_2 of Q such that $Q_1 \cap Q_2 = 1$ and $|Q : Q_1 Q_2|$ is finite. By Lemma 20, the inclusion $\iota : Q_1 \times Q_2 \to Q$ induces a homeomorphism $\iota^* : S(Q) \overset{\sim}{\to} S(Q_1 \times Q_2)$ sending Σ_A onto $\Sigma_{\tilde{A}}$, where $\tilde{A}$ denotes A as $k(Q_1 \times Q_2)$-module. Now $S(Q_1 \times Q_2, Q_1) \subseteq \Sigma_{\tilde{A}}$ if, and only if, for every non-zero homomorphism

$$\tilde{v} = (0, v_2) : Q_1 \times Q_2 \to \mathbb{R},$$

the module A is finitely generated over $k(Q_1 \times Q_2)_{\tilde{v}}$. Since $(Q_1 \times Q_2)_{\tilde{v}} = Q_1 \times (Q_2)_{v_2}$, the hypothesis on A therefore implies that A is finitely generated over $(kQ_1)(Q_2)_{v_2}$ for every $[v_2] \in S(Q_2)$, whence A must be finitely generated over kQ_1 by Theorem 26.

Finitely generated normal subgroups with infinite cyclic quotient.   Suppose $r_0(Q) > 1$. If A is a finitely generated *tame* kQ-module, the connected space $S(Q)$ is the union of the open subsets Σ_A and $-\Sigma_A$, and so $\Sigma_A \cap -\Sigma_A$ is a non-empty open subset. Since discrete points are dense in $S(Q)$, it follows that Σ_A contains a pair of antipodal points $\{[v], [-v]\}$ or, equivalently, Σ_A contains the 0-dimensional great subsphere $S(Q, \ker v)$. From Corollary 27 (or from a direct argument using Lemma 19) we conclude that A is finitely generated over $k(\ker v)$.

As an application we have

COROLLARY 28 ([22], Theorem)   *Let G be a finitely generated nilpotent-by-abelian group. If $r_0(G^{ab}) > 1$ and if $(G')^{ab}$ is a tame $\mathbb{Z}G$-module, then G contains a finitely generated normal subgroup N with infinite cyclic quotient.*

Proof.   By the above reasoning, $A = (G')^{ab}$ is finitely generated over some $\mathbb{Z}(N/G')$, where G/N is infinite cyclic. Choose generators

$\{g_1 G', \ldots, g_m G'\}$ of N/G' and a finite set $\mathscr{A} \subseteq G'$ such that $\mathscr{A} G''$ generates A as a $\mathbb{Z}(N/G')$-module. Then the conjugates $\{g^{-1} ag \,|\, a \in \mathscr{A} \text{ and } g \in \langle g_1, \ldots, g_m \rangle\}$ generate the nilpotent group G' modulo G'' and so generate G'. It follows that

$$N = \langle g_1, \ldots, g_m, \mathscr{A} \rangle$$

is finitely generated.

Remarks.

(i) The example $G = \langle a, t; t^{-1} a t = a^m \rangle$ with $|m| > 1$ shows that Corollary 28 is false without the hypothesis that $r_0(G^{ab}) > 1$.

(ii) Let $G = \langle a, s, t; a^t = aa^s, [s, t] = 1 = [a, a^s] \rangle$ be the group of Theorem A and set $t_1 = st$. Then

$$G = \langle a, s, t_1; a^{t_1} = a^s a^{s^2}, [s, t_1] = 1 = [a, s^s] \rangle$$

and $N = \langle a, a^s, t_1 \rangle$ is a finitely generated normal subgroup with infinite cyclic factor group. The new presentation reveals that $G' = \langle a^{t_1^j}, a^{st_1^j} \,|\, j \in \mathbb{Z} \rangle$ is free abelian of rank $\aleph_0$.

6.4   Finite generation of tensor powers

Let A be a finitely generated kQ-module and $m \geqslant 2$ an integer. We say A is *m-tame* if for every $k \leqslant m$ all k-point subsets of $\Sigma_A^c = S(Q) \setminus \Sigma_A$ are contained in an open hemisphere of $S(Q)$. Note that A is m-tame if it is $(m+1)$-tame; moreover, A is 2-tame precisely if Σ_A^c does not contain a pair of antipodal points or, equivalently, if A is tame.

Consider the tensor product $A_1 \otimes_k A_2 \otimes_k \ldots \otimes_k A_m$ of m finitely generated kQ-modules $A_1, A_2, \ldots, A_m$. This can be viewed as a kQ^m-module via the canonical isomorphism $k(Q^m) \xrightarrow{\sim} \otimes^m kQ$, and as such it is finitely generated over $k(Q^m)$. Let

$$(v_1, v_2, \ldots, v_m) \colon Q^m \to \mathbb{R}$$

be a non-zero homomorphism. If $[v_i] \in \Sigma_{A_i}$ for some $i \in \{1, 2, \ldots, m\}$, there exists by Lemma 19 an element $\lambda_i \in C(A_i)$ for which $(v_i)(\lambda_i) > 0$. Then $\lambda_i' = 1 \otimes \ldots \otimes 1 \otimes \lambda_i \otimes 1 \otimes \ldots \otimes 1$ centralizes $A_1 \otimes_k \ldots \otimes_k A_m$ and $(v_1, \ldots, v_m)(\lambda_i') = v_i(\lambda_i) > 0$. We conclude that

$$(31) \quad \{[(v_1, \ldots, v_m)] \in S(Q^m) \,|\, [v_i] \in \Sigma_{A_i} \text{ for some } 1 \leqslant i \leqslant m\} \subseteq \Sigma_{A_1 \otimes \ldots \otimes A_m}.$$

Suppose now that A is a finitely generated m-tame kQ-module. Then $\otimes^m A$, the m-fold tensor power of A, is finitely generated as $k(Q^m)$-module. Let $\Delta: Q \to Q^m$ be the diagonal homomorphism $q \mapsto (q, \ldots, q)$. Viewing $\otimes^m A$ as a kQ-module via Δ, Corollary 27 (applied to the pair $(Q^m, \Delta Q)$) shows $\otimes^m A$ is finitely generated if, and only if, $S(Q^m, \Delta Q) \subseteq \Sigma_{\otimes^m A}$. Now $[(v_1, \ldots, v_m)] \in S(Q^m, \Delta Q)$ precisely if $v_1 + v_2 + \ldots + v_m$ is zero. If this is so and if $v_{j_1}, \ldots, v_{j_k}$ denote the non-zero components of $(v_1, \ldots, v_m)$, then the k-point set $\{[v_{j_1}], \ldots, [v_{j_k}]\}$ is not contained in an open hemisphere. Since A is m-tame we conclude that at least one point of this k-point set is in Σ_A, whence $[(v_1, \ldots, v_m)]$ lies in $\Sigma_{\otimes^m A}$ by (31). This argument establishes

THEOREM 29 ([16], p. 377)   *Let Q be a finitely generated abelian group, k a commutative ring, A a finitely generated kQ-module and $m \geqslant 2$. If A is m-tame the m-fold tensor power $\otimes^m A$ is finitely generated as a kQ-module with diagonal action.*

Remark.   Suppose $Q = \langle t \rangle$ is infinite cyclic and $A = \mathbb{Z}Q/(2t-3)\mathbb{Z}Q$. Then (Lemma 23) Σ_A^c equals $S(Q)$, and so A is not 2-tame, despite the fact that all tensor powers $\otimes^m A$ are finitely generated $\mathbb{Z}Q$-modules under diagonal action; indeed $\otimes^m A \cong \mathbb{Z}Q/(2^m t - 3^m)\mathbb{Z}Q$. This example shows that the converse of Theorem 29 is false in general. It is true however, if k is a field:

THEOREM 30 (Bieri and Groves [16], Theorem C)   *If Q, A and m are as in Theorem 29, and k is a field, the following statements are equivalent:*

(i)   *A is m-tame.*
(ii) *The m-fold tensor power $\otimes^m A$ is finitely generated as a kQ-module with diagonal action.*
(iii) *The i-fold exterior powers $\wedge^i A$ are finitely generated as kQ-modules with diagonal action for $i = 2, 3, \ldots, m$.*

The proof is significantly harder than that of Theorem 29, cf. [16].

Applications to finitely presented soluble groups.   An easy consequence of Theorem 29 is

COROLLARY 31   *Suppose G is a finitely generated group, M a normal subgroup with abelian quotient $Q = G/M$ and $m \geqslant 2$. If M^{ab} is an m-tame $\mathbb{Z}Q$-module (via conjugation) then $G/\gamma_{m+1}M$ satisfies max-n.*

Proof. The lower central factors $\gamma_j M/\gamma_{j+1}M$ are images of the $\mathbb{Z}(\Delta Q)$-modules $\otimes^j M^{ab}$. For $2 \leqslant j \leqslant m$ these are finitely generated by Theorem 29.

As an *illustration* consider the matrix group $H = H(l, \mathbb{Z}[1/p], \langle p \rangle)$, with $l \geqslant 3$, defined by (8). Let M consist of the lower unitriangular matrices. Then $Q = H/M$ is free abelian on the $(l-2)$ cosets

$$q_2 = \mathrm{diag}(1, p, 1, \ldots, 1) \cdot M, \ldots, q_{l-1} = \mathrm{diag}(1, \ldots, 1, p, 1) \cdot M.$$

Next M^{ab} is the direct sum of $(l-1)$ modules $A_1, \ldots, A_{l-1}$, which are formed by the elements

$$\{1 + rE_{i+1,i} \mid r \in \mathbb{Z}[1/p]\}.$$

The geometric invariants $\Sigma^c_{A_i}$ can be computed by using formula (18) (or, more easily, by formula (34) below). The outcome is that $\Sigma^c_{A_i} = \{[v_i]\}$, where

$$v_1(q_j) = \begin{cases} -1 & \text{if } j = 2 \\ 0 & \text{if } j > 2 \end{cases}$$

$$v_i(q_j) = \begin{cases} 1 & \text{if } j = i \\ -1 & \text{if } j = i+1 \quad \text{for } 1 < i < l-1 \\ 0 & \text{otherwise} \end{cases}$$

$$v_{l-1}(q_j) = \begin{cases} 1 & \text{if } j = l-1 \\ 0 & \text{otherwise.} \end{cases}$$

Then formula (24) yields $\Sigma^c_A = \{[v_1], [v_2], \ldots, [v_{l-1}]\}$.

Now $v_1 + v_2 + \ldots + v_{l-1} = 0$ and so M^{ab} is not $(l-1)$-tame. But every $(l-2)$-point subset is contained in an open hemisphere H_q (cf. p. 282); indeed for $\Sigma^c_A \setminus \{[v_i]\}$ one can take

$$q = q_2^{-2} q_3^{-3} \ldots q_i^{-i} q_{i+1}^{l-1} \ldots q_{l-1}^{i+1},$$

and so M^{ab} is $(l-2)$-tame. These findings and Corollary 31 give us again the result that $H/\gamma_{l-1}M$, which coincides with $H/\zeta H$, satisfies max-n (cf. p. 277).

Our second application combines Theorem 29 and the method of proving Theorem C.

COROLLARY 32 ([20], Theorem 5.7) *Suppose G is a finitely generated group and M is a normal subgroup with abelian factor group Q. If $A = M^{ab}$ is a tame $\mathbb{Z}Q$-module, every quotient of $G/\gamma_3 M$ satisfies* max-n *and is residually finite.*

Proof. An easy argument shows that it is enough to prove the result for $G/\gamma_3 M$. We therefore assume that $\gamma_3 M = 1$. As A is 2-tame, Corollary 31 shows that G has max-n. By a result of Hall ([34], Theorem 1) it follows that $\gamma_2 M$ is a residually finite $\mathbb{Z}Q$-module. Thus to prove G is residually finite it will suffice to verify that a group $\bar{G}$ of the form $\bar{G} = G/L$ is residually finite, whenever L is a normal subgroup of G of finite index in $\gamma_2 M$. Henceforth write G for $\bar{G}$; so $\gamma_2 M$ is finite. Its centralizer G_1 has finite index in G and contains M. If $Q_1 = G_1/M$, then M^{ab} is a tame $\mathbb{Z}Q_1$-module, by Lemma 20. If the argument of the proof of Theorem C, as given in subsection 5.4, is applied to the centre-by-metabelian group G_1 one sees that $B = M/\zeta M$ is finitely generated abelian (for M^{ab} is a tame $\mathbb{Z}Q_1$-module). Because the alternating bimultiplicative form $\beta\colon B \times B \to \gamma_2 M$, induced by commutation, is non-degenerate and $\gamma_2 M$ is finite, B must have finite exponent and so is finite. Consequently G_1, and so G itself, is metabelian-by-finite and hence residually finite by Hall's result.

Remark. It is pointed out in [31] and [57] that there are mistakes in [7] and [48] where claims conflicting with the corollary are made.

6.5 Results based on the connection with valuations

Let Q be a finitely generated abelian group, k a commutative ring, I an ideal of kQ and let A be the k-algebra kQ/I. The theme of this subsection will be the computation of Σ_A^c by means of generalized valuations of A.

The basic result. Let Γ be an ordered abelian group, ∞ an additional element larger than every element of Γ and set $\Gamma_\infty = \Gamma \cup \{\infty\}$. The usual conventions for ∞ then make $\langle \Gamma_\infty, + \rangle$ into an ordered monoid. If $P \lhd A$ is a prime ideal, we let L denote the field of fractions $ff(A/P)$ of the domain A/P and $L^\times$ the multiplicative group of L.

A *valuation* $w\colon L \to \Gamma_\infty$ of L with values in Γ_∞ is a function which restricts to a homomorphism on $L^\times$ and satisfies

$$(32) \quad w(0) = \infty, \text{ and } w(x+y) \geqslant \min(w(x), w(y)) \quad \text{for all } (x, y) \in L^2.$$

The composite $\tilde{w}: A \xrightarrow{\;\pi\;} A/P \xrightarrow{\;w\;} \Gamma_\infty$ is called a *generalized valuation*. Let $v \in \mathrm{Hom}(Q, \mathbb{R})$ and suppose there exists an order preserving injective homomorphism $\sigma: v(Q) \to \Gamma$ so that the diagram

(33)
$$
\begin{array}{ccc}
Q \longrightarrow & \mathrm{im}\, v \longhookrightarrow & \mathbb{R} \\
\downarrow{\scriptstyle \kappa} & \downarrow{\scriptstyle \sigma} & \\
A \longrightarrow & \Gamma_\infty &
\end{array}
$$

commutes (here $\kappa(q) = q + I$). In this situation we shall say $\tilde{\omega}$ *extends* v.

THEOREM 33 (cf. [13], Theorem 2; [21], Theorem 2.1) *A non-zero homomorphism* $v: Q \to \mathbb{R}$ *represents a point of* Σ_A^c *if, and only if, there exists an ordered abelian group* Γ *and a generalized valuation* $\tilde{w}: A \to \Gamma_\infty$ *which extends* v *and satisfies* $\tilde{w}(k + I/I) \geqslant 0$.
    Further, if A is a domain, $[v] \in \Sigma_A^c$ *if, and only if, there exists a valuation w on $ff(A)$ so that w extends v and $w(k + I/I) \geqslant 0$.*

Proof. Let $kQ_v^+ = \{\lambda \in kQ_v \mid v(\lambda) > 0\}$. Then kQ_v^+ is an ideal in kQ_v and by Lemma 19 $[v] \in \Sigma_A$ if, and only if, there exists $\mu \in I$ with $1 + \mu$ in kQ_v^+. Therefore $[v]$ is in Σ_A^c precisely if I is disjoint from the multiplicatively closed set $S = 1 + kQ_v^+$. Now if $I \cap S = \emptyset$, every ideal P containing I and maximal with respect to $P \cap S = \emptyset$ must be a prime ideal. We conclude that $[v] \in \Sigma_A^c$ if, and only if, there exists a prime ideal $P \supseteq I$ such that

$$
P_v = (kQ_v^+ + P)/P
$$

is a proper ideal of the domain

$$
A_v = (kQ_v + P)/P.
$$

Assume now v can be extended to a generalized valuation $\tilde{w}: A = kQ/I \to kQ/P \to \Gamma_\infty$ with $\tilde{w}(k + I/I) \geqslant 0$. Let σ satisfy (33). The axioms of a valuation imply that $\sigma(v(\lambda)) \leqslant w(\lambda + P)$ for all $\lambda \in kQ$. Consequently A_v is contained in the valuation ring $V = \{x \in ff(kQ/P) \mid w(x) \geqslant 0\}$ and P_v is contained in the valuation ideal $M = \{x \in ff(kQ/P) \mid w(x) > 0\}$, whence $P_v \neq A_v$ and so $[v]$ is in Σ_A^c. Conversely, if $P_v \neq A_v$, the basic extension theorem for valuations (see, e.g., [36], pp. 35, 36) yields a valuation ring $V \subseteq ff(kQ/P) = L$ with maximal ideal M and such that

$$
k + I/I \twoheadrightarrow k + P/P \leqslant A_v \leqslant V \quad \text{and} \quad P_v \leqslant M.
$$

The constructions of Γ and of $w: L \to \Gamma_\infty$, as well as the verification that $\tilde{w} = w \circ \pi$ extends v are standard: cf. [21], p. 176.

Remark. If kQ is noetherian, the radical ideal $\sqrt{I}$ is a finite intersection $P_1 \cap \ldots \cap P_m$ of prime ideals, and Lemma 22 shows that

$$\Sigma^c_{kQ/I} = \Sigma^c_{kQ/P_1} \cup \ldots \cup \Sigma^c_{kQ/P_m}.$$

When applying Theorem 35 we can therefore always restrict attention to valuations of the fields $ff(kQ/P_i)$, $i = 1, 2, \ldots, m$.

Modules of finite Prüfer rank. Theorem 33 and the fact that valuations of Dedekind domains are well understood, give the useful qualitative result Corollary 35 below. First we discuss a special case:

Example 34. Let Q be a finitely generated abelian group, acting on the field of rational numbers $\mathbb{Q}$ by a homomorphism $\kappa: Q \to \mathbb{Q}^\times$ and define A to be the cyclic $\mathbb{Z}Q$-module $1 . \mathbb{Z}Q \leqslant \mathbb{Q}$. Since the valuations of $\mathbb{Q}$ are just the p-adic valuations $w_p: \mathbb{Q} \to \mathbb{Z}_\infty$, for p prime, the last assertion of Theorem 33 gives

$$\Sigma^c_A = \{[w_p \circ \kappa] \mid w_p \circ \kappa \neq 0\}.$$

More explicitly, select in Q a maximal $\mathbb{Z}$-linearly independent subset $\{q_1, \ldots, q_n\}$ and compute the prime factorizations

$$|\kappa(q_j)| = \prod_p p^{m_{jp}}.$$

If $\mathscr{P}$ denotes the set of primes p such that $m_{j_p} \neq 0$ for at least one j, then Σ^c_A has at most $|\mathscr{P}|$ elements. Let e_j be the jth standard vector of $\mathbb{R}^n$ and θ^* the homeomorphism $\mathbb{S}^{n-1} \xrightarrow{\sim} S(Q)$ induced by $\theta(q_j) = e_j$. Then

$$(34) \qquad (\theta^*)^{-1}(\Sigma^c_A) = \left\{ \frac{m_p}{\|m_p\|} \,\middle|\, p \in \mathscr{P} \right\} \subseteq \mathbb{S}^{n-1},$$

where $m_p = (m_{1_p}, \ldots, m_{n_p})$. Notice that every point of Σ^c_A is discrete. Conversely, if $\Sigma \subseteq S(Q)$ is a finite subset made up of discrete points, there exists a homomorphism $\kappa: Q \to \mathbb{Q}^\times$ such that the derived $\mathbb{Z}Q$-module A satisfies $\Sigma^c_A = \Sigma$.

The special feature of Example 34 is that Σ^c_A is a finite set of discrete points. This is shared by a wider class of modules:

COROLLARY 35 *If A is a finitely generated $\mathbb{Z}Q$-module whose additive group has finite Prüfer rank, then Σ_A^c is a finite set.*

Proof. If $P_1, \ldots, P_m$ are the prime ideals of $\mathbb{Z}Q$ minimal over $\mathrm{Ann}(A)$, then $\Sigma_A^c = \Sigma_{\mathbb{Z}Q/P_1}^c \cup \ldots \cup \Sigma_{\mathbb{Z}Q/P_m}^c$. The hypothesis that A has finite Prüfer rank implies the same for each $\mathbb{Z}Q/P_i$. Hence $\mathbb{Z}Q/P_i$ is either a finite field, or its field L_i of fractions is an algebraic number field with ring of integers $\mathfrak{O}_i$. Every valuation of L_i is discrete and associated with a non-zero prime ideal $\mathfrak{p}$ of $\mathfrak{O}_i$. Moreover, if $q \in Q$, the factorization of the fractional ideal $\mathfrak{O}_i(q + P_i)$ involves only finitely many prime ideals. As Q is finitely generated, the result follows.

Integral elements. Let q be an element of Q and A a finitely generated kQ-module with annihilator ideal I. We say q is *integral with respect to A* if $q + I \in kQ/I$ is integral over $(k + I)/I$; i.e. if there exists a monic polynomial $f(X) \in k[X]$ with $f(q) \in I$. Elements of finite order are always integral and are of little interest. If q is an *element of infinite order* which is integral with respect to A, it may be deduced from Lemma 19 that Σ_A contains the open hemisphere $H_{q^{-1}}$, whence Σ_A^c is contained in the closed hemisphere $\bar{H}_q = \{[v] \in S(Q) \mid v(q) \geqslant 0\}$. The converse of this statement need not be true for non-noetherian rings (see Example 39 below), but for noetherian rings it must:

THEOREM 36 ([21], Theorem 4.8) *Assume k is noetherian. If q is an element of infinite order in Q, the following statements are equivalent:*

(i) *q is integral with respect to A.*
(ii) $\Sigma_A^c \subseteq \bar{H}_q$.
(iii) $\mathrm{dis}\,\Sigma_A^c = \Sigma_A^c \cap \mathrm{dis}\,S(Q) \subseteq \bar{H}_q$.

Proof. (i) $\Rightarrow$ (ii) has been discussed above, and (ii) $\Rightarrow$ (iii) is clear. To prove the implication (iii) $\Rightarrow$ (i), let $P_1, \ldots, P_m$ be the minimal prime ideals over I, the annihilator ideal of A. Then $\sqrt{I} = P_1 \cap \ldots \cap P_m$ and q is integral with respect to kQ/I if, and only if, it is integral with respect to each kQ/P_i. On the other hand, $\Sigma_A^c = \Sigma_{kQ/P_1}^c \cup \ldots \cup \Sigma_{kQ/P_m}^c$ and so $\mathrm{dis}\,\Sigma_A^c$ is contained in $\bar{H}_q$ if, and only if, all $\mathrm{dis}\,\Sigma_{kQ/P_i}^c$ are contained in $\bar{H}_q$. Both facts taken together show that we may assume I is prime. Let us write $P = I$ and, without loss of generality, take $A = kQ/P$.

Assume q is *not* integral with respect to A. Then a point $[v]$ of $\Sigma_{kQ/P}^c$ which does not lie in $\bar{H}_q$ can be constructed as follows. Set $R = (k[q^{-1}] + P)/P$, $K = ff(R)$ and write $x = q^{-1} + P$. Then x is not a unit of R. Let M be a minimal prime ideal over xR. Since R is noetherian, M

has height 1 by Krull's Principal Ideal Theorem (see e.g. [36], p. 104) and so the local ring $S = R_M$ is 1-dimensional and noetherian; its integral closure $S' \subseteq K$ will still be noetherian (see e.g. [36], p. 61), whence S' is a Dedekind domain. Since $x \in S'$ and is not a unit, there exists a discrete valuation $w_1 \colon K \twoheadrightarrow \mathbb{Z}_\infty$ with $w_1(x) > 0$ and $w(S') \geqslant 0$. Because $L = ff(A)$ is finitely generated over K, therefore L is finite over a purely transcendental extension of K. Hence the discrete valuation w_1 on K can be extended to a *discrete* valuation $w \colon L \to \mathbb{R}_\infty$. The composite $v = w \circ \kappa \colon Q \to L^\times \to \mathbb{R}$ is then a rank 1 homomorphism with $v(q^{-1}) = w(x) > 0$ and $w(k + P/P) \geqslant 0$. It represents by Theorem 35 a point $[v]$ in $\Sigma^c_{kQ/P}$ but outside $\bar{H}_q$.

If in the theorem dis Σ^c_A is empty, then every q will be integral with respect to A so that kQ/I will be finitely generated over k. Clearly A will also be so. This finishes the missing (iii) $\Rightarrow$ (i) on p. 287.

Further examples. Let F be a field, Q a free abelian group of rank n and $A = FQ$. Let k be an F-subalgebra of A. To avoid confusion, we shall write the elements of A as functions $Q \to F$; in particular, χ_q will denote the characteristic function of $\{q\}$. Define $\kappa \colon kQ \twoheadrightarrow A = FQ$ to be the k-algebra epimorphism induced by the inclusion $k \leqslant FQ$ and $q \mapsto \chi_q$, and use κ to turn A into a cyclic kQ-module.

If $q \in Q \setminus \{1\}$ is an element with $\chi_q \in k$ the element $q - \chi_q$ is integral with respect to A and hence $\Sigma^c_A \subseteq \bar{H}_q$. In this way one gets the following upper bound for Σ^c_A:

$$(35) \qquad \Sigma^c_A \subseteq \cap \{\bar{H}_q \mid \chi_q \in k\}$$
$$= \{[v] \in S(Q) \mid v(q) \geqslant 0 \text{ for all } q \text{ with } \chi_q \in k\}.$$

On the other hand, every non-zero homomorphism $v \colon Q \to \mathbb{R}$ extends to a valuation $v \colon FQ \to \mathbb{R}_\infty$, defined by the minimum formula (12), and this, in turn, extends uniquely to a valuation $w \colon L = ff(FQ) \to \mathbb{R}_\infty$. Theorem 35 states that $[v] \in \Sigma^c_A$ if $w(k) \geqslant 0$, i.e. if $k \leqslant FQ_v$, and so one obtains the lower bound

$$(36) \qquad \{[v] \in S(Q) \mid k \leqslant FQ_v\} \subseteq \Sigma^c_A.$$

In the special case, where k *is the F-subalgebra of FQ generated by a submonoid Q_0 of Q*, the bounds (35) and (36) coincide and yield

$$(37) \qquad \Sigma^c_A = \{[v] \in S(Q) \mid Q_0 \leqslant Q_v\} = \cap \{\bar{H}_q \mid q \in Q_0 \setminus \{1\}\}.$$

Remarks.

(i) Suppose $k = F[\mu] \leqslant FQ$ is generated by the element $\mu\colon Q \to F$. Then $\lambda = \sum \mu(q)\,.\,q - \mu\,.\,1 \in kQ$ annihilates A; in point of fact $\kappa\colon kQ \twoheadrightarrow A$ induces an isomorphism $kQ/\lambda\,.\,kQ \overset{\sim}{\to} A$. So Σ_A^c can be computed explicitly by Lemma 23.

(ii) Clearly Q_v contains a submonoid Q_0 if, and only if, it contains a set of monoid generators J of Q_0. Hence (37) implies that *every intersection* $\cap \{\bar{H}_q \mid q \in J \setminus \{1\}\}$ *of rationally defined closed hemispheres occurs as* Σ_A^c for a suitable domain k and a suitable cyclic kQ-module A: it suffices to take Q_0 to be the monoid generated by J, to set $k = FQ_0$ and to define $\kappa\colon kQ \twoheadrightarrow A = FQ$ as above.

Example 37.   Let $\{q_1, \ldots, q_n\}$ be a basis of Q and define $\theta\colon Q \overset{\sim}{\to} \mathbb{Z}^n$ to be the isomorphism taking q_j to the jth standard basis vector e_j of $\mathbb{Z}^n \subset \mathbb{R}^n$. Choose $0 \leqslant m \leqslant n$ and let Q_0 be the monoid generated by $\{q_1, \ldots, q_m, q_{m+1}^{\pm 1}, \ldots, q_n^{\pm 1}\}$. Then formula (37) translates to

$$(38) \quad (\theta^*)^{-1}(\Sigma_A^c) = \{(x_1, \ldots, x_m, 0, \ldots, 0) \in \mathbb{S}^{n-1} \mid x_1 \geqslant 0, \ldots, x_m \geqslant 0\}.$$

Notice that $L = f\!f(FQ)$ is algebraic over the noetherian domain $FQ_0 \leqslant FQ$ but that (38) may nevertheless be infinite, in contrast to the situation treated in Corollary 35.

Example 38.   Let $v\colon Q \to \mathbb{R}$ be a non-zero homomorphism and put $Q_0 = Q_v$. Then $\Sigma_A^c = \{[v]\}$. If v is not discrete, we have a situation where $\varnothing = \mathrm{dis}\,\Sigma_A^c \neq \Sigma_A^c$. This shows that in Proposition 21, (iii) need not imply (i) for a non-noetherian ring k.

Example 39.   Let $\{q_1, q_2\}$ be a basis of Q and take Q_0 to be the monoid

$$\{q_1^{m_1}\,.\,q_2^{m_2} \mid \text{either } m_1 > 0, \text{ or } m_1 = 0 \text{ and } m_2 \geqslant 0\}.$$

Then $\Sigma_A^c = \{[v]\}$, where $v\colon Q \to \mathbb{R}$ takes q_1 to 1 and q_2 to 0. Consequently, $\Sigma_A^c \subseteq \bar{H}_{q_1} \cap \bar{H}_{q_2} \cap \bar{H}_{q_2^{-1}}$. Now q_1 and q_2 are in Q_0 and so they are integral over $k = FQ_0$; but q_2^{-1} is not integral over k. This example illustrates that the characterization of integral elements given by Theorem 36 fails for non-noetherian domains.

On the geometric structure of Σ_A^c.   We shall review the qualitative results obtained so far about Σ_A^c and compare them with recent results of Bieri and Groves [17].

Let Q be a finitely generated abelian group, k a commutative ring and A a finitely generated kQ-module. We may assume Q is free abelian because of Lemma 20. Let $I = \mathrm{Ann}_{kQ}(A)$. If $\sqrt{I}$ is the intersection $P_1 \cap \ldots \cap P_m$ of finitely many prime ideals, as happens when k is noetherian, then (cf. p. 288)

$$\Sigma^c_A = \Sigma^c_{kQ/P_1} \cup \ldots \cup \Sigma^c_{kQ/P_m}.$$

Thus one may restrict attention to prime ideals I if k is noetherian.

When $I = \mu kQ$ and k is a domain, there is an explicit formula for $\Sigma^c_{kQ/I}$ given by Lemma 23: it implies that $\Sigma^c_{kQ/I}$ is a finite union

(39)
$$\bigcup_{d \in D} \left(\bigcap_{q \in J(d)} \bar{H}_q \right)$$

of convex, rationally defined spherical polyhedra. The index sets D and $J(d)$ are all finite. As any set of the form (39) has the property that its discrete points are dense, it follows that dis $\Sigma^c_{kQ/I}$ is dense in $\Sigma^c_{kQ/I}$.

When $I = \mu kQ$, the structure of the domain k has little effect on Σ^c_A, for the geometric invariant depends only on supp μ and on those corners of μ whose coefficients are units. But in general the structure of k does influence Σ^c_A greatly. For example, suppose we consider only modules A with $k \leqslant A \leqslant ff(k)$. Then every intersection of $\bar{H}_q$'s arises for a suitable domain k (see Remark (ii) on p. 302), whereas Σ^c_A is a finite set of discrete points if $k = \mathbb{Z}$ (by Example 34).

These considerations suggest sharp qualitative results only exist for severely restricted classes of domains. That results of this kind do exist at all has recently been demonstrated by Bieri and Groves. They proved

THEOREM 40 [17]   *Assume k is a Dedekind domain and P a prime ideal in kQ with $k \cap P = 0$. Set $A = kQ/P$ and $L = ff(A)$. Then there exists a finite set $\mathscr{P}$ of prime ideals $\mathfrak{p}$ of k, including 0, for which the following statements are true:*

(i)
$$\Sigma^c_A = \bigcup \{\Sigma^c_{A_\mathfrak{p}} \mid \mathfrak{p} \in \mathscr{P}\},$$

where the localized domains $A_\mathfrak{p} \cong k_\mathfrak{p} Q/P_\mathfrak{p}$ are to be considered as modules over $k_\mathfrak{p} Q$;

(ii) *if $v: Q \to \mathbb{R}$ is a non-zero homomorphism, $[v]$ is in $\Sigma^c_{A_\mathfrak{p}}$ if and only if* *there exists a valuation* $w\colon L \to \mathbb{R}_\infty$ *which extends* v *and which is such that its restriction to* k *is the* $\mathfrak{p}$*-adic valuation;*

(iii) Σ_A^c *is a finite union of convex, rationally defined spherical polyhedra.*

COROLLARY 41 [17]   *Theorem B1 and Theorem 40 imply Theorem B2.*

Proof. We have to show that a finitely generated $\mathbb{Z}Q$-module A, for which Σ_A^c does not contain a pair of antipodal discrete points, does not contain an antipodal pair at all. Put differently, we have to establish that $\Sigma_A^c \cap -\Sigma_A^c \neq \emptyset$ implies that $\operatorname{dis}(\Sigma_A^c \cap -\Sigma_A^c) \neq \emptyset$. By Theorem 40(iii) and formula (39), $\Sigma_A^c \cap -\Sigma_A^c$ is a finite union of finite intersections $\cap \{\bar{H}_q \mid q \in J\}$. The corollary thus follows from the fact that every non-empty intersection of the form $\cap \{\bar{H}_q \mid q \in J\}$ contains a discrete point. ∎

7.   Finitely Presented Centre-by-Metabelian Groups II

In this section we shall discuss Theorem D. As a matter of fact there is a slightly more general result:

Let G be a finitely generated centre-by-metabelian group; if $(G')^{ab}$ is a tame $\mathbb{Z}G^{ab}$-module, then G is finitely presentable.

For $(G')^{ab}$ is then 2-tame, so G satisfies max-n by Corollary 31. Hence G'', being central in G, is finitely generated and the assertion follows from Theorem D (the metabelian case) and Lemma 9(i).

Returning to Theorem D, a familiar Lemma 20 argument shows that the theorem will follow from

THEOREM D1 ([20], Theorem 3.1)   *Let Q be a free abelian group of rank n and A a finitely generated tame $\mathbb{Z}Q$-module. Then every extension of A by Q can be finitely presented.*

7.1   An infinite presentation for G

Consider the extension $A \lhd G \xrightarrow{\pi} Q$, with Q free abelian of rank n and A a finitely generated $\mathbb{Z}Q$-module. Choose a finite set of generators $\mathscr{A} \cup \mathscr{T}$ of G having the properties that $\mathscr{T} = \{t_1, \ldots, t_n\}$ maps under π onto a basis of Q and that $\mathscr{A}$ is contained in A and includes every commutator $a_{ij} = [t_i, t_j]$ $(1 \leqslant i < j \leqslant n)$. (The involvement of the commutators is needed for technical reasons.) Let $F = F(\mathscr{A} \cup \mathscr{T})$ be the free group on $\mathscr{A} \cup \mathscr{T}$, $F(\mathscr{T})$ the free group on $\mathscr{T}$ and $\rho: F \twoheadrightarrow G$ be the obvious epimorphism.

Our presentation will have three types of defining relators. First are those which ensure that Q is abelian,

$$\mathscr{R}_Q = \{a_{ij}^{-1}[t_i, t_j] \mid 1 \leqslant i < j \leqslant n\}$$

(here $a_{ij} \in \mathscr{A}$ and $[t_i, t_j] \in F(\mathscr{T})$, so $\mathscr{R}_Q \subseteq F$); next the infinite set of commutators

$$\mathscr{K} = \{[a, b^w] \mid a, b \text{ in } \mathscr{A} \text{ and } w \text{ in } F(\mathscr{T})\}.$$

The relators $\mathscr{R}_Q \cup \mathscr{K}$ define a metabelian group $\tilde{G}$, where $\tilde{A} \lhd \tilde{G} \twoheadrightarrow Q$. Now $\ker(\tilde{G} \twoheadrightarrow G)$ is contained in the finitely generated $\mathbb{Z}Q$-module $\tilde{A}$ and so is finitely generated by Hilbert's Basis Theorem. Thus there exists a finite set $\mathscr{R}_A \subseteq \langle \mathscr{A} \rangle^F$ whose image under $F \twoheadrightarrow \tilde{G}$ generates $\ker(\tilde{G} \twoheadrightarrow G)$ and hence

$$\rho_*: \langle \mathscr{A} \cup \mathscr{T}; \mathscr{R}_A \cup \mathscr{R}_Q \cup \mathscr{K} \rangle \overset{\sim}{\to} G.$$

7.2 Auxiliary relations codifying tameness

We shall prove the finite presentability of our group G as follows: Using the tameness of A, we find a set of auxiliary relators $\mathscr{C}_\Lambda$; then establish that almost all relators of $\mathscr{K}$ are redundant in the presence of $\mathscr{C}_\Lambda \cup \mathscr{R}_Q$ and hence *a fortiori* in the presence of $\mathscr{R}_A \cup \mathscr{C}_\Lambda \cup \mathscr{R}_Q$.

By formula (18), Σ_A has a canonical cover by open subsets

$$O_\lambda = \{[v] \in S(Q) \mid v(\lambda) > 0\},$$

where λ runs over the centralizer $C(A)$ of A in $\mathbb{Z}Q$. (Note that $O_q = H_q$ for $q \in Q$.) Since $\Sigma_A \cup -\Sigma_A = S(Q)$ by the tameness of A, and since $S(Q)$ is compact there exists a finite set $\Lambda \subseteq C(A)$ such that

$$S(Q) = \bigcup_{\lambda \in \Lambda} (O_\lambda \cup -O_\lambda).$$

Let $\theta: Q \overset{\sim}{\to} \mathbb{Z}^n$ be the isomorphism taking $\pi(t_j)$ to the jth canonical basis vector e_j of $\mathbb{Z}^n \subset \mathbb{R}^n$. Define $\mathscr{F}$ to be the finite collection of the subsets

$$L_\lambda = \theta(\text{supp } \lambda) \quad \text{and} \quad -L_\lambda = -\theta(\text{supp } \lambda)$$

of $\mathbb{Z}^n$ with $\lambda \in \Lambda$. Then $\mathscr{F}$ satisfies the assumptions of Lemma 25. Consequently there exists a natural number m_0 with the property that for every x in $\mathbb{Z}^n$ with $\|x\|^2 = m+1 > m_0$ there exists $\varepsilon = \pm 1$ and $\lambda \in \Lambda \subseteq C(A)$ such that $x + \varepsilon L_\lambda$ is in the ball $\mathbb{B}_m$.

Denote by $\mathscr{S}$ the subset $\{t_1^{l_1} t_2^{l_2} \ldots t_n^{l_n} \mid (l_1, l_2, \ldots, l_n) \in \mathbb{Z}^n\}$ of $F(\mathscr{F})$ and let $^\wedge : \mathscr{S} \rightarrowtail Q$ be the bijection determined by $F \xrightarrow{\rho} G \xrightarrow{\pi} Q$ (cf. subsection 7.1). We now define the auxiliary relators as

$$\mathscr{C}_\Lambda = \left\{ a^{-1} \prod_{s \in \mathscr{S}} (a^{\lambda(\hat{s})})^s \mid a \in \mathscr{A} \text{ and } \lambda \in \Lambda \right\}.$$

Here $\lambda(\hat{s})$ means the coefficient of $\hat{s}$ in the element λ of $\mathbb{Z}Q$ (the value of $\hat{s}$ under the function $\lambda : Q \to \mathbb{Z}$) and the products, which are of course finite, are taken with respect to some definite ordering.

7.3 Proof for split extensions

The proof of Theorem D1 is based on a geometric induction, paralleling that used for Theorem 26. We shall restrict our argument to the split extension case. In the general case some fairly delicate commutator calculations are needed to cope with local geometric problems (cf. [20], pp. 448–451), and these are likely to obscure the gist of the argument.

Suppose then that $G = Q \ltimes A$. Here we can choose $\mathscr{F}$ so that $[t_i, t_j] = 1$ for all relevant i, j, whence $\mathscr{R}_Q$ is now

$$\mathscr{R}_Q = \{[t_i, t_j] \mid 1 \leqslant i < j \leqslant n\}.$$

Let $\mathscr{S}_m$ be the subset of $\mathscr{S}$ whose image under $\theta^\wedge$ lies in $\mathbb{B}_m$ and set

$$\mathscr{K}_m = \{[a, b^s] \mid a, b \text{ in } \mathscr{A} \text{ and } s \text{ in } \mathscr{S}_m\}.$$

Then modulo $\mathscr{R}_Q$, our previous $\mathscr{K}$ is $\bigcup_{m \geqslant 1} \mathscr{K}_m$ (because G is split) and if

$$G_m = \langle \mathscr{A} \cup \mathscr{F}; \mathscr{R}_A \cup \mathscr{C}_\Lambda \cup \mathscr{R}_Q \cup \mathscr{K}_m \rangle,$$

then $\varinjlim_m G_m = G$.

We claim that the identity on $\mathscr{A} \cup \mathscr{F}$ induces an isomorphism $G_{m_0} \rightarrowtail G$, thereby proving Theorem D1 in this case. Here m_0 is as in subsection 7.2. Our claim will follow if we can show that the identity on $\mathscr{A} \cup \mathscr{F}$ induces isomorphisms $G_m \rightarrowtail G_{m+1}$ for every $m \geqslant m_0$.

So let $m \geqslant m_0$ and consider an element s' in $\mathscr{S}$ with $\theta^\wedge(s') = \theta(\hat{s'})$ of norm $\sqrt{m+1}$ (i.e. $s' \in \mathscr{S}_{m+1} \backslash \mathscr{S}_m$). By the definitions of $\mathscr{F}$ and m_0, there is a sign $\varepsilon = \pm 1$ and an element λ in $\Lambda \subseteq C(A)$ with $\theta^\wedge(s') + \varepsilon L_\lambda \subseteq \mathbb{B}_m$. Now for any s in $\mathscr{S}$, $\lambda(\hat{s}) \neq 0$ if, and only if, $\theta(\hat{s}) \in \theta(\operatorname{supp}\lambda) = L_\lambda$ and so $\lambda(\hat{s}) \neq 0$ implies $\theta^\wedge(s's^\varepsilon) = \theta(\hat{s'}) + \varepsilon\theta(\hat{s}) \in \mathbb{B}_m$. Since $\mathscr{S}$ is multiplicatively closed modulo the relations $\mathscr{R}_Q$, we conclude

$$\lambda(\hat{s}) \neq 0 \Rightarrow s's^\varepsilon \in \mathscr{S}_m \quad (\operatorname{mod} \mathscr{R}_Q).$$

Hence, for any a, b in $\mathscr{A}$ and any s in $\mathscr{S}_{m+1} \backslash \mathscr{S}_m$,

(40)
$$\text{(i) if } \varepsilon = 1, \text{ then } [a, (b^{\lambda(\hat{s})})^{ss'}] = 1 \quad \text{in} \quad G_m;$$
$$\text{(ii) if } \varepsilon = -1, \text{ then } [a^{\lambda(\hat{s})}, b^{s's^{-1}}] = 1 \quad \text{in} \quad G_{\hat{m}}.$$

Suppose $\varepsilon = 1$. The identity $[u, vw] = [u, w][u, v]^w$ and the defining relation $b = \prod_s (b^{\lambda(\hat{s})})^s$ show that $[a, b^s]$, as an element in G_m, is a product of conjugates of $[a, (b^{\lambda(\hat{s})})^{ss'}]$ and so is 1 in G_m by (40)(i).

If $\varepsilon = -1$, the identity $[uv, w] = [u, w]^v [v, w]$ and the defining relation $a = \prod_s (a^{\lambda(\hat{s})})^s$ give $[a, b^s]$ as a product of conjugates of $[(a^{\lambda(\hat{s})})^s, b^s] = [a^{\lambda(\hat{s})}, b^{s's^{-1}}]^s$ and so is 1 in G_m by (40)(ii).

Thus all defining relations of G_{m+1} hold in G_m. Induction therefore yields $G_{m_0} \rightsquigarrow \varinjlim G_m \cong G$.

7.4 Some examples

Example 42. The Baumslag–Remeslennikov method revisited. Let Q be free abelian of rank $n = 2g$ with basis $\{s_1, t_1, \ldots, s_g, t_g\}$, and suppose A is a finitely generated $\mathbb{Z}Q$-module annihilated by elements of the form

(41)
$$\mu_i = t_i - p_i(s_i),$$

with $p_i(X)$ special (cf. (6)), one for each pair (s_i, t_i). We shall verify that A is a tame $\mathbb{Z}Q$-module and use the proof of Theorem D1 to find a finite presentation.

(I) *Reduction to the case $g = 1$.* Suppose $\mathbb{Z}^n \subset \mathbb{R}^n$ can be written as an orthogonal direct sum $\mathbb{Z}^{n_1} \perp \mathbb{Z}^{n_2} \perp \ldots \perp \mathbb{Z}^{n_g}$ of g sublattices $\mathbb{Z}^{n_i} \subset \mathbb{R}^{n_i}$, the decomposition being such that for each i there exists a finite collection $\mathscr{F}^{(i)}$ of finite sets $L^{(i)} \subset \mathbb{Z}^{n_i}$ which satisfy the hypotheses of Lemma 25 for $\mathbb{S}^{n_i - 1} \subset \mathbb{R}^{n_i}$. Let $C_i > 0$ and $D_i > 0$ be the numbers produced in the proof of Lemma 25. For every x in $\mathbb{Z}^n$ there exists $1 \leqslant i \leqslant g$ so that the

norm of the ith orthogonal component $x^{(i)}$ in $\mathbb{Z}^{n_i}$ satisfies $\|x^{(i)}\| \geqslant$
$g^{-1/2}\|x\|$. The proof of Lemma 25 reveals then that the collection
$\mathscr{F} = \bigcup_i \mathscr{F}^{(i)}$ satisfies the hypotheses of the lemma and that m_0 can be
chosen to be the integral part of $g \max_i \{(D_i^2/2C_i)^2\}$.

In the situation of Example 42 each $n_i = 2$. The construction of the
subsets $\mathscr{F}^{(i)}$ is explained below.

(II) *Finding $\mathscr{F}$, C and D in the case $g = 1$.* Let Q be free abelian on $\{s, t\}$
and consider $A = \mathbb{Z}Q/\mu\mathbb{Z}Q$, where μ is the element

$$\mu = t - p(s),$$

and $p(X)$ is a special polynomial (cf. p. 274) of degree d. The corners of
μ are $1, s^d, t$ and since their coefficients are units in $\mathbb{Z}$, we obtain three
centralizing elements

(42)
$$\begin{aligned} \lambda_1 &= 1 + \mu \\ \lambda_{s^d} &= 1 + \mu s^{-d} \\ \lambda_t &= 1 - \mu t^{-1}. \end{aligned}$$

The proof of Lemma 23(i) implies that

$$\Sigma_A = C_1 \cup C_{s^d} \cup C_t = O(\lambda_1) \cup O(\lambda_{s^d}) \cup O(\lambda_t).$$

It is easily seen that Σ_A^c is made up of the three points $[v_1]$, $[v_2]$ and
$[v_3]$ given by $v_1(s) = 1$, $v_1(t) = 0$; $v_2(s) = 0$, $v_2(t) = 1$; $v_3(s) = -1$, $v_3(t) = -d$.
Obviously, Σ_A^c does not contain a pair of antipodal points.

To compute C and D, let $\theta: Q \xrightarrow{\sim} \mathbb{Z}^2$ be the isomorphism sending s to
$(1, 0)$ and t to $(0, 1)$ and take $\Lambda = \{\lambda_1, \lambda_{s^d}, \lambda_t\}$. Then $\mathscr{F}$ consists of all
$\pm L$, where L is the θ-image of the support of λ in Λ; and $\mathbb{B}_{d^2+1}$ is the
smallest ball containing the elements of $\mathscr{F}$, whence $D^2 = d^2 + 1$. The
number

$$C = \min_u \left(\max_L \min_y \{\langle u, y \rangle \mid y \in L \in \mathscr{F}, u \in \mathbb{S}^1\} \right)$$

works out to be $((d+1)^2 + 1)^{-1/2}$, and so m_0 can be taken to be the
integral part $f(d)$ say, of $(D^2/2C)^2 = \frac{1}{4}(d^2+1)^2((d+1)^2+1)$.

Returning to the original situation of Example 42, where A is annihilated by elements (41), we conclude from the reasoning in (I) that A is tame and that one may take

(43) $$m_0 = g \max_i \{f(d_i) \mid 1 \leqslant i \leqslant g\}$$

in the given proof of Theorem D1.

(III) *An illustration.* Let us apply the previous considerations to the example of Theorem A where $g = 1$ and $\mu = t - (1 + s)$. Then formula (43) yields $m_0 = 5$ and so $G = Q \,\complement\, (\mathbb{Z}Q/\mu \,.\, \mathbb{Z}Q)$ has the presentation

$$\langle a, s, t; \, \mathscr{R}_A \cup \mathscr{R}_Q \cup \mathscr{C}_\Lambda \cup \mathscr{K}_5 \rangle.$$

Here $\mathscr{R}_A = \{(a^t)^{-1} a a^s\}$, $\mathscr{R}_Q = \{[s, t]\}$, while $\mathscr{C}_\Lambda$ consists of the three auxiliary relators (corresponding to (42))

(44) $\quad a^{-1} a^t (a^{-1})^s, \quad a^{-1} (a^{-1})^{s^{-1}} a^{s^{-1}t} \quad$ and $\quad a^{-1} a^{t^{-1}} a^{st^{-1}}$

derived from $a^t = a a^s$, whereas $\mathscr{K}_5$ is made up of the 21 commutator relators

(45) $$[a, a^{s^{l_1} t^{l_2}}] \quad \text{for} \quad l_1^2 + l_2^2 \leqslant 5.$$

The relators (44) were derived from $a^t = a a^s$ in the free group on $\{a, s, t\}$ and can therefore be discarded. The set $\mathscr{K}_5$ is extravagantly large; indeed one can check painlessly that the geometric procedure used in subsection 7.3 to derive new commutator relators, implies that all relators of $\mathscr{K}_5$ are consequences of $a^t = a a^s$ and $[a, a^s] = 1$. Thus we find that $Q \,\complement\, (\mathbb{Z}Q/(t - 1 - s)\mathbb{Z}Q)$ has the presentation

$$\langle a, s, t; \, a^t = a a^s, [s, t] = 1 = [a, a^s] \rangle.$$

Example 43.   Let $Q = \langle s, t \rangle$ be free abelian of rank 2 and choose, as in Illustration 24,

$$\mu = m s^{-1} t^{-1} + s^{-1} + t^{-1} + n + st,$$

with $|m| > 1$. The group $G = Q \,\complement\, (\mathbb{Z}Q/\mu\mathbb{Z}Q)$ has a finite presentation, but we cannot obtain it by the method of Baumslag–Remeslennikov. If, however, we proceed as in the previous example, we find that

$C = (10)^{-1/2}$, $D^2 = 8$ and $m_0 = (D^2/2C)^2 = 160$ and arrive at the finite presentation

$$\langle a, s, t; 1 = (a^m)^{s^{-1}t^{-1}} a^{s^{-1}} a^{t^{-1}} a^n a^{st} = [s, t] = [a, a^{s^h t^l}],$$
$$\text{where} \quad h^2 + l^2 \leqslant 160 \rangle.$$

The displayed presentation has approximately 500 relations; but this number can be reduced considerably by geometric arguments.

8. Some Open Problems

Our understanding of what makes a finitely generated infinite soluble group finitely related is still very limited, but the results obtained indicate where to look for further understanding. Quite naturally the problems suggested by the known results are of two kinds, reflecting the two kinds of original discovery.

The problems of the first kind are concerned with the *construction of new examples*. On a naive level we might simply ask whether a finitely generated soluble group with such and such a property can be finitely related. We know that finitely presented soluble groups can have an infinitely generated centre and need not be residually finite (H. Abels [1]); they may have an insoluble word problem (Kharlampovič [37]), and every finitely generated, relatively free $\mathfrak{N}_c\mathfrak{A}$-group can be embedded in one (Thomson [57]). But what about the other properties?

Problem E.   Can a finitely presented soluble group contain a subgroup which is free soluble of rank 2 and derived length 3 (Baumslag [8])? Can it contain the wreath product $S_3 \wr \langle t \rangle$ of the symmetric group on 3 letters and an infinite cyclic group (Baumslag)? Can it contain the additive group of rationals?

Positive solutions to questions as exemplified by Problem E would stress the diversity of finitely presented soluble groups, but may not by themselves deepen our understanding. The next three problems indicate questions on a more systematic level, the answers to which are likely to lead to greater insight.

Problem F. (F. B. Cannonito [25]).   *Which finitely generated $\mathfrak{N}_2\mathfrak{A}$-groups embed in finitely presented $\mathfrak{N}_2\mathfrak{A}$-groups?*

Problem G. (D. J. S. Robinson). *Which finitely generated soluble mini-max groups G embed in finitely presented soluble groups $\tilde{G}$? When can $\tilde{G}$ be chosen to be a minimax group?*

Problem H. (Baumslag [8]). *Can every finitely generated abelian-by-polycyclic group be embedded in a finitely presented abelian-by-polycyclic group?*

We comment briefly on these. Clearly there are only countably many finitely generated subgroups of finitely presented groups; in point of fact, every such subgroup must be recursively presentable (cf. [39], p. 214 ff.).

Finitely generated metabelian and more generally, abelian-by-poly-cyclic groups satisfy max-n [33], a fact from which one can infer that every such group can be recursively presented. Thus the necessary condition is automatically satisfied for these groups. Theorem A* tells us there is no further restriction on the finitely generated subgroups of finitely presented *metabelian groups*.

The situation is entirely different for finitely generated centre-by-metabelian groups, of which there exist continuously many. But Theorem C forces finitely generated subgroups of finitely presented centre-by-metabelian groups to be abelian-by-polycyclic; and Theorem 13 tells us that there is no further restriction.

Problem F deals with the larger class of groups whose derived group is nilpotent of class 2. Certain necessary conditions can be inferred from Corollary 32, for example residual finiteness. It seems a difficult enterprise to spell out explicitly what follows from the assumption that $(G')^{ab}$ is a tame $\mathbb{Z}G^{ab}$-module.

Problem G is of interest because of the following facts (see [26] for more details). Every finitely generated soluble minimax group G has a divisible abelian normal subgroup D such that G/D is (torsion-free)-by-finite. Hence G/D embeds in a constructible soluble group (see Theorem 10(iv)). On the other hand, Abels' example $H(4, \mathbb{Z}[1/p], \langle p \rangle)$ (p. 277) has a finitely presented central quotient $H/\langle z \rangle$ which contains a Prüfer group of type p^{∞} and thus is not (torsion-free)-by-finite. Finally one has to bear in mind that finitely generated soluble minimax groups cannot in general be recursively presented and so are not usually embeddable in finitely presented groups.

We turn next to the problem of finding analogues of Theorem B and Theorem B2. The goal is indicated by

Problem J.   *Find further characteristic properties of finitely presented infinite soluble groups.*

Note that both Theorem B and Theorem B2 were derived from properties of HNN-extensions or of generalized free products. The multiplicator, however, was never used. The experience obtained so far indicates that it is the wrong kind of invariant in the context under discussion: it is both less informative and harder to compute than the geometric invariant.

Finally there is the question of finding analogues of Corollary BD, describing the finitely related groups of a suitable class of finitely generated soluble groups. The solution of the following problems might be especially revealing:

Problem K.   *Characterize the finitely related $\mathfrak{N}_2\mathfrak{A}$-groups.*

Problem L.   *Characterize the finitely related $\mathfrak{A}\mathfrak{N}_2$-groups.*

References

[1] Abels, H., An example of a finitely presented solvable group, *in* 'Homological Group Theory' (ed. C. T. C. Wall), *London Math. Soc. Lecture Note Series* **36**, 205–211, Cambridge Univ. Press (1979).

[2] Artin, E., Theorie der Zöpfe, *Abh. Math. Sem. Univ. Hamburg* **4** (1925), 47–72.

[3] Baumslag, G., Wreath products and finitely presented groups, *Math. Z.* **75** (1961), 22–28.

[4] Baumslag, G., 'Finitely Presented Groups', *Internat. Conf. Theory of Groups* (Canberra 1965), Gordon and Breach, New York, 1967, 37–50.

[5] Baumslag, G., A finitely presented metabelian group with a free abelian derived group of infinite rank, *Proc. Amer. Math. Soc.* **35** (1972), 61–62.

[6] Baumslag, G., Subgroups of finitely presented metabelian groups, *J. Austral. Math. Soc.* **16** (1973), 98–110.

[7] Baumslag, G., A finitely presented solvable group that is not residually finite, *Math. Z.* **133** (1973), 125–127.

[8] Baumslag, G., 'Finitely Presented Metabelian Groups', *Springer Lecture Notes in Math.* **372** (1974), 65–74.

[9] Baumslag, G., Bieri, R., Constructible solvable groups, *Math. Z.* **151** (1976), 249–257.

[10] Baumslag, G., Strebel, R., Some finitely generated, infinitely related metabelian groups with trivial multiplicator, *J. Algebra* **40** (1976), 46–62.

[11] Baumslag, G., Gildenhuys, D., Strebel, R., Algorithmically insoluble problems about finitely presented soluble groups and Lie algebras (preprint 1983).

[12] Bergman, G. M., A weak Nullstellensatz for valuations, *Proc. Amer. Math. Soc.* **28** (1971), 32–38.

[13] Bergman, G. M., The logarithmic limit-set of an algebraic variety, *Trans. Amer. Math. Soc.* **157** (1971), 459–469.

[14] Bieri, R., 'Homological Dimensions of Discrete Groups', *Queen Mary College Mathematics Notes*, 2nd ed., London (1981).

[15] Bieri, R., A connection between the integral homology and the centre of a rational linear group, *Math. Z.* **170** (1980), 263–266.

[16] Bieri, R., Groves, J. R. J., Metabelian groups of type $(FP)_\infty$ are virtually of type (FP), *Proc. London Math. Soc.* (3) **45** (1982), 365–384.

[17] Bieri, R., Groves, J. R. J., On the geometry of the set of characters induced by valuations, *J. reine angew. Math.* **347** (1984), 168–195.

[18] Bieri, R., Strebel, R., Almost finitely presented soluble groups, *Comment. Math. Helv.* **53** (1978), 258–278.

[19] Bieri, R., Strebel, R., Soluble groups with coherent group rings, *in* 'Homological Group Theory' (ed. C. T. C. Wall), *London Math. Soc. Lecture Note Series* **36**, 235–240, Cambridge Univ. Press (1979).

[20] Bieri, R., Strebel, R., Valuations and finitely presented metabelian groups, *Proc. London Math. Soc.* (3) **41** (1980), 439–464.

[21] Bieri, R., Strebel, R., A geometric invariant for modules over an abelian group, *J. reine angew. Math.* **322** (1981), 170–189.

[22] Bieri, R., Strebel, R., On the existence of finitely generated normal subgroups with infinite cyclic quotients, *Arch. Math.* **36** (1981), 401–403.

[23] Bieri, R., Strebel, R., A geometric invariant for nilpotent-by-abelian-by-finite groups, *J. Pure Appl. Algebra* **25** (1982), 1–20.

[24] Blackburn, N., Some homology groups of wreathe products, *Illinois J. Math.* **16** (1972), 116–129.

[25] Cannonito, F. B., On varietal analogs of Higman's embedding theorem, *Contemp. Math.* (to appear).

[26] Cannonito, F. B., Robinson, D. J. S., The word problem for finitely generated soluble groups of finite rank, *Bull. London Math. Soc.* **58** (1984), 43–46.

[27] Dehn, M., Über die Topologie des dreidimensionalen Raumes, *Math. Ann.* **69** (1910), 137–168.

[28] Eckmann, B., Müller, H., Poincaré-duality groups of dimension two, *Comment. Math. Helv.* **55** (1980), 510–520.

[29] Groves, J. R. J., Varieties of soluble groups and a dichotomy of P. Hall, *Bull. Austral. Math. Soc.* **5** (1971), 391–400.

[30] Groves, J. R. J., Soluble groups in which every finitely generated subgroup is finitely presented, *J. Austral. Math. Soc.* (A) **26** (1978), 115–125.

[31] Groves, J. R. J., Finitely presented centre-by-metabelian groups, *J. London Math. Soc.* (2) **18** (1978), 65–69.

[32] Gupta, C. K., Gupta, N. D., On the linearity of free nilpotent-by-abelian groups, *J. Algebra* **24** (1973), 293–302.

[33] Hall, P., Finiteness conditions for soluble groups, *Proc. London Math. Soc.* (3) **4** (1954), 419–436.

[34] Hall, P., On the finiteness of certain soluble groups, *Proc. London Math. Soc.* (3) **9** (1959), 595–622.

[35] Hall, P., The Frattini subgroup of finitely generated groups, *Proc. London Math. Soc.* (3) **11** (1961), 327–352.

[36] Kaplansky, I., 'Commutative Rings' (rev. ed.), The Univ. Chicago Press (1974).

[37] Kharlampovič, O. G., A finitely presented solvable group with unsolvable word problem, *Izv. Akad. Nauk SSSR Ser. Mat.* **45** (1981), 852–873 (Russian); *Math. USSR Izvestiya* **49** (1982), 151–169 (English translation).

[38] Kilsch, D., On minimax groups which are embeddable into constructible groups, *J. London Math. Soc.* (2) **18** (1978), 472–474.

[39] Lyndon, R. C., Schupp, P. E., 'Combinatorial Group Theory', Springer-Verlag (1977).

[40] Magnus, W., Über *n*-dimensionale Gittertransformationen, *Acta Math.* **64** (1934), 353–367.

[41] Magnus, W., On a theorem of Marshall Hall, *Ann. Math.* **40** (1939), 764–768.

[42] Magnus, W., Karrass, A., Solitar, D., 'Combinatorial Group Theory', Interscience Publishers (1966).

[43] Neumann, B. H., Some remarks on infinite groups, *J. London Math. Soc.* **12** (1937), 120–127.

[44] Nielsen, J., Die Isomorphismengruppen der freien Gruppen, *Math. Ann.* **91** (1924), 169–209.

[45] Nielsen, J., Die Gruppe der dreidimensionalen Gittertransformationen, *Kgl. Danske Vidensk. Selsk., Mat. Fys. Meddels.* **12** (1924), 1–29.

[46] Passman, D. S., 'The Algebraic Structure of Group Rings', Wiley-Interscience (1977).

[47] Remeslennikov, V. N., Finite approximability of metabelian groups, *Algebra i Logika* **7** (1968), 106–113 (Russian); *Algebra and Logic* **7** (1968), 268–272 (English translation).

[48] Remeslennikov, V. N., An example of a finitely defined solvable group without the maximality condition for normal subgroups, *Mat. Zametki* **12** (1972), 287–293 (Russian); *Math. Notes* **12** (1972), 606–609 (English translation).

[49] Remeslennikov, V. N., On finitely presented groups, *Proc. Fourth All-Union Symposium on the Theory of Groups*, 164–169, Novosibirsk (1973).

[50] Robinson, D. J. S., 'A Course in the Theory of Groups', Springer (1982).

[51] Robinson, D. J. S., Strebel, R., Some finitely presented soluble groups which are not nilpotent by abelian by finite, *J. London Math. Soc.* (2) **26** (1982), 435–440.

[52] Šmel'kin, A. L., On soluble products of groups, *Sibirsk. Mat. Ž.* **6** (1965), 212–220.

[53] Strebel, R., On one-relator soluble groups, *Comment. Math. Helv.* **56** (1981), 123–131.

[54] Strebel, R., On finitely related abelian by nilpotent groups (preprint 1981).

[55] Strebel, R., Subgroups of finitely presented centre-by-metabelian groups, *J. London Math. Soc.* (2) **28** (1983), 481–491.

[56] Strebel, R., On quotients of groups having finite homological type, *Arch. Math.* **41** (1983), 419–426.

[57] Thomson, M. W., Subgroups of finitely presented solvable linear groups, *Trans. Amer. Math. Soc.* **231** (1977), 133–142.

[58] Wehrfritz, B. A. F., 'Infinite Linear Groups', Springer-Verlag (1973).

[59] Wirtinger, W., Über die Verzweigungen der Funktionen von zwei Veränderlichen, *Jahresber. Deutschen Math. Ver.* **14** (1905), 517.

The Algebra of Partitions

I. G. MACDONALD

1. Introduction 315
2. The algebra of symmetric functions 316
3. Hall polynomials and the Hall algebra 321
4. Hall–Littlewood symmetric functions 324
5. GL_n over a finite field 328
6. GL_n over a local field 330
7. Concluding remarks 332
References 333

1. Introduction

Philip Hall published little on combinatorial questions. Nevertheless his work in this area has had a profound influence, mainly through his discovery of the polynomials and the algebra which now bear his name.

Let us begin by recalling some definitions. A *partition* is any (finite or infinite) sequence $\lambda = (\lambda_1, \lambda_2, \ldots)$ of non-negative integers such that $\lambda_1 \geqslant \lambda_2 \geqslant \ldots$, and such that $|\lambda| = \sum \lambda_i$ is finite (so that only finitely many of the λ_i are non-zero). We shall find it convenient not to distinguish between two partitions which differ only by a string of zeros at the end: thus for example $(2,1)$, $(2,1,0)$, $(2,1,0,0,\ldots)$ are to be regarded as the same partition. The number of non-zero terms λ_i is called the *length* $l(\lambda)$ of the partition λ. A partition λ may be described graphically by its *diagram* $D(\lambda)$, which is the set of lattice points (i,j) in $\mathbb{Z}^2$ such that $1 \leqslant j \leqslant \lambda_i$. We adopt the convention (as with matrices) that the first coordinate (the row index) increases as one goes downwards, and the second coordinate (the column index) increases as one goes from left to right. If we reflect the diagram $D(\lambda)$ in the main diagonal, we obtain the diagram of a partition $\lambda' = (\lambda_1', \lambda_2', \ldots)$ called the *conjugate* of λ. Thus λ_i' is the number of integers $j \geqslant 1$ such that

GROUP THEORY: essays for Philip Hall
ISBN 0-12-304880-X

$\lambda_j \geqslant i$; in particular, $\lambda_1' = l(\lambda)$, and dually $\lambda_1 = l(\lambda')$. It is clear that $|\lambda'| = |\lambda|$, and that $\lambda'' = \lambda$.

Now let p be a prime number. A finite abelian p-group G is a direct sum of cyclic subgroups, of orders $p^{\lambda_1}, p^{\lambda_2}, \ldots, p^{\lambda_r}$ say, where we may suppose that $\lambda_1 \geqslant \lambda_2 \geqslant \ldots \geqslant \lambda_r > 0$. The partition $\lambda = (\lambda_1, \lambda_2, \ldots, \lambda_r)$ determines G up to isomorphism, and is called the *type* of G. (The conjugate partition λ' enters the picture as follows: for each $i \geqslant 1$, λ_i' is the rank of the elementary p-group $p^{i-1}G/p^i G$.) Thus the isomorphism classes of finite abelian p-groups are naturally indexed by partitions. All this is of course elementary and familiar: what more is there to be said on this subject?

Hall showed that, nevertheless, there are problems worthy of study in this domain. In a short note [2] in 1938 he proved the following 'rather curious' result. Fix a prime p, and for each partition λ let G_λ be a finite abelian p-group of type λ. Then

$$(1.1) \qquad \sum_\lambda |G_\lambda|^{-1} = \sum_\lambda |\mathrm{Aut}\, G_\lambda|^{-1}$$

where the sum on either side is over all partitions λ, that is to say over all isomorphism classes of finite abelian p-groups, and $\mathrm{Aut}\, G_\lambda$ is the group of automorphisms of G_λ. A little later, in 1940 [3], Hall stated a number of identities for finite abelian 2-groups, of which the following are typical. Let $v(G_\lambda)$ denote the number of composition series of G_λ. Then

$$(1.2) \qquad \sum_{|\lambda|=n} \frac{v(G_\lambda)}{|\mathrm{Aut}\, G_\lambda|} = 1,$$

$$(1.3) \qquad \sum_{|\lambda|=n} \frac{v(G_\lambda)^2}{|\mathrm{Aut}\, G_\lambda|} = n!$$

(When $p \neq 2$, the right-hand sides of (1.2) and (1.3) must be multiplied by $(p-1)^{-n}$.) As we shall see later, these formulas and others like them are simple corollaries of the formalism of the Hall algebra.

2.   The Algebra of Symmetric Functions

Before we introduce the Hall algebra, it will be as well to review briefly the theory of symmetric functions [8].

Let Λ denote the ring of symmetric functions in infinitely many independent variables $x_1, x_2, \ldots$, with integer coefficients. Λ is a graded ring, namely $\Lambda = \bigoplus_{n \geqslant 0} \Lambda_n$, where Λ_n consists of the homogeneous symmetric functions of degree n. For each partition $\lambda = (\lambda_1, \lambda_2, \ldots)$ the *monomial symmetric function* m_λ is defined to be the formal infinite sum

$$m_\lambda = \sum x_1^{\lambda_1} x_2^{\lambda_2} \cdots$$

taken over all distinct monomials obtainable from $x_1^{\lambda_1} x_2^{\lambda_2} \cdots$ by permutations of all the variables. For example, $m_{(2,1)} = \sum_{i \neq j} x_i^2 x_j$. The m_λ form a $\mathbb{Z}$-basis of Λ, and for each $n \geqslant 0$ the m_λ with $|\lambda| = n$ form a $\mathbb{Z}$-basis of Λ_n.

In particular, when $\lambda = (1^n)$ we obtain the *elementary symmetric functions*

$$m_{(1^n)} = e_n = \sum x_1 x_2 \ldots x_n \quad (n \geqslant 1)$$

and the fundamental theorem on symmetric functions states that Λ is freely generated as a polynomial ring by the e_n. In other words, if for each partition λ we define

$$e_\lambda = e_{\lambda_1} e_{\lambda_2} \cdots,$$

then the e_λ form another $\mathbb{Z}$-basis of Λ.

Next, the *complete homogeneous symmetric functions* h_n are defined by

$$h_n = \sum_{|\lambda| = n} m_\lambda \quad (n \geqslant 1)$$

so that h_n is the sum of all distinct monomials of degree n in the x_i. The e_n and the h_n have the generating functions

(2.1) $$E(t) = \sum_{n=0}^{\infty} e_n t^n = \prod_{i=1}^{\infty} (1 + x_i t),$$

(2.2) $$H(t) = \sum_{n=0}^{\infty} h_n t^n = \prod_{i=1}^{\infty} (1 - x_i t)^{-1}$$

(where $e_0 = h_0 = 1$), from which it follows that

(2.3) $$\sum_{r=0}^{n} (-1)^r e_r h_{n-r} = 0 \quad (n \geqslant 1).$$

It follows from these relations that the ring homomorphism $\omega\colon \Lambda \to \Lambda$ defined by $\omega(e_n) = h_n$ for each $n \geqslant 1$ is an *involution*, and therefore Λ is also freely generated as a polynomial ring by the h's: in other words, the products $h_\lambda = h_{\lambda_1} h_{\lambda_2} \ldots$, for all partitions λ, form a third $\mathbb{Z}$-basis of Λ.

Another class of symmetric functions consists of the *power-sums* p_n and their products, where

$$p_n = m_{(n)} = \sum x_1^n \quad (n \geqslant 1).$$

Their generating function is

$$P(t) = \sum_{n=1}^{\infty} p_n t^{n-1} = \sum_{i \geqslant 1} \frac{x_i}{1 - x_i t}$$

so that we have

(2.4)
$$P(t) = \frac{d}{dt} \log H(t)$$

and likewise

(2.5)
$$P(-t) = \frac{d}{dt} \log E(t).$$

The power-sum products $p_\lambda = p_{\lambda_1} p_{\lambda_2} \ldots$ (we define p_0 to be 1) form a $\mathbb{Q}$-basis of $\Lambda_\mathbb{Q} = \Lambda \otimes_\mathbb{Z} \mathbb{Q}$, but do not form a $\mathbb{Z}$-basis of Λ. From (2.4) and (2.5) it follows that $\omega(p_n) = (-1)^{n-1} p_n$, so that

$$\omega(p_\lambda) = \varepsilon_\lambda p_\lambda$$

where $\varepsilon_\lambda = (-1)^{|\lambda| - l(\lambda)}$; hence the p_λ are eigenfunctions for the involution ω.

This linear-algebra approach to the algebra of symmetric functions is due to Hall [4], who also observed that it is possible to introduce in a significant way a scalar product on Λ: the bilinear form on $\Lambda \times \Lambda$ defined by

(2.6)
$$\langle h_\lambda, m_\mu \rangle = \delta_{\lambda\mu}$$

is *symmetric*; and this in turn leads to 'what may be regarded as the central fact in the algebra of symmetric functions' (*loc. cit.*), namely the existence of an *orthonormal* $\mathbb{Z}$-basis (s_λ) of Λ, indexed by the set of all partitions, and such that

$$(2.7) \qquad \langle s_\lambda, s_\mu \rangle = \delta_{\lambda\mu}.$$

The s_λ are the so-called *Schur functions* or *S-functions*, which may be defined in various equivalent ways (see, e.g., [8], chapter I).

For the moment let us work with a finite set of variables $x_1, \ldots, x_n$ and define, for each sequence $\alpha = (\alpha_1, \ldots, \alpha_n)$ of non-negative integers,

$$a_\alpha = \det(x_i^{\alpha_j})_{1 \leqslant i, j \leqslant n}$$

so that, for example, if we put $\delta = (n-1, n-2, \ldots, 1, 0)$, then a_δ is the Vandermonde determinant, equal to the product $\prod_{i<j}(x_i - x_j)$. Clearly a_α will merely change sign if two of the α_i are interchanged, and in particular will vanish if two are equal. Up to sign, we may therefore assume that $\alpha_1 > \alpha_2 > \ldots > \alpha_n \geqslant 0$, and write $\alpha_i = \lambda_i + n - i$ for $1 \leqslant i \leqslant n$, that is to say $\alpha = \lambda + \delta$, where $\lambda = (\lambda_1, \ldots, \lambda_n)$ is a *partition* (of length $\leqslant n$). The determinant $a_\alpha = a_{\lambda+\delta}$ is divisible by the Vandermonde determinant a_δ in the polynomial ring $\mathbb{Z}[x_1, \ldots, x_n]$, and the quotient

$$s_\lambda(x_1, \ldots, x_n) = a_{\lambda+\delta}/a_\delta$$

is a *symmetric* polynomial, homogeneous of degree $|\lambda|$. It is easily seen that on passing to $n+1$ variables we have $s_\lambda(x_1, \ldots, x_n, 0) = s_\lambda(x_1, \ldots, x_n)$, and it follows that there is a uniquely defined element s_λ of Λ which reduces to $s_\lambda(x_1, \ldots, x_n)$ when $x_{n+1}, x_{n+2}, \ldots$ are all set equal to 0, for any $n \geqslant l(\lambda)$.

The S-functions s_λ can of course be expressed in terms of each of the bases (e_μ), (h_μ), (m_μ), (p_μ) previously introduced. We consider them in turn.

(i) If $\lambda' = (\lambda'_1, \lambda'_2, \ldots)$ is the conjugate of the partition λ, we have

$$(2.8) \qquad s_\lambda = \det(e_{\lambda'_i - i + j})_{1 \leqslant i, j \leqslant l(\lambda')}.$$

(ii) The corresponding formula expressing s_λ in terms of the h's is

$$(2.9) \qquad s_\lambda = \det(h_{\lambda_i - i + j})_{1 \leqslant i, j \leqslant l(\lambda)},$$

where $e_0 = h_0 = 1$, and $e_i = h_i = 0$ for $i < 0$.

Since the involution ω interchanges the e's and the h's, it follows from (2.8) and (2.9) that

$$(2.10) \qquad\qquad \omega(s_\lambda) = s_{\lambda'}.$$

(iii) To express s_λ as a linear combination of monomial symmetric functions we introduce the following notion. By a *tableau of shape* λ we mean any mapping τ of the diagram $D(\lambda)$ into the positive integers such that $\tau(i,j)$ increases in the weak sense along the rows of $D(\lambda)$, and in the strict sense down the columns. The *weight* of τ is defined to be the sequence $(\alpha_r)_{r \geqslant 1}$, where α_r is the number of lattice points (i,j) in $D(\lambda)$ such that $\tau(i,j) = r$; it may or may not be a partition. With this explained, we have

$$(2.11) \qquad\qquad s_\lambda = \sum_\mu K_{\lambda\mu} m_\mu$$

where $K_{\lambda\mu}$ is the number of tableaux τ of shape λ and weight μ.

(iv) The expression of the S-functions s_λ in terms of the power-sum products p_μ involves the characters of the symmetric groups S_n. First of all, the p_μ form an orthogonal (but not orthonormal) basis of $\Lambda_{\mathbb{Q}}$ for the scalar product (2.6): we have

$$(2.12) \qquad\qquad \langle p_\mu, p_\nu \rangle = \delta_{\mu\nu} z_\mu$$

where (if $|\mu| = n$) $n!/z_\mu$ is the number of elements in S_n whose cycle-type is the partition μ: in other words, z_μ is the order of the centralizer in S_n of such an element. We then have

$$(2.13) \qquad\qquad s_\lambda = \sum_\mu z_\mu^{-1} \chi_\mu^\lambda p_\mu,$$

$$(2.13') \qquad\qquad p_\mu = \sum_\lambda \chi_\mu^\lambda s_\lambda$$

where the matrix $(\chi_\mu^\lambda)_{|\lambda|=|\mu|=n}$ is the character table of S_n. More precisely, the function χ^λ on S_n which takes the value χ_μ^λ at elements of cycle-type μ is an irreducible character of S_n, and all the irreducible characters occur in this way, without repetition. In particular, $\chi^{(n)}$ is the trivial character ($\chi_\mu^{(n)} = 1$ for all partitions μ of n) and $\chi^{(1^n)}$ is the sign character ($\chi_\mu^{(1^n)} = (-1)^{n-l(\mu)}$).

Finally, we should mention the 'Littlewood–Richardson rule' for multiplying two S-functions. If μ, ν are partitions, the product $s_\mu s_\nu$ can of course be written as a linear combination of S-functions, say

$$s_\mu s_\nu = \sum_\lambda c^\lambda_{\mu\nu} s_\lambda,$$

where $c^\lambda_{\mu\nu} = \langle s_\lambda, s_\mu s_\nu \rangle$ by (2.7). These coefficients $c^\lambda_{\mu\nu}$ are in fact non-negative integers. Clearly $c^\lambda_{\mu\nu} = 0$ unless $|\lambda| = |\mu| + |\nu|$, by consideration of degrees; also $c^\lambda_{\mu\nu} = 0$ unless $D(\lambda) \supset D(\mu) \cup D(\nu)$. To explain the result of Littlewood and Richardson, we need to generalize slightly the notion of a tableau introduced above. If λ, μ are partitions such that $D(\lambda) \supset D(\mu)$ (i.e. $\lambda_i \geqslant \mu_i$ for all $i \geqslant 1$) a *tableau of shape* $\lambda - \mu$ is any mapping τ of $D(\lambda) - D(\mu)$ into the positive integers, which satisfies the same conditions as before. A tableau τ is an *LR-tableau* if it satisfies the following condition: if the numbers $\tau(i,j)$ are read from right to left in successive rows, starting from the top (i.e. $\tau(1, \lambda_1), \ldots,$ $\tau(1, \mu_1+1), \tau(2, \lambda_2), \ldots, \tau(2, \mu_2+1), \ldots$) the resulting sequence of integers, say $(\alpha_1, \alpha_2, \ldots, \alpha_N)$, has the property that for $r = 1, 2, \ldots, N$ and each positive integer $s < \max(\alpha_1, \ldots, \alpha_N)$, the number of occurrences of s in $(\alpha_1, \ldots, \alpha_r)$ is not less than the number of occurrences of $s+1$.

With this explained, the coefficient $c^\lambda_{\mu\nu}$ is equal to the number of LR-tableaux of shape $\lambda - \mu$ and weight ν (and also to the number of LR-tableaux of shape $\lambda - \nu$ and weight μ).

3.   Hall Polynomials and the Hall Algebra

Let p be a fixed prime number, let λ, μ, ν be partitions and let G be a finite abelian p-group of type λ. Denote by $g^\lambda_{\mu\nu}(p)$ the number of subgroups H of G such that H has type μ and G/H has type ν. Clearly this number will be zero unless $|\lambda| = |\mu| + |\nu|$. More generally, given partitions λ and $\mu^1, \ldots, \mu^r$, let $g^\lambda_{\mu^1 \ldots \mu^r}(p)$ denote the number of chains of subgroups

$$1 = H_0 < H_1 < \ldots < H_r = G$$

in G of type λ such that H_i/H_{i-1} has type μ^i for $i = 1, \ldots, r$. Hall showed that $g^\lambda_{\mu\nu}(p)$ (and more generally $g^\lambda_{\mu^1 \ldots \mu^r}(p)$) is a polynomial function of p with integer coefficients. These polynomials $g^\lambda_{\mu\nu}$ are the 'Hall polynomials'.

Furthermore, Hall was able to determine the degree and leading coefficient of $g^\lambda_{\mu\nu}$. If $\lambda = (\lambda_1, \lambda_2, \ldots)$ is any partition, let

(3.1) $$n(\lambda) = \sum_{i \geqslant 1} (i-1)\lambda_i$$

and recall from section 2 that $c_{\mu\nu}^{\lambda}$ denotes the coefficient $\langle s_\lambda, s_\mu s_\nu \rangle$ of s_λ in $s_\mu s_\nu$. Then Hall's result [4] is

(3.2)   (i) If $c_{\mu\nu}^{\lambda} > 0$, the Hall polynomial $g_{\mu\nu}^{\lambda}$ has degree $n(\lambda) - n(\mu) - n(\nu)$ and leading coefficient $c_{\mu\nu}^{\lambda}$;

(ii) If $c_{\mu\nu}^{\lambda} = 0$, then $g_{\mu\nu}^{\lambda} = 0$ identically.

This result has since been sharpened by T. Klein [6]. Consider as above a finite abelian p-group G of type λ and a subgroup H of type μ such that G/H has type ν. From these data we can construct a tableau $\tau = \tau(H)$ of shape $\lambda' - \nu'$ and weight μ', as follows. For each integer $r \geqslant 0$, let λ^r denote the type of $G/p^r H$, so that $\lambda^0 = \nu$ and $\lambda^r = \lambda$ for $r \geqslant \mu_1$. We have $D(\lambda^r) \supset D(\lambda^{r-1})$, and $|\lambda^r| - |\lambda^{r-1}| = \mu'_r$ for $r \geqslant 1$, and we define $\tau(i,j)$ to be equal to r if and only if $(j, i) \in D(\lambda^r) - D(\lambda^{r-1})$. In fact $\tau(H)$ is an *LR-tableau* of shape $\lambda' - \nu'$ and weight μ'.

For each *LR*-tableau τ of shape $\lambda' - \nu'$ and weight μ', let $g_\tau(p)$ denote the number of subgroups H of G such that $\tau(H) = \tau$. Then Klein's result is that $g_\tau(p)$ is a *monic* polynomial function of p, with integer coefficients and degree $n(\lambda) - n(\mu) - n(\nu)$. From this it follows that the Hall polynomial $g_{\mu\nu}^{\lambda}(p)$ is expressed as a sum of monic 'Klein polynomials' g_τ:

$$g_{\mu\nu}^{\lambda} = \sum_\tau g_\tau$$

where τ runs through the set of all *LR*-tableaux of shape $\lambda' - \nu'$ and weight μ'. The number of such tableaux is $c_{\mu'\nu'}^{\lambda'}$, which is equal to $c_{\mu\nu}^{\lambda}$ by virtue of (2.10).

Next, Hall used the polynomials $g_{\mu\nu}^{\lambda}$ to construct an algebra which reflects the lattice structure of finite abelian p-groups. Let $H(p)$ be a free $\mathbb{Z}$-module with basis $(u_\lambda(p))$ indexed by the set of all partitions, and define a bilinear multiplication in $H(p)$ by

(3.3) $$u_\mu(p)u_\nu(p) = \sum_\lambda g_{\mu\nu}^{\lambda}(p)u_\lambda(p)$$

for all pairs of partitions μ, ν. (The sum on the right of (3.3) is finite, since as remarked earlier we have $g_{\mu\nu}^{\lambda}(p) = 0$ unless $|\lambda| = |\mu| + |\nu|$.) It is not hard to show that this multiplication makes $H(p)$ into a commutative, associative ring with identity element, and that $g_{\mu^1 \ldots \mu^r}^{\lambda}(p)$ is the coefficient of $u_\lambda(p)$ in the product $u_{\mu^1}(p) \ldots u_{\mu^r}(p)$. In particular, the number $v(G_\lambda)$ of composition series of G_λ (§1) is the coefficient of $u_\lambda(p)$ in $u_1(p)^{|\lambda|}$.

The algebra $H(p)$ is the *Hall algebra* (for the prime p). It may be identified with an algebra of symmetric functions, as follows. For each $n \geq 1$, let $v_n(p) = \sum_{|\lambda|=n} u_\lambda(p)$; then the ring homomorphism $\Lambda \to H(p)$ defined by $h_n \mapsto v_n(p)$ for all n extends to an isomorphism of the $\mathbb{Q}$-algebra $\Lambda_{\mathbb{Q}}$ onto $H_{\mathbb{Q}}(p) = H(p) \otimes_{\mathbb{Z}} \mathbb{Q}$. We shall identify $H_{\mathbb{Q}}(p)$ with $\Lambda_{\mathbb{Q}}$ via this isomorphism, so that now

$$(3.4) \qquad h_n = \sum_{|\lambda|=n} u_\lambda(p),$$

and each $u_\lambda(p)$ is a well-defined symmetric function (with rational coefficients), homogeneous of degree λ. For example, it is not hard to show that

$$(3.5) \qquad \sum_{r=0}^{n} (-1)^r p^{r(r-1)/2} u_{(1^r)}(p) h_{n-r} = 0$$

and comparison of (3.5) with (2.3) shows that

$$(3.6) \qquad e_r = p^{r(r-1)/2} u_{(1^r)}(p) \quad (r \geq 1).$$

To establish (3.5), in view of (3.4) and the definition (3.3) of multiplication in $H(p)$, it is enough to show that

$$(3.7) \qquad \sum_{\mu, r} (-1)^r p^{r(r-1)/2} g^{\lambda}_{(1^r)\mu}(p) = 0$$

for each partition λ. Now $\sum_\mu g^{\lambda}_{(1^r)\mu}(p)$ is the number of elementary subgroups of order p^r in a fixed p-group G of type λ. All the elementary subgroups of G lie in the socle, which is a vector space of dimension $l(\lambda)$ over the field of p elements, and (3.7) is therefore a consequence of Möbius inversion in vector spaces over finite fields [9].

We should observe at this point that the Hall polynomials and the Hall algebra can be defined in a slightly more general context. Let $\mathfrak{o}$ be a discrete valuation ring with maximal ideal $\mathfrak{p}$ and finite residue field $k = \mathfrak{o}/\mathfrak{p}$. An $\mathfrak{o}$-module M is *finite* (i.e. has a finite number of elements) if and only if it is killed by a power of $\mathfrak{p}$, or again if and only if it has finite length (i.e. satisfies both the ascending and the descending chain conditions). Since $\mathfrak{o}$ is a principal ideal domain, every finite $\mathfrak{o}$-module M is a direct sum of cyclic modules $M_1, \ldots, M_r$, where $M_i \cong \mathfrak{o}/\mathfrak{p}^{\lambda_i}$ say, and the partition $\lambda = (\lambda_1, \ldots, \lambda_r)$ (called the *type* of M) determines M up to isomorphism. The number of submodules N of M such that N has

type μ and M/N has type ν is then $g^\lambda_{\mu\nu}(q)$, where $g^\lambda_{\mu\nu}$ is the Hall polynomial defined above, and q is the number of elements in the residue
field k (hence is a prime power). In place of the Hall algebra $H(p)$ we
have $H(q)$, depending on the ring $\mathfrak{o}$ only through the cardinality of its
residue field, and *all* the results stated so far remain true in this more
general context (with p replaced by q throughout).

The original case of finite abelian p-groups is that in which $\mathfrak{o}$ is the
ring of p-adic integers. Another case of interest is the following: let V
be a finite-dimensional vector space over a finite field k, and let θ be a
nilpotent endomorphism of V. Then V may be regarded as a finite
module over the ring $\mathfrak{o} = k[[t]]$ of formal power series with coefficients
in k, by defining $tv = \theta v$ for v in V.

Remark. It has recently been realized that Hall was anticipated by
more than half a century by E. Steinitz, who in 1900 defined what we
have called the Hall polynomials and the Hall algebra, recognized
their connection with S-functions, and conjectured Hall's theorem
(3.2). Steinitz's note [11] is a summary of a lecture given at the annual
meeting of the DMV in Aachen in 1900; it gives neither proofs nor
indications of method, and remained forgotten until brought to light
recently by K. Johnsen [5].

4. Hall–Littlewood Symmetric Functions

The problem of finding an explicit expression for the symmetric
function $u_\lambda(q)$ for any partition λ was solved by D. E. Littlewood [7],
inspired by work of Schur [10], to which we shall have occasion to
return later. Let $x_1,\ldots,x_n,t$ be independent indeterminates, let $\lambda = (\lambda_1,\ldots,\lambda_n)$ be a partition of length $\leqslant n$, and consider the expression

$$(4.1) \qquad R_\lambda(x_1,\ldots,x_n;t) = \sum x_1^{\lambda_1}\ldots x_n^{\lambda_n} \prod_{i<j} \frac{x_i - tx_j}{x_i - x_j}$$

where the sum on the right is extended over all $n!$ terms obtained from
that written by permuting the x's in all possible ways. It is not difficult
to see that R_λ is in fact a *symmetric polynomial* in $x_1,\ldots,x_n$, homo-
geneous of degree $|\lambda|$, and that when expressed as a linear combination
of S-functions, say

$$R_\lambda = \sum_\mu a_{\lambda\mu}(t)s_\mu$$

the coefficients $a_{\lambda\mu}(t)$ are polynomials in t with integer coefficients. The highest common factor of the $a_{\lambda\mu}(t)$ (for fixed λ) is the polynomial $v_{\lambda}(t)$ defined as follows: for each $i \geqslant 0$, let m_i denote the number of indices $j = 1, \ldots, n$ such that $\lambda_j = i$; then

$$v_{\lambda}(t) = \prod_{i \geqslant 0} \prod_{j=1}^{m_i} \frac{1-t^j}{1-t}.$$

We now define

(4.2) $$P_{\lambda}(x_1, \ldots, x_n; t) = \frac{1}{v_{\lambda}(t)} R_{\lambda}(x_1, \ldots, x_n; t).$$

The polynomials P_{λ} (unlike the R_{λ}) have the property that

$$P_{\lambda}(x_1, \ldots, x_n, 0; t) = P_{\lambda}(x_1, \ldots, x_n; t)$$

and therefore, as in the case of S-functions, for each partition λ there is a well-defined element $P_{\lambda}(t) \in \Lambda[t] = \Lambda \otimes_{\mathbb{Z}} \mathbb{Z}[t]$ which reduces to $P_{\lambda}(x_1, \ldots, x_n; t)$ when $x_{n+1}, x_{n+2}, \ldots$ are set equal to 0, for any $n \geqslant l(\lambda)$. These symmetric functions $P_{\lambda}(t)$ may be called *Hall–Littlewood symmetric functions*, or *HL-functions* [8]. They form a $\mathbb{Z}[t]$-basis of $\Lambda[t]$. In particular, when $\lambda = (1^n)$ we have

(4.3) $$P_{(1^n)}(t) = e_n$$

independent of t.

The $P_{\lambda}(t)$ serve to interpolate between the monomial functions m_{λ} and the S-functions s_{λ}, since

(4.4) $$P_{\lambda}(0) = s_{\lambda}, \quad P_{\lambda}(1) = m_{\lambda}$$

for any partition λ. In the particular case $t = -1$ they were introduced much earlier by Schur [10], in connection with the projective representations of symmetric groups. Their relevance to the Hall algebra is that for any partition λ we have (Littlewood [7])

(4.5) $$u_{\lambda}(q) = q^{-n(\lambda)} P_{\lambda}(q^{-1})$$

($n(\lambda)$ was defined in (3.1)). Since $P_{(1^n)}(q^{-1}) = e_n$ (4.3), this result includes (3.6) as a special case.

Since the $P_\lambda(t)$ form a $\mathbb{Z}[t]$-basis of $\Lambda[t]$, the product of two *HL*-functions can be written as a linear combination of *HL*-functions, say

$$(4.6) \qquad P_\mu(t)P_\nu(t) = \sum_\lambda f^\lambda_{\mu\nu}(t)P_\lambda(t)$$

with coefficients $f^\lambda_{\mu\nu}(t)$ in $\mathbb{Z}[t]$. From (4.5), (4.6) and (3.3) it follows that the $f^\lambda_{\mu\nu}(t)$ are related to the Hall polynomials by

$$(4.7) \qquad g^\lambda_{\mu\nu}(q) = q^{n(\lambda)-n(\mu)-n(\nu)} f^\lambda_{\mu\nu}(q^{-1}).$$

Many of the formal properties of *S*-functions have counterparts for *HL*-functions. In particular, there is a symmetric bilinear form $\langle u,v\rangle_q$ on $\Lambda_\mathbb{Q}$ such that

$$(4.8) \qquad \langle u_\lambda(q), u_\mu(q)\rangle_q = \delta_{\lambda\mu} a_\lambda(q)^{-1}$$

where $a_\lambda(q)$ is the order of the group of automorphisms of a finite o-module of type λ. For this scalar product we have also

$$(4.9) \qquad \langle P_\lambda, P_\mu\rangle_q = \delta_{\lambda\mu} y_\lambda(q)$$

where

$$(4.10) \qquad y_\lambda(q) = z_\lambda \prod_{i\geq 1} (q^{\lambda_i}-1)^{-1}.$$

The power sum products p_μ can be expressed in terms of the symmetric functions $u_\lambda(q)$, say

$$(4.11) \qquad p_\mu = \sum_\lambda Q^\lambda_\mu(q)u_\lambda(q).$$

The coefficients $Q^\lambda_\mu(q)$ occurring in (4.11) are called *Green's polynomials*: $Q^\lambda_\mu(q)$ is a polynomial function of q, of degree $n(\lambda)-n(\mu)$, with leading coefficient χ^λ_μ (this follows from (4.4), (4.5) and comparison of (4.11) with (2.13')).

From (4.8), (4.9) and (4.11) it follows that the $Q^\lambda_\mu(q)$ satisfy orthogonality relations analogous to those for the characters of the symmetric groups S_n, namely

$$(4.12) \qquad \sum_\lambda a_\lambda(q)^{-1} Q^\lambda_\mu(q)Q^\lambda_\nu(q) = \delta_{\mu\nu} y_\mu(q).$$

These orthogonality relations were discovered by J. A. Green [1], and play a crucial role in his determination of the irreducible characters of the general linear groups over finite fields.

To end this section we shall indicate how the identities (1.1)–(1.3) from [2] and [3] follow from the formalism of HL-functions as we have sketched it. If the variables $x_1, x_2, \ldots$ are specialized by

$$x_i \mapsto q^{-i} \quad (i \geqslant 1)$$

it is not hard to show that

(4.13) $$u_\lambda(q) \mapsto a_\lambda(q)^{-1}$$

for each partition λ. Hence from (3.4) (with p replaced by q) we have

$$\sum_{|\lambda|=n} a_\lambda(q)^{-1} = h_n(q^{-1}, q^{-2}, \ldots)$$

and therefore, summing over all partitions λ,

$$\begin{aligned}
\sum_\lambda a_\lambda(q)^{-1} &= \sum_{n \geqslant 0} h_n(q^{-1}, q^{-2}, \ldots) \\
&= \prod_{i \geqslant 1} (1 - q^{-i})^{-1} \quad \text{by} \quad (2.2) \\
&= \sum_\lambda q^{-|\lambda|}.
\end{aligned}$$

In other words, we have

$$\sum_{(M)} |\operatorname{Aut} M|^{-1} = \sum_{(M)} |M|^{-1}$$

as convergent infinite series, where on either side the sum is over all isomorphism classes of finite $\mathfrak{o}$-modules. When $\mathfrak{o} = \mathbb{Z}_p$ this reduces to the identity (1.1).

Next, as in section 1, let $v(M_\lambda)$ denote the number of composition series of a finite $\mathfrak{o}$-module M_λ of type λ. Then $v(M_\lambda)$ is the coefficient of u_λ in $u_1^n = p_1^n$, where $n = |\lambda|$, so that

(4.14) $$\sum_{|\lambda|=n} v(M_\lambda) u_\lambda = p_1^n.$$

Since $p_1(q^{-1}, q^{-2}, \ldots) = \sum_{i \geqslant 1} q^{-i} = (q-1)^{-1}$, it follows from (4.13) and (4.14) that

$$\sum_{|\lambda|=n} \frac{v(M_\lambda)}{|\text{Aut } M_\lambda|} = \frac{1}{(q-1)^n},$$

which reduces to (1.2) when $q = 2$ and $\mathfrak{o} = \mathbb{Z}_2$.

Finally, it follows from (4.11) that $v(M_\lambda) = Q^\lambda_{(1^n)}(q)$, and therefore

$$\sum_{|\lambda|=n} \frac{v(M_\lambda)^2}{|\text{Aut } M_\lambda|} = \sum_{|\lambda|=n} a_\lambda(q)^{-1} Q^\lambda_{(1^n)}(q)^2$$

which by (4.12) is equal to $y_{(1^n)}(q)$, i.e. we have

$$\sum_{|\lambda|=n} \frac{v(M_\lambda)^2}{|\text{Aut } M_\lambda|} = \frac{n!}{(q-1)^n}$$

which reduces to (1.3) when $q = 2$ and $\mathfrak{o} = \mathbb{Z}_2$.

All the identities for abelian groups stated by Hall in [3] may be derived in the same way.

5. GL_n over a Finite Field

Undoubtedly the most striking application of Hall's theory is to the determination of the irreducible characters of the general linear groups over a finite field [1]. The full story is too complex to be summarized here; we shall merely indicate how the Hall algebra comes into the picture, and refer to [1] or [8] for full details.

Let k be a finite field and let $G_n = GL_n(k)$ denote the group of all non-singular $n \times n$ matrices over k (when $n = 0$, we make the convention that G_0 is a group with one element). Let Φ denote the set of all irreducible monic polynomials $f \in k[t]$, with the exception of the polynomial t. Consideration of the Jordan normal form for matrices over finite fields shows that the conjugacy classes of the group G_n are naturally parametrized by functions $\mu : \Phi \to \mathscr{P}$, where $\mathscr{P}$ is the set of all partitions, satisfying

(5.1) $$\|\mu\| = \sum_{f \in \Phi} d(f) |\mu(f)| = n$$

where $d(f)$ is the degree of $f \in \Phi$. Let c_μ denote the conjugacy class corresponding to μ.

The mechanism of induction from parabolic subgroups plays a key role in the representation theory of the groups G_n. Let m, n be positive integers, and let $G_{m,n}$ denote the subgroup of all $g \in G_{m+n}$ of the form

$$g = \begin{pmatrix} g_{11} & g_{12} \\ 0 & g_{22} \end{pmatrix}$$

where $g_{11} \in G_m$ and $g_{22} \in G_n$. If u_1 (resp. u_2) is a class function on G_m (resp. G_n), then $u(g) = u_1(g_{11})u_2(g_{22})$ is a class function on $G_{m,n}$, and we define $u_1 \circ u_2$ to be the class function on G_{m+n} induced by u. If u_1, u_2 are *characters*, so also is $u_1 \circ u_2$.

Let A_n denote the space of all (complex-valued) class functions on G_n. Then the 'induction product' $u_1 \circ u_2$ just defined gives $A = \bigoplus_{n \geqslant 0} A_n$ the structure of a commutative and associative $\mathbb{C}$-algebra. The algebra A carries a natural scalar product, for which the individual spaces A_n are mutually orthogonal, and which on A_n is the usual scalar product of class functions:

(5.2)
$$\langle u, v \rangle = \frac{1}{|G_n|} \sum_{g \in G_n} u(g)\overline{v(g)}.$$

For each function $\mu: \Phi \to \mathscr{P}$ satisfying (5.1), let u_μ denote the characteristic function of the conjugacy class c_μ of G_n. Clearly these u_μ form a basis of the vector space A_n, and the algebra structure of A is therefore determined by the coefficients $g^\lambda_{\mu\nu}$ in the equations

(5.3)
$$u_\mu \circ u_\nu = \sum_\lambda g^\lambda_{\mu\nu} u_\lambda$$

for all $\mu, \nu: \Phi \to \mathscr{P}$ such that $\|\mu\|$ and $\|\nu\|$ are finite. Now these coefficients $g^\lambda_{\mu\nu}$ are products of Hall polynomials: namely

(5.4)
$$g^\lambda_{\mu\nu} = \prod_{f \in \Phi} g^{\lambda(f)}_{\mu(f)\nu(f)}(q^{d(f)})$$

where q is the number of elements in the finite field k. From (5.3) and (5.4) it follows that

(5.5)
$$A \cong H \otimes_{\mathbb{Z}} \mathbb{C},$$

where H denotes the tensor product (over $\mathbb{Z}$) of the Hall algebras $H(q^{d(f)})$ for all $f \in \Phi$; the isomorphism (5.5) is such that $u_\lambda \in A$ corresponds

to $\bigotimes_{f \in \Phi} u_{\lambda(f)}(q^{d(f)})$. Moreover, the scalar product (4.8) on the Hall algebra determines a scalar product on H, for which $\langle u_\lambda, u_\mu \rangle = \delta_{\lambda\mu} a_\lambda$, where

$$a_\lambda = \prod_{f \in \Phi} a_{\lambda(f)}(q^{d(f)})$$

is the order of the centralizer of any element in the conjugacy class c_λ. With respect to this scalar product and the scalar product (5.2) on A, the isomorphism (5.5) is an isometry.

6. GL_n over a Local Field

We have seen in the previous section that the Hall algebra is an essential ingredient of the representation theory of GL_n over a finite field. In this section we shall try to indicate its relevance to the representation theory of GL_n over a (non-archimedean) local field F. For the sake of simplicity we shall here take F to be the field $\mathbb{Q}_p$ of p-adic numbers, for some prime number p, although the following results will be valid, with appropriate modifications, for any non-archimedean local field [8].

The absolute value on F ($= \mathbb{Q}_p$) defines a topology, for which F is a locally compact topological field, and the ring $\mathfrak{o} = \mathbb{Z}_p$ of p-adic integers is an open compact subring of F. Let $G = GL_n(F)$ be the group of all invertible $n \times n$ matrices over F, and let G^+ denote the subset of G consisting of all (x_{ij}) in G with entries $x_{ij} \in \mathfrak{o}$. G^+ is closed under multiplication, but is not a *subgroup* of G. Let $K = GL_n(\mathfrak{o}) = G^+ \cap (G^+)^{-1}$, the group of all $n \times n$ matrices over $\mathfrak{o}$ with determinant a unit of $\mathfrak{o}$.

We may regard G as an open subset of matrix space $M_n(F) = F^{n^2}$. As such it inherits a topology for which it is a locally compact topological group. Since $\mathfrak{o}$ is compact and open in F, it follows that G^+ is compact and open in G, and that K is a compact open subgroup of G.

The group G possesses a (left- and right-invariant) Haar measure, unique up to a scalar factor. It will be convenient to normalize this measure so that the subgroup K has measure 1.

Let $H(G, K)$ denote the ring of all compactly supported continuous functions $f: G \to \mathbb{Z}$ such that $f(k_1 x k_2) = f(x)$ for all $x \in G$ and all $k_1, k_2 \in K$, with multiplication defined as follows: for $f, g \in H(G, K)$,

$$(fg)(x) = \int_G f(xy^{-1})g(y)\, dy$$

where dy is the normalized Haar measure. Also let $H(G^+, K)$ denote the subset of $H(G, K)$ consisting of the functions which vanish outside G^+. Then $H(G, K)$ is a commutative, associative ring, called the *Hecke ring* of G relative to K, and $H(G^+, K)$ is a subring of $H(G, K)$.

Each function $f \in H(G, K)$ is constant on each double coset KxK in G. These double cosets are compact, open and mutually disjoint. Since f has compact support, it takes non-zero values on only finitely many double cosets KxK, and hence the characteristic functions of these double cosets form a $\mathbb{Z}$-*basis* of $H(G, K)$. The characteristic function of K is the identity element of $H(G, K)$.

Consider a double coset KxK, where $x \in G$. By multiplying x by a suitable power of p we can bring x into G^+. The theory of elementary divisors for matrices over a principal ideal domain now shows that by pre- and post-multiplying x by suitable elements of K we can reduce x to a diagonal matrix. Multiplying further by a diagonal matrix belonging to K will produce a diagonal matrix whose entries are powers of p, and finally conjugation by a permutation matrix will arrange the exponents in descending order. It follows from this discussion that the double coset KxK has a unique representative of the form $p^\lambda = \text{diag}(p^{\lambda_1}, \ldots, p^{\lambda_n})$, where $\lambda_1 \geqslant \ldots \geqslant \lambda_n$; and $\lambda_n \geqslant 0$ if and only if $x \in G^+$. Thus the double cosets of K in G^+ are indexed by *partitions* $\lambda = (\lambda_1, \ldots, \lambda_n)$ of length $\leqslant n$.

Let φ_λ denote the characteristic function of the double coset $Kp^\lambda K$. Thus in particular $\varphi_{(1^n)}$ is the characteristic function of pK, and the characteristic function of $p^r . Kp^\lambda K$ (where $r \in \mathbb{Z}$) is $\varphi_\lambda \varphi_{(1^n)}^r$. From this it follows without difficulty that

$$(6.1) \qquad H(G, K) = H(G^+, K)[\varphi_{(1^n)}^{-1}]$$

and we may therefore concentrate attention on $H(G^+, K)$.

The φ_λ, for all partitions λ of length $\leqslant n$, form a $\mathbb{Z}$-basis of $H(G^+, K)$, and in fact we have

$$(6.2) \qquad \varphi_\mu \varphi_\nu = \sum_\lambda g_{\mu\nu}^\lambda(p)\varphi_\lambda$$

where the coefficients $g_{\mu\nu}^\lambda(p)$ are Hall polynomials. It follows that the Hecke ring $H(G^+, K)$ is just the Hall algebra $H(p)$ 'truncated': to be precise, if $H_n(p)$ denotes the quotient of $H(p)$ by the ideal generated by the $u_\lambda(p)$ such that $l(\lambda) > n$, then the mapping

$$(6.3) \qquad H(G^+, K) \to H_n(p)$$

defined by $\varphi_\lambda \mapsto u_\lambda(p)$ (for all partitions λ of length $\leqslant n$) is an *isomorphism* of rings. Hence, from (4.5), the linear mapping

$$H(G^+, K) \rightarrow \mathbb{Q}[x_1, \ldots, x_n]^{S_n}$$

defined by

$$\varphi_\lambda \mapsto p^{-n(\lambda)} P_\lambda(x_1, \ldots, x_n; p^{-1})$$

is an isomorphism of the Hecke ring $H(G^+, K)$ onto a subring of $\mathbb{Q}[x_1, \ldots, x_n]^{s_n}$. Thus we have an explicit realization of $H(G^+, K)$ as an algebra of symmetric polynomials in n variables; and by virtue of (6.1) this extends to $H(G, K)$, since $\varphi_{(1^n)} \mapsto p^{-n(n-1)/2} x_1 \ldots x_n$.

Knowing the structure of $H(G, K)$, it is now an easy exercise for us to compute explicitly the zonal spherical functions on G relative to K. We shall not go into this matter here, since it would take us too far afield; the details may be found in [8].

7. Concluding Remarks

We have seen in the preceding sections that the Hall–Littlewood symmetric functions $P_\lambda(t)$ are significant for various values of the parameter t. When $t = 0$ they reduce to the Schur functions s_λ; when $t = 1$ they become the monomial symmetric functions m_λ; and when $t = 1/q$ (q a prime power) they arise naturally in the representation theory of the general linear groups over both finite fields and local fields. When $t = -1/q$, they play an analogous part in the representation theory of unitary groups over finite fields.

There is at least one other value of t which is of interest, namely $t = -1$. In Schur's account [10] of the projective representations of the symmetric group S_n, he introduced certain symmetric functions Q_λ (defined only for partitions with all parts distinct). In our notation, Q_λ is $2^{l(\lambda)} P_\lambda(-1)$. Schur's work antedates by many years everything we have been discussing, with the exception of the forgotten note [11] of Steinitz.

It would take too long to explain in detail the part played by these symmetric functions in Schur's theory, and we give only a brief indication. The irreducible projective representations of S_n can be lifted to irreducible ordinary representations of a certain double covering $\tilde{S}_n$ of S_n. If we now define polynomials $X_\mu^\lambda(t) \in \mathbb{Z}[t]$ by the equations

$$p_\mu = \sum_\lambda X_\mu^\lambda(t) P_\lambda(t)$$

(so that $X_\mu^\lambda(0) = \chi_\mu^\lambda$, and $q^{n(\lambda)} X_\mu^\lambda(q^{-1}) = Q_\mu^\lambda(q)$, then part of the character table of $\tilde{S}_n$ is the matrix

$$(2^{e(\lambda, \mu)} X_\mu^\lambda(-1))$$

in which λ runs through the partitions of n with all parts distinct, μ runs through the partitions of n with all parts odd, and $e(\lambda, \mu)$ denotes the integral part of $\frac{1}{2}(l(\mu) - l(\lambda))$. Here λ indexes the irreducible representations of $\tilde{S}_n$ which do not factor through S_n, and μ indexes suitably chosen conjugacy classes of elements of $\tilde{S}_n$ whose images in S_n have μ as cycle-type.

References

[1] Green, J. A., The characters of the finite general linear groups, *Trans. Amer. Math. Soc.* **80** (1955), 402–447.

[2] Hall, P., A partition formula connected with abelian groups, *Comm. Math. Helv.* **11** (1938), 126–129.

[3] Hall, P., On groups of automorphisms, *J. reine angew. Math.* **182** (1940), 194–204.

[4] Hall, P., The algebra of partitions, *Proc. 4th Canadian Math. Congress*, (*Banff* 1957), 147–149.

[5] Johnsen, K., On a forgotten note by Ernst Steinitz in the theory of Abelian groups, *Bull. London Math. Soc.* **14** (1982), 353–355.

[6] Klein, T., The Hall polynomial, *J. Algebra* **12** (1969), 61–78.

[7] Littlewood, D. E., On certain symmetric functions, *Proc. London Math. Soc.* **43** (1961), 485–498.

[8] Macdonald, I. G., 'Symmetric Functions and Hall Polynomials', Oxford Univ. Press (1979).

[9] Rota, G.-C., On the foundations of combinatorial theory I: Theory of Möbius functions, *Z. Wahrscheinlichkeitstheorie* **2** (1964), 340–368.

[10] Schur, I., Über die Darstellung der symmetrischen und der alternierenden Gruppen durch gebrochene lineare Substitutionen, *J. reine angew. Math.* **139** (1911), 155–250.

[11] Steinitz, E., Zur Theorie der Abel'schen Gruppen, *Jahresbericht der DMV* **9** (1901), 80–85.

Index of Notation

Special symbols

$H \wr K$	restricted wreath product of H and K, 100
$H \wr_{\mathfrak{V}} K$	verbal wreath product of H and K, 107
$H \bar{\wr} K$	unrestricted wreath product of H and K, 197
$B \lrcorner A$	split extension of B by A, 22
$A \llcorner B$	split extension of B by A, 175
$\mho^1(G)$	subgroup of the p-group G generated by all the pth powers, 64
$\nabla_G(H)$	$\nabla_G(H)/H = \Delta(N_G(H)/H)$, 244
$\Sigma \searrow H$	the system Σ reduces into H, 40
$\lesssim$	28
$\underset{\lambda \in \Lambda}{\text{Cr }} G_\lambda$	cartesian product, 187
$\underset{\lambda \in \Lambda}{\text{Dr }} G_\lambda$	direct product
$\text{Dr } H^\Lambda$	direct power, 195
$\underset{\lambda \in \Lambda}{\text{Fr }} G_\lambda$	free product, 175
$\underset{\lambda \in \Lambda}{*} G_\lambda$	free product, 266
$H * K$	free product, 197
$H *_{G_0} K$	amalgamated free product, 259
$\text{Wr } H^\Lambda$	wreath power, 179
$I^\dagger$	238
$M^{\otimes r}$	rth tensor power of M, 22
$\otimes^r M$	rth tensor power of M, 295
$\wedge^r M$	rth exterior power of M, 295

Open face

$\mathbb{A}$	finite adele ring, 129
$\mathbb{A}_\infty$	ring of finite integral adeles, 122
$\mathbb{B}_m$	$\{x \in \mathbb{Z}^n; \langle x, x \rangle \leqslant m\}$, 291
$\mathbb{C}$	complex numbers
$\mathbb{F}_p$	field with p elements
$\mathbb{F}_p^\times$	multiplicative group $\mathbb{F}_p \setminus \{0\}$
$\mathbb{P}$	set of all prime numbers, 14
$\mathbb{P}^*$	set of all positive powers of prime numbers, 29
$\mathbb{Q}$	rational numbers
$\mathbb{Q}_p$	p-adic rationals, 79
$\mathbb{R}$	real numbers
$\mathbb{S}^{n-1}$	unit sphere in $\mathbb{R}^n$, 282
$\mathbb{Z}$	integers
$\mathbb{Z}_p$	p-adic integers, 79

Upper case Roman

$\mathrm{Ann}_R(M)$	annihilator of the R-module M, 274
$\mathrm{Ann}(M)$	annihilator of M, 291
$\mathrm{Aut}(G)$	automorphism group of G, 15
$\mathrm{Aut}_c(G)$	group of central automorphisms of G, 63
$BLF(f)$	Baer local formation defined by f, 23
$C_G(X)$	centralizer of X in G, 4
$\mathrm{Core}_G(H)$	$\bigcap_{x \in G} H^x$, 19
$\mathrm{Cov}_{\mathfrak{X}}(G)$	set of covering $\mathfrak{X}$-subgroups of G, 17
C_π	with exactly one conjugacy class of Hall π-subgroups, 54
$D(\lambda)$	diagram of the partition λ, 315
$D_H(G)$	set of G-orbital elements of H, 219
$D_n(F)$	nth dimension subgroup of F, 84
$D_n(R, G)$	nth dimension subgroup of G over R, 86
D_π	with C_π and such that every π-subgroup is contained in some Hall π-subgroup, 54
E_π	with at least one Hall π-subgroup, 8, 54
$F(G)$	Fitting radical of G, 12
$F(\mathfrak{X})$	free group on $\mathfrak{X}$, 264
$\mathrm{Frat}\, H$	Frattini subgroup of H, 172
$\mathrm{Gal}(\ )$	Galois group, 125
$\mathrm{Gasch}_\Omega(G)$	set of Gaschütz Ω-subgroups of G, 30

$G(p)$	set of elements of infinite p-height, 98
G^R	group of R points of $\mathfrak{G}$, 122
H_G	*see* $\mathrm{Core}_G(H)$, 271
$H(G, K)$	Hecke ring of G relative to K, 330–1
$H(l, r, \mathbb{Q})$	Abel's matrix group, 277
$H(p)$	Hall algebra for the prime p, 323
H_q	$\{[v] \in S(Q) \mid v(q) > 0\}$, 282
$H_{\mathbb{Q}}(p)$	$H(p) \otimes_{\mathbb{Z}} \mathbb{Q}$, 323
$\mathrm{Inj}_{\mathfrak{F}}(G)$	set of $\mathfrak{F}$-injectors of G, 36
$\mathrm{Inn}(G)$	inner automorphism group of G, 15
$J(R)$	Jacobson radical of R, 211
$LF(f)$	local formation defined by f, 20
$\mathrm{Locksec}(\mathfrak{F})$	Lockett section of $\mathfrak{F}$, 45
$M(G)$	multiplicator of G, 265
$M(\Lambda, F)$	McLain's group, 167
$NCF(m)$	68
$N_G(X)$	normalizer of X in G, 4
$O(G)$	odd order radical of G, 44
$O_p(G)$	largest normal p-subgroup of G, 8
$O_{p'p}(G)$	largest normal $p'p$-subgroup of G, 20
$O^p(G)$	p-residual of G, 26
$O^{\pi}(G)$	π-residual of G, 24
$\mathrm{Priv}\,R$	set of primitive ideals of R, 247
$\mathrm{Proj}_{\mathfrak{H}}(G)$	set of $\mathfrak{H}$-projectors of G, 27
$P_{\lambda}(t)$	Hall–Littlewood functions, 325
$\mathrm{Soc}(G)$	socle of G, 27
$S(Q)$	valuation sphere of Q, 262
$\mathrm{Syl}_p(G)$	set of Sylow-p-subgroups of G, 41
$\mathrm{Syl}_{\Omega}(G)$	set of generalized Sylow Ω-subgroups of G, 31
$\mathrm{Tr}(n, R)$	lower triangular $n \times n$ matrices over R, 276
$U_n(R)$	upper unitriangular $n \times n$ matrices over R, 145
$Z(G)$	centre of G, 8
$Z_{\infty}(G)$	hypercentre of G, 15

Lower case Roman

$b(\mathfrak{H})$	boundary of $\mathfrak{H}$, 26
$\mathrm{cl}(X)$	class of the nilpotent group X, 77
$c(G)$	class number of G, 137
$c'(G)$	weak class number of G, 132
$\mathrm{dis}\,S(Q)$	set of discrete points of $S(Q)$, 262

e_λ elementary symmetric function, 317
$ff(R)$ field of fractions of R, 297
$h(G)$ Hirsch number of G, 215
h_n complete homogeneous symmetric function, 317
$i_G(H)$ isolator of H in G, 237
$k_\mathfrak{p}$ local ring of k at $\mathfrak{p}$, 303
p' set of all primes different from p
m_λ monomial symmetric function, 317
$l(\lambda)$ length of the partition λ, 315
p_n power-sum, 318
$r_0(A)$ torsion-free rank of A, 261
s_λ Schur function, 319
supp λ support of λ, 219
$u(\mathfrak{A})$ unipotent radical of $\mathfrak{A}$, 136

Upper case Greek

$\Delta(G)$ the FC-centre of G, 210
$\Delta^+(G)$ the periodic subgroup of $\Delta(G)$, 112, 210
Δ^p 248
$\Lambda(\mathfrak{X})$ the Lausch group of $\mathfrak{X}$, 50
$\Phi(G)$ the Frattini subgroup of G, 17
$\Phi_A(B)$ intersection of the maximal A-subgroups of B, 21
Σ_A geometric invariant of the module A, 261
Σ_A^c complement of Σ_A in $S(Q)$, 287
$\Omega_k(G)$ set of elements in the p-group G, of order dividing p^k,
                 8, 66

Lower case Greek

$\gamma_n(G)$ nth term of the lower central series of G, 62
$\bar{\gamma}_n(G)$ $\bar{\gamma}_n(G)/\gamma_n(G)$ is the torsion subgroup of $G/\gamma_n(G)$, 87
$\zeta_i(G)$ ith term of the upper central series of G, 180
λ' the partition conjugate to λ, 315
$|\lambda|$ the sum of the parts of λ, 315
$\pi(G)$ the set of primes p for which G has an element of order p,
                 14
$\pi(\mathfrak{X})$ $\bigcup_{G \in \mathfrak{X}} \pi(G)$, 17
π' the set of primes not in π, 14

π_H projection map, 210

$\omega^*(G)$ marginal subgroup corresponding to word ω, 182

Script

$\mathscr{C}(G)$ lattice of all subgroups of $G^{\mathbb{Q}}$, 132

$\mathscr{D}_k(G)$ Lie ring associated with dimension subgroups of G over k, 87

$\mathscr{F}_H$ the Fitting set $\mathscr{F}$ of H, 34

$\mathscr{F}(G)$ the set of isomorphism types of finite quotients of G, 130

$\mathscr{F}_p(G)$ the set of isomorphism types of finite p-quotients of G, 130

$\mathscr{G}_k(G)$ graded ring associated with the powers $k\mathfrak{g}^n$, 87

$\mathscr{G}_{\mathbb{Z}}(G)$ graded ring corresponding to the $\mathfrak{g}^n$ in $\mathbb{Z}G$, 88

$\mathscr{L}(R)$ Lie ring associated with the free group R, 110

$\mathscr{L}_G(k)$ 123

$\mathscr{R}_{k/\mathbb{Q}}$ restriction of scalars functor, 147

$\mathscr{U}(L)$ universal envelope of L, 88

Upper case Fraktur

Classes of groups

$\mathfrak{A}$ finite abelian groups, 17; abelian groups, 100, 160

$\mathfrak{C}_\pi$ finite π-groups, 17

$\mathfrak{F}$ finite groups, 160

$\mathfrak{F}_p$ finite p-groups, 100

$\mathfrak{F}^*, \mathfrak{F}_*$ upper and lower limits of Locksec($\mathfrak{F}$), 45

$\mathfrak{L}_A$ countable locally finite groups whose centre contains A, 191

$\mathfrak{N}$ nilpotent groups

$\mathfrak{N}_0$ torsion-free nilpotent groups, 100

$\mathfrak{N}_p$ nilpotent groups of finite p-power exponent, 100

$\mathfrak{N}^r$ r-step nilpotent groups, 21

$\mathfrak{Q}^\pi$ π-perfect groups, 25

$\mathfrak{P}$ polycyclic groups, 208

$\mathfrak{S}$ soluble groups, 17

$\mathfrak{S}_\pi$ finite soluble π-groups, 17

$\mathfrak{U}$ supersoluble groups, 19

$\mathfrak{X}\mathfrak{Y}$ $\mathfrak{X}$-by-$\mathfrak{Y}$ groups, 160

Other uses

$\mathfrak{A}_L$	algebraic automorphism group of L, 124
$\mathfrak{A}_G$	$\mathfrak{A}_L$, where $L = \mathscr{L}_G(\mathbb{Q})$
$\mathfrak{B}_L$	image of $\mathfrak{A}_L$ in $\mathfrak{A}_{L/L'}$, 136
$\mathfrak{B}_G$	$\mathfrak{B}_L$, where $L = \mathscr{L}_G(\mathbb{Q})$
$\mathfrak{E}_k(K)$	$\mathrm{Aut}\,(k \otimes_{\mathbb{Q}} K)$, 147
$G_{\mathfrak{X}}$	$\mathfrak{X}$-radical of G, 32
$G^{\mathfrak{X}}$	$\mathfrak{X}$-residual of G, 17
$\mathfrak{G}(R)$	group of R points of $\mathfrak{G}$, 122
$\mathfrak{B}(G)$	the $\mathfrak{B}$-residual of G, 129

Lower case Fraktur

$\mathfrak{g}, \mathfrak{h}, \mathfrak{x}, \ldots$ augmentation ideals of $\mathbb{Z}G, \mathbb{Z}H, \mathbb{Z}X, \ldots$

Operators on classes

17	R_0
24	D_0
32	N_0
99	R
161	$\langle A, B \rangle$, H, L, P, Q, S, S_n
162	$\acute{P}, \grave{P}, \hat{P}, \hat{P}_n, \hat{P}_{S_n}, \acute{S}, \grave{S}, \hat{S}$
168	$N, \acute{N}$
176	$\acute{P}_{S_n}, \acute{P}_n$

These are defined on the pages indicated except for N_0 and R_0, which are defined by saying that a class $\mathfrak{X}$ is N_0-closed if in any group G the product of two normal $\mathfrak{X}$-subgroups of G is in $\mathfrak{X}$, and R_0-closed if $G/(H \cap K)$ is in $\mathfrak{X}$ whenever the quotient groups G/H and G/K are both in $\mathfrak{X}$.

Subject Index

Abelian-by-polycyclic groups
  Frattini subgroups of, 226–228
  and max-n, 209
  and residual finiteness, 215–225
abnormal subgroup, 39
absolute field, 216
adele ring
  finite, 129
  finite integral, 122
A-groups, 9
algebraic automorphism group,
    124–125
algebraic groups
  as algebraic automorphism groups
    of nilpotent Lie algebras, 125,
    136–137
  associated with a $\mathcal{T}$-group, 142
  Borel's theorem on, 129
  Hasse principle for, 126
Anderson's local approach to
    injectors, 34
Andreev correspondence, 106
annihilator free ideals, 229–231
AR-property, *see* Artin–Rees property
Artin–Rees property in group rings,
    223–226
ascendant subgroups, 162
  groups generated by abelian ones,
    168
augmentation ideals, 83–90
  conditions for $\mathfrak{q}^{\omega} = 0$, 102
  over a field, 86–90
  of finitely generated nilpotent
    groups, 89
  and residual nilpotence, 102–109
automorphism classes, groups with
    finite, 187–188
automorphism groups
  faithfully stabilizing a series,
    174–176
  of free groups, 258

groups with finite, 182–184
automorphisms
  of finite p-groups, 80–83
  fixed point free, 8, 81–83

Baer
  function, 23
  groups, 167
  local formation, 23
  nilpotent groups, 172
basic sequence, 86
Baumslag–Remeslennikov method,
    260, 273–275, 307–309
Berger's theorem on $\mathfrak{F}_{*}$-radicals, 52
Bergman's theorem, 219, 257
  statement, 221
binary form, 141, 153–157
  and $\mathcal{D}^{*}$-groups, 151
  Pfaffian, 148
  projective automorphisms of, 141
  projective equivalence of, 141
boundary of a Schunck class, 26
Burnside, W.
  views on group theory, 2
Burnside's $p^{a}q^{b}$-theorem
  and Hall's theorem, 14
  and Sylow systems, 8

Carter subgroups, 16
  as nilpotent covering subgroups, 19
central extension, 265
centralizer, and Σ_{A}, 284, 287
centralizers of involutions, 193–194
  Brauer–Fowler theorem, 193
  finite, 193
central relations, *see* relations
centre-by-finite groups, 180–181
  and finite automorphism groups,
    183

341

centre-by-metabelian groups, 280–287
    embedding in finitely presented,
        200, 275
chain conditions, 164
characteristically simple group,
        McLain's, 166–167
characters
    Brauer's characterization, 10
    of higher relation modules, 111
    irreducible ones of GL_n, 328–332
    in the Odd Order Paper, 10–12
classes of groups, 160–161
    product of, 160
classification
    of $\mathscr{D}^*$-groups, 150, 152–153
    effective, 126
    and finite p-group theory, 63
    of finite simple groups, 2; use in
        resolving open questions, 54
    of k-forms, 124
    of nilpotent Lie algebras, 124,
        135–136
    of torsion-free nilpotent groups,
        121–158; of irreducible
        $\mathscr{T}$-groups, 142; of the groups
        G^R, 142; of certain $\mathscr{T}_2$-groups,
        148–153
class number of a $\mathscr{T}$-group, 137
closure operations, 160–161
    partial ordering of, 161
    unary, 161
coclass, see p-groups
coherent group, 272–273
cohomology, see Galois cohomology,
        multiplicator
commensurability, see $\mathscr{T}$-groups
commutators
    basic, 85–86; sequence, 86
    collection process, 63–64
    Hall–Petresco formula, 64
complement basis, see Sylow
        complements
complements, partial, see supplements
complete groups
    finite supersoluble of odd order, 116
    as semi-direct products of p-groups,
        116
completion
    q-adic of G, 123
    Mal'cev, and augmentation

ideals, 102; of a torsion-free
        nilpotent group, 122
    profinite, of a torsion-free nilpotent
        group, 122
    pro-p, of a torsion-free nilpotent
        group, 122
composition factors, as fossils, 1
composition series, see series
conjugacy, of Hall systems, 15
conjugacy classes, join of pronormal,
        39
constructible groups
    embedding in, 200–201, 259–260
    soluble, 270–273
control, 228
controller of an ideal, 229
convex polyhedra, 283
Conway's group, 11
coordinate isomorphism, 281
corner, 288
covering groups of Hall, 185–187
covering subgroups, 17–20
    and Carter subgroups, 19
    and Hall subgroups, 19
    and projectors, 25
    and saturated formations, 17–19
    and Schunck classes, 25
    supersoluble, 19
C-theorems
    and Carter, Hall and Sylow
        subgroups, 16
    for Fitting class injectors, 34
    implied by E-theorem, 55
    do not imply D-theorems, 57

D-classes, 25
    sublattice of lattice of Schunck
        classes, 28
degree of commutativity
    of an $NCF(m)$-group, 68
    of a p-group of maximal class, 66
dense sets of linear transformations,
        113
$\mathscr{D}^*$-groups, 149–153
    and binary forms, 151
    central decomposition, 149
    classification of, 150, 152–153
    indecomposable, 149
descendant subgroup, 162

descent, theory of, 124ff.
dimension subgroups
  over a commutative ring, 86; over a field, 86–87
  conjecture, 90
  of a free group, 84–85
  integral, 90–97
  of nilpotent and residually nilpotent groups, 97, 99
  Rips' example, 90
  Sjogren's work, 91–97
  transfinite, 97–99; in nilpotent groups, 97, 99; higher, 99
discrete points, 262, 282
discriminated, 101
division rings and polycyclic-by-finite group rings, 251
D-theorems
  and Carter's theorem, 16
  implied by E-theorems, 55

eccentric factors, of a plinth series, 246
eccentric length, of a polycyclic group, 246
embedding
  in constructible groups, 200–201, 272
  of countable groups in 2-generator groups, 163, 194, 195; Hall's construction, 195–196
  in finitely generated groups, 194–197
  in finitely presented groups, 198–201, 260, 274–277
  in a normal product of two free groups, 170
  in simple groups, 194, 197–198
  subnormal, 196–197
  theorems, 194–201
  in universal locally finite groups, 189
embedding, normal, 48
endomorphism dimension, finite, 216
Engel groups
  need not be locally nilpotent, 172
  with max-ab or min-ab, 172
ergodic conjecture, 242

E-theorems, 8–10
  and Carter, Hall and Sylow subgroups, 16
  conjectures about $E_{2'}$, 55
  for Fitting class injectors, 34
  in the Odd Order Paper, 10
evaluation map, 282
existentially closed groups, 190
extension, central, 265
extra special
  2-groups, 12
  p-groups, see p-groups

factors
  characteristically central, hypercentral, 49
  chief, and system normalizers, 15
  lower central, of a free group, 86, 110–112; of a free polynilpotent group, 109
faithfulness, of relative higher relation modules, 112–114
FC-center, 210
FC-groups, 184–188
  central quotient residually finite, 184
  countable periodic residually finite, 185
  covering groups of Hall, 185–186
  sections of direct products of finite groups, 185–187
FC-hypercentral, see hyper-Δ, 230
Feit, Walter, 10
fertile subgroup, 4
  characteristic, 5
finite-by-nilpotent groups, 181
  and the AR-property, 224
finitely presented groups
  abelian-by-nilpotent, 278
  Baumslag–Remeslennikov method, 260, 273–275, 307–309
  centre-by-metabelian, 261, 275, 280–287, 304–310
  linear (Thomson's theorem), 277
  need not have max-n, 277
  metabelian, 260–264; images of, 264; but not constructible, 260, 273–274; embeddings in, 260, 274–275

finitely presented groups—*contd.*
  nilpotent-by-abelian, 293
  nilpotent-by-cyclic, 278–280
  open problems, 310–312
  and the word problem, 199
finiteness conditions, 163–165
  on commutators and conjugates,
    165, 180–188
  effect on generalized soluble
    groups, 178–179
  finite conjugacy classes, *see*
    FC-groups
  $2^{\aleph_0}$ finitely generated groups, 163,
    258
  see also finite rank
finiteness theorem
  of Borel and Serre, 125
  of Pickel, 128
finite rank, 164–165
  abelian section, 165
  abelian subgroup, 165
  Prüfer, 164
Fischer–Griess group, 3
Fischer's characterization of
    pronormal subgroups, 40
Fischer subgroups, 33
Fitting classes, 32
  Dark's example, 33
  as dual of formation, 33
  and finite *p*-groups, 116
  normal, 43–45; classification, 44, 45,
    48; $\mathfrak{D}(p)$ and $\mathfrak{Y}$, 43, 44;
    smallest, 45
  subgroup closed, 43
Fitting groups, 167
Fitting pairs, 48
  of Laue, Lausch and Pain, 51
  universal, 50
Fitting sets, 32–37
  dualizations, 37
  radicals and injectors, 34, 36
Fitting subgroup, 32, 231
  generalized, 11
formations, 17
  Baer local, 23
  and covering subgroups, 17
  and *D*-classes, 25
  dual, 33
  function, 20
  local, 20–23; form a Fitting class, 33

saturated, 17, 20–24; solubly
    saturated, 24
FP_2, groups of type, 268
Frattini argument, and pronormality,
    37
Frattini subgroup
  in finitely generated soluble
    groups, 196, 226–228
  in *n*-step nilpotent groups, 227
free groups
  automorphism groups of, 258
  and formal power series, 83–86
  normal product of, 170
  and residual nilpotence, 170
free product with amalgamation, 259
Frobenius
  groups, 5
  kernel, 8
Fuchsian groups, 258

Galois cohomology
  finiteness theorem of Borel and
    Serre, 125
  and *k*-forms, 125
  and $\mathscr{T}$-groups, 125
Gaschütz Ω-subgroups, 30
  and Hall subgroups, 31
  and work of Heineken and Harman,
    31
generalized
  free product, 259
  nilpotent groups, 165–175
  periodic elements, 100
  soluble groups, 175–180; simple,
    179–180
  valuation, 298
geometric invariant Σ_A, 262, 281–285
  its complement Σ_A^c in $S(Q)$, 287,
    302–304
  connection with centralizer, 284
Glauberman's *ZJ*-theorem
  and *E*-theorems, 10
  and weak closure, 12
Goldie rank, 250
graded ring associated with a group
    ring, Quillen's work, 87ff.
Green's polynomials, 326
Gruenberg groups, 167
Grün's complex, 8

Hall, Philip
  dimension subgroups, 86
  early work, 13–16
  embeddings theorems, 194–198
  E_π-conjectures, 8, 55
  FC-groups, 185–187
  Frattini subgroups, 196, 207, 227
  ideas on construction of soluble
      groups, 38
  irreducible modules, 216–218
  letter to J. G. Thompson, 3–7
  max-n, 207
  normal product of locally soluble
      groups, 175
  partitions, 315–333
  problem on verbal and marginal
      subgroups, 182
  question about finitely related
      groups and max-n, 227
  residual finiteness, 207ff.
  simple groups with ascending
      series, 180
  theorem on automorphism groups,
      316, 327
  theory of basic commutators, 85–86
  universal locally finite group,
      188–189
  version of the Nullstellensatz,
      226–227
  work on finite-by-nilpotent groups,
      188–189
  work on soluble groups, 8
  $\bar{Z}$-groups, 170
Hall
  algebra, 321–324
  polynomials, 321–324
  subgroups, 14; and Carter
      subgroups, 16; and Gaschütz
      and Sylow Ω-subgroups, 31;
      and injectors and projectors,
      34
  systems, 14–16; reducing into a
      subgroup, 16, 39; and
      pronormality, 40
Hall–Hartley theorems on stability
      groups, 173–175
Hall–Higman paper, 7, 12
Hall–Kulatilaka theorem, 192
Hall–Littlewood symmetric functions,
      324–328

Hasse principle, 126
  counterexamples, 146
Hecke ring, 331
Higman, Neumann, Neumann
  embedding theorems, 194–195;
      generalizations, 195–196
Hirsch number of a polycyclic group,
      215
Hirsch–Plotkin theorem, 165–166
HL-functions, see Hall–Littlewood
      symmetric functions
HNN-extensions, 259–260
  ascending, 260, 268
  and cyclic extensions, 266–268
hyperabelian groups, 176
hypercentral groups, 168
  and stability groups, 174
hypercentrality in group rings, 231
hypercentre, as intersection of
      system normalizers, 15
hyper-Δ and hyper-$D(W)$ groups, and
      control, 230–234
hypoabelian groups, 176
hypocentral groups, 170

ideals
  centrally generated, 223
  faithful, almost faithful, 238; almost
      N-faithful, 244
  polycentral, 223
  prime, 208; image standard, 243
  primitive, 208
idempotents, 248–249, 250–251
  and primitive group rings, 213
infiltration of a sequence, 94
  shuffle, 94
injectors, 32–37
  Anderson's local approach, 34
  E- and C-theorems, 33
  and Hall and Sylow subgroups, 34
  and projectors, 34
  relative to Fitting sets, 34
  smallest Fitting class giving rise to
      class of, 36
intersection theorem and generalized
      intersection theorem, 229
intersection theorem for powers of $\mathfrak{g}$,
      98
isolated subgroup, 138–140, 220

isolator of a subgroup, 237
isomorphism problem for $\mathscr{T}$-groups, 128
isomorphisms connected with $\mathscr{T}$-groups, 123ff.
  comparison of, 130–134
  of k-forms, 125
  of lattice groups, 128

Jacobson radical, in polycyclic group rings, 248, *see also* primitive group rings
J-subgroup, 3–8
  discovery of, 7

k-forms of a Lie algebra
  classification problem, 124
  correspondence with Galois cohomology set, 125
Klein polynomials, 322
Kuroš and Černikov's survey article, 160

lattice groups, 127
  their isomorphism classes, 128; effective procedure for deciding isomorphism, 128
Lausch group, 45, 48–52
  construction, 50
  lattice isomorphic with Locksub, 51
  structure of, 52
Leech lattice, 11
Lie algebras
  action of finite groups on, 115–116
  Andreev correspondence, 106
  associated with lower p-series, 115
  nilpotent, *see* nilpotent Lie algebras
  and Šmelkin's work on residual nilpotence, 105
  verbal wreath product of, 107
Lie rings, associated with
  a free group, 81
  dimension subgroups, 87
  higher relation modules, 110ff.
  a p-group, 81
linear groups
  characters, 328–332
  and finite presentability, 276–277

realizable as automorphism groups of Frattini quotient of a p-group, 116
lith of a monolithic module, 215
Littlewood, *see* Hall–Littlewood
locally finite groups, 163, 188–194
  and centralizers of involutions, 193–194
  existentially closed, 190–192
  with finite Prüfer rank, 193
  infinite abelian subgroups of, 192–193
  with min-ab, 192
  with min-p, 193
  p-groups, 165; with no non-trivial ascendant subgroup, 168
  simple linear, 193
  simple but not locally-(finite simple), 192
  universal, 188–192
locally free groups
  embedding in finitely presented groups, 199
locally nilpotent groups, 165–169
  countable ones are Gruenberg groups, 168
  with finite abelian subgroup rank, 166
  Hirsch–Plotkin theorem, 165–166
  McLain's groups, 166–167
  maximal subgroups of, 166
  with max-n, 166
  with min-n, 166
  principal factors of, 166
  subnormal structure of, 166
locally polycyclic-by-finite groups embed in finitely presented groups, 199
locally soluble groups, 175
  normal product of, 175
  principal factors of, 175
  are $\overline{SI}$-groups, 176
Lockett section, 45, 46–48
  lower limit of, 47
Locksub, 50
  lattice isomorphic with Lausch group, 51
log subgroup, 219
LR-tableau, 321

McLain's characteristically simple
    group, 166–167
  and normal product of locally
    soluble groups, 175
Magnus power series method in free
    groups, 83–84
Magnus variety, 108
Mal'cev correspondence for $\mathcal{T}$-groups
    and nilpotent Lie algebras,
    124
marginal subgroup, 182
max-n
  and Abelian-by-polycyclic groups,
    209
  groups with, 208–209
  and tame modules, 284, 295–296
  and wreath products, 209
metabelian groups, *see also* finitely
    presented groups embedding in
    finitely presented ones, 200
module
  finite induced, 247
  finitely stable, 99
  of finite Prüfer rank and the
    geometric invariant, 299–300
  injective, 251
  integral elements and the geometric
    invariant, 300–301
  irreducible, over group ring of
    polycyclic group, 208
  locally finite, 251
  monolithic, 215
  projective, over a group ring of a
    polycyclic group, 250–251
  rationally irreducible, 219
  relation, *see* relation module
  residually monolithic, 216
  residually simple, 228
  stable, 99
  tame, 263, 294–297; and finite
    presentability of metabelian
    groups, 304–307
  tensor powers, 113; and modules for
    GL_n, 140, 142; finite generation
    of, 294–297; and finitely
    presented soluble groups,
    295–297
multiplicator, 265–266
  of wreath product, 265

N-groups, 2
$NCF(m)$-groups, 68
nilpotent groups, 61–120
  dimension subgroups, 97; the ωth
    of these, 98; generalized, 159
  form an N_0-closed class, 32
  generalized, *see* generalized
    nilpotent groups
  m^cth powers in, 98
  and primitive group rings, 214
  torsion-free, *see* $\mathcal{T}$-groups
  torsion-free, of class two, 63; degree
    of commutativity, 68; the class
    $NCF(m)$, 68; *see also* $\mathcal{T}_2$-groups
nilpotent Lie algebras
  and $\mathcal{T}$-groups, 123ff.
  classification, 135
  2-generator over $\mathbb{Q}$ with $\mathfrak{A}_L$
    unipotent, 137
  Hasse principle, 126
  Mal'cev correspondence, 124
nio(G), 237
normalizer condition, 168
normally embedded subgroup, *see*
    strongly pronormal subgroup
Nullstellensatz and polycyclic group
    rings, 217, 218–222
  Hall's version, 226–227
numerology,
  and Conway's group, 11
  and Fischer–Griess group, 3, 11

Odd Order Paper
  E-theorems in, 10
  groups of symplectic type, 11–12
  and Hall, 7
  self-normalizing cyclic subgroups,
    10
  signalizers in, 9
  tamely embedded subsets, 10
  '3 against 2' argument, 10
OOP, *see* Odd Order Paper
operator group, Ω-, 162
orbital, 219
  ideals, 229ff.
  prime ideals, 239–243
orbitally sound
  conditions to be, 237, 245

orbitally sound—*contd.*
  groups, 234–239
  nio(G), 237
outer commutator words and Hall's
  problems, 182

p-adic uniserial groups, 79
partitions
  and abelian p-groups, 316
  algebra of, 315–333
  diagram of, 315
  length of, 315
p'-closed subgroup, 4, 6
perfect groups
  and closure operations, 161
  and self-normalizing subgroups, 10
  S_3 not subnormal in, 53
periodic group which is not locally
  finite, 163
p-groups, finite, 62–83
  associated Lie ring, 81–82
  asymptotic value of number of
    groups of order p^n, 63
  automorphisms, 80–83; outer, 80;
    with boundedly many fixed
    points, 81–82
  breadth, 80; class-breadth
    conjecture, 80
  with characteristic Abelian
    subgroups all cyclic, 11
  coclass, 76–80; 3-groups of coclass
    2, 77; associated graph, 77–78;
    of pro-p groups, 79
  derived lengths, bounds for, 77, 79,
    82
  extra special, crucial role in
    p-length paper, 12; and
    sporadic simple groups, 12
  foundations of systematic study
    by Hall, 62
  and Fitting classes, 116
  of maximal class, 66–76; 2-groups,
    66; 3-groups, 77; degree of
    commutativity, 66; exceptional,
    67–68; the subgroup G_1, 66, its
    derived length, 70, groups with
    G_1 of class at most 2, 70,
    71–76; isomorphism problem,
    76; metabelian, 71

  and partitions, 316, 322–324
  power structure, 64–66
  regular, 64; analogue of theory in
    arbitrary finite p-groups,
    64–66
  and simple groups, 62–63
  solutions of $x^{p^k} = a$ in, 65
  of symplectic type, 11–12
p-groups, infinite but finitely
  generated, 163
  as automorphism groups of regular
    trees, 163
  simple, 192
p-height, elements of infinite, 98
  commute with p-elements, 99
  generalized, 100
π-perfect groups, 25
  as Schunck classes with normal
    projectors, 44
plinth
  length, 243–244
  series, 234
  socle, 248
plinths, 219
  Bergman's theorem on, 221
  rational, 234
p-nilpotent groups, and the
  AR-property, 224
p-normal groups, 3–8
  Thompson's paper on, 3; Hall's
    letter on this, 3–7
polycyclic group
  coherent, 272–273
  constructible, 260, 272–273
polycyclic group rings, 207–256
  cyclic modules, 217
  irreducible modules, 216–218, 222
  monolithic modules, 225
  prime ideals, 238–239, 243–245
  prime length, 243–244
  primitive ideals and finite induced
    modules, 247
  primitive length, 246
  primitivity, 246
  Noetherian, 208
  semiprimitivity, 248
polynilpotent groups
  free ones, 103, 109; their lower
    central factors, 109
power-sums, 318

p-reducible p-core, 8
presentation
  free, and residual nilpotence,
    103–109
  recursive, and embeddability in
    finitely presented groups, 199,
    201
prime ideals, 208, 228
  in polycyclic group rings, 238–239,
    243–245
  source of, 245
  standard, 243; image standard, 243
  vertex of, 245
prime length, 243–245
primitive group
  and boundaries, 26
  conjugacy of complements in, 19
  and Schunck classes, 24
primitive group rings, 209–215
primitive ideals, 208, 247
  in polycyclic group rings, 246, 247
Priv R, 247
product, permutable
  of soluble subgroups of co-prime
    order, 54
  of Sylow subgroups, 14
projectors, 25
  arithmetical properties of, 29–32
  covering subgroup definition of, 17
  dual theory, 33
  and Hall and Sylow subgroups, 34
  and injectors, 34
  of a join of Schunck classes, 41
pronormal subgroups, 37–41
  conjugacy classes of, 39
  and the Frattini argument, 32
  persist in intermediate groups, 37
  and reducibility of Hall systems, 40
  and self-normalizing subgroups, 39
  strongly pronormal subgroups, 41
  and subgroup-closed Fitting sets, 42
pro-p completion, 122
pro-p group, 79
  coclass of, 79
  soluble of finite coclass, 79
p-series, lower central, 110
  of a free group, 115; actions of
    finite groups on the factors, 116
p-soluble group, of p-length at most
    two, 6, 7

radicable hull, 122
  of $G^{\mathbb{Z}_p}$ and $G^{\mathbb{A}_\infty}$, 130
radical, 32
  behaviour in direct products, 43, 45
  calculation of the $\mathfrak{F}_*$-, 45, 52
  relative to a Fitting set, 34
radical groups, 177–178
  are $\overline{\overline{SN}}$, 178
rationally defined
  convex polyhedra, 283
  great subsphere, 282
  hemisphere, 302
relation modules
  faithfulness and residual
    nilpotence, 104
  higher, 110–112; characters of, 111;
    contain a large free module,
    111
  relative higher, 112–114; and free
    modules, 113, 114
relations, groups defined by central,
    138–142
relatively free groups
  centre of $F/\gamma_{c+1}(R)$, 111
  finitely related, 269–270
representation, stable, 99
residual, 17
  behaviour in direct products, 43
residually, 99
  central groups, 171–172; with min-n,
    171; need not be Z-groups, 171
  commutable groups, 178
  finite groups, 215–218, 225
  finite p-groups, 103
  nilpotent groups, 170; faithfully
    stabilizing an ascending series,
    174; dimension subgroups of,
    97, 99
  simple modules, 228
  soluble groups, 176
residual nilpotence, 99–109
  and abelian-by-polycyclic groups,
    225
  and augmentation ideals, 102–103
  and free presentations, 103–109
  of F/R', 103
  of $C_\infty \wr G$, 104
  of $F/R'R^p$, 105
  and free products, 109
  and wreath products, 99–102

rigid groups, 143–148

Rips and dimension subgroups, 90

R-operator group, see R-powered group

R-powered group, 122

Schunck classes, 24–28
  boundaries of, 26–27
  form a complete lattice, 28
  and covering subgroups, 25
  join of, 28
  with normal projectors, 44
Schur
  functions, 319ff.
  multiplicator, see multiplicator
self-normalizing subgroups
  cyclic, 16
  nilpotent, 16
serial subgroups, 162
  groups in which every subgroup is serial, 172
series, 161–163
  terms, factors, ascending, descending, Ω-series, normal, subnormal, 162; composition, 162–163; Ω-composition, principal, chief, 163
  with abelian factors, see SN-groups
  lower central, 83–90
  groups with a central series, see Z-groups; with a descending central series of length ω, 170
S-functions, see Schur functions
signalizers, 8–10
  introduction, 9
  p-signalizers, 11
SI-groups, 176
  $\underline{\text{are}}$ residually commutable, 178
$\overline{SI}$-groups, 176
  need not be $\overline{SN}$, 177
simple groups
  absolutely, 163
  Brauer–Fowler theorem on, 193
  Chevalley groups and $E_{p,q}$, 55
  classification of, 2
  embeddings in, 197–198
  evolution from solvable groups, 2
  finitely generated p-group, 192
  Fischer–Griess F_1, 3

locally finite, 192; linear, 193
minimal simple, A-group, 9; of odd order, 11
their rarity, 2
$\overline{\overline{SN}}$-groups, 177, 180
sporadic, 3; connexion with extra special 2-groups, 12
S-isomorphism of $\mathscr{T}$-groups, 123
Sjogren, J. A., work on dimension subgroups, 91–97
Šmelkin, A. L., work on residual nilpotence, 105–108
$\underline{SN}$-groups, 175–177
SN-groups, 176
  with a non-cyclic free subgroup, 176–177
$\overline{\overline{SN}}$-groups, 177
  simple ones, 177, 180
soluble groups, finite, 13–60
  the early years, 13–16
  characterizations by Sylow complements, 13; by permutable nilpotent subgroups, 14
  their construction using supplements, 38
  geological analogy, 1
  Hall's $E_{p,q}$ conjecture, 8, 55
  of p-length two, 7
solvable groups, see soluble groups
source of a prime ideal, 245
space groups
  p-adic uniserial, 79
  p-uniserial, 80
special polynomial, 274
stability groups, 173–175
  of an ascending series, 174
  of a descending series, 174–175
  of a series of finite length, 174
  and wreath products, 173
stable representations, 99
Steinitz, E. and Hall's combinatorial work, 324
subnormal subgroups
  groups with every subgroup subnormal, 168
  join problem, 197
subsoluble groups, 176
Šunkov, V. P., work on locally finite groups, 192, 193

supersoluble groups, 19
  finite complete, 116
  do not form a Fitting class, 33
  and saturated formations, 33
  supplements, 37
    Hall's use of them, 38ff.
Suzuki's *CA*-groups theorem, 10
Sylow basis, 14, 15
Sylow complements, 13, 14
  complement basis, 14
Sylow subgroups
  groups with them all abelian, 9
  and Carter subgroups, 16
  generalized, 31; criterion for
    existence of, 31
  and injectors and projectors, 34
  permutable, 1
  permuting in pairs, 14
Sylow systems
  starting point of, 8
  now called Hall systems, 14
symmetric functions
  algebra of, 316–321; involution ω,
    318; scalar product, 318
  complete homogeneous, 317
  elementary, 317
  monomial, 317
symmetric group of degree 3
  and perfect groups, 53
  and the restriction map of
    cohomology, 156
system normalizers
  cover and avoidance properties, 15
  intersection, 15
  minimal elements of subabnormal
    chains, 39
  relative, 38

tableau, 320
  LR-, 321
tamely embedded subsets, 10
Tarski
  p-groups, 164
  monsters and group rings, 211
tensor power of a module, *see* module
$\mathscr{T}$-groups, 121–158
  and algebraic groups, 122; the
    groups G^R, 122–123
  of class 2, *see* $\mathscr{T}_2$-groups

class number of, 137
commensurability of, 126;
  commensurability problem
    effectively soluble, 127
comparison of isomorphisms,
  130–134
defined by ring extensions, 142–148
with the same finite quotients, 126,
  129
irreducible, 142
isomorphism problem effectively
  soluble, 128
lattice groups, 128; isomorphism
  problem effectively soluble, 128
Mal'cev's correspondence for, 124
and nilpotent Lie algebras, 124
$\mathbb{Q}$-isomorphism classes of, 124;
  $\mathbb{Q}_p$-classes formed from them,
  126
$\mathbb{Q}_p$-isomorphic for all p, but not
  commensurable, 141
rigid, 143–148
weak class number of, 132
$\mathscr{T}_2$-groups, *see also* $\mathscr{D}^*$-groups
  of Hirsch length 6, 148
  of Hirsch length at most 5, 151
  with small centre, 148–153
  strongly isolated, 143, 147
Thompson, J. G.
  Hall's letter to, 3–7
  J-subgroup, 7
  paper on p-normal groups, 3ff.
  thesis of, 7
  transitivity theorem of, 9
torsion groups, *see* periodic groups
trace, of a Fitting set, 34
transitivity theorem, 9

uniqueness theorems, 10–11
unique product group, 249
unit sphere, 262
universal locally finite groups,
  180–192

valuations, 297
  generalized, 298
  and the geometric invariant,
    281–285, 298–299

valuations—*contd.*
  and log subgroups, 220
valuation sphere $S(Q)$, 262, 282
varieties
  Andreev correspondence, 106
  of Lie type, 107
  Magnus, 108
  soluble, and finite presentability,
    269–270
verbal subgroup, 182
vertex of a prime ideal, 245

weak class number, 132
  characterization of, 132, 134
  bounded above by class number,
    137
weak closure, 3–8
  of normal elementary Abelian
    subgroup, 12
  Wielandt's theorems on, 5
Wielandt, H.
  1940 theorems, 5; their relevance to
    groups with fixed point free
    automorphisms, 8
  1960 theorem used in proof of Hall's
    theorem, 13–14
  theorem on product of two nilpotent
    subgroups, 14, 54

word problem, 194
  finitely generated groups with it
    soluble, Higman's theorem, 199
  and finitely presented groups, 278
wreath power
  in construction of simple groups, 179
wreath products
  and embedding theorems, 197
  and finite-by-nilpotent groups, 181
  finitely related, 266
  infinitely related, 265
  and max-n, 209
  multiplicator of, 265
  and normal product of locally
    soluble groups, 175
  and primitive group rings, 212–214
  and residual nilpotence, 99–102
  and stability groups, 173
  verbal, 107

Zalesskii subgroup of a polycyclic-by-
    finite group, 234
Zalesskii type subgroups, 228–234
ZD-groups, *see* hypocentral groups
zero divisors in group rings, 249
Z-groups, 169–172
$\bar{Z}$-groups, 169–172
  with non-cyclic free subgroups, 170

Contributors

Grunewald, F., Prof.   Mathematisches Institut, Universität Bonn, D-5300, Bonn 1, West Germany.

Hartley, B., Prof.   Department of Mathematics, University of Manchester, Manchester M13 9PL, England.

Hawkes, T. O., Dr.   Mathematics Institute, University of Warwick, Coventry CV4 7AL, England.

Macdonald, I. G., Prof.   Department of Pure Mathematics, Queen Mary College, London E1 4NS, England.

Passman, D. S., Prof.   Department of Mathematics, University of Wisconsin, Madison, Wisconsin 53706, United States of America.

Robinson, D. J. S., Prof.   Department of Mathematics, University of Illinois, Urbana, Illinois 61801, United States of America.

Segal, D., Dr.   All Souls College, Oxford OX1 4AL, England.

Strebel, R., Dr.   Mathematisches Institut, Universität Heidelberg, D-6900, Heidelberg, West Germany.

Thompson, J. G., Prof.   Department of Pure Mathematics and Mathematical Statistics, University of Cambridge, Cambridge CB2 1SB, England.